I0029383

Energy Pricing Policies for Inclusive Growth in Latin America and the Caribbean

DIRECTIONS IN DEVELOPMENT
Energy and Mining

Energy Pricing Policies for Inclusive Growth in Latin America and the Caribbean

Guillermo Beylis and Barbara Cunha

WORLD BANK GROUP

© 2017 International Bank for Reconstruction and Development / The World Bank
1818 H Street NW, Washington, DC 20433
Telephone: 202-473-1000; Internet: www.worldbank.org

Some rights reserved

1 2 3 4   20 19 18 17

This work is a product of the staff of The World Bank with external contributions. The findings, interpretations, and conclusions expressed in this work do not necessarily reflect the views of The World Bank, its Board of Executive Directors, or the governments they represent. The World Bank does not guarantee the accuracy of the data included in this work. The boundaries, colors, denominations, and other information shown on any map in this work do not imply any judgment on the part of The World Bank concerning the legal status of any territory or the endorsement or acceptance of such boundaries.

Nothing herein shall constitute or be considered to be a limitation upon or waiver of the privileges and immunities of The World Bank, all of which are specifically reserved.

Rights and Permissions

This work is available under the Creative Commons Attribution 3.0 IGO license (CC BY 3.0 IGO) http://creativecommons.org/licenses/by/3.0/igo. Under the Creative Commons Attribution license, you are free to copy, distribute, transmit, and adapt this work, including for commercial purposes, under the following conditions:

Attribution—Please cite the work as follows: Beylis, Guillermo, and Barbara Cunha. 2017. *Energy Pricing Policies for Inclusive Growth in Latin America and the Caribbean*. Directions in Development. Washington, DC: World Bank. doi:10.1596/978-1-4648-1111-1. License: Creative Commons Attribution CC BY 3.0 IGO

Translations—If you create a translation of this work, please add the following disclaimer along with the attribution: *This translation was not created by The World Bank and should not be considered an official World Bank translation. The World Bank shall not be liable for any content or error in this translation.*

Adaptations—If you create an adaptation of this work, please add the following disclaimer along with the attribution: *This is an adaptation of an original work by The World Bank. Views and opinions expressed in the adaptation are the sole responsibility of the author or authors of the adaptation and are not endorsed by The World Bank.*

Third-party content—The World Bank does not necessarily own each component of the content contained within the work. The World Bank therefore does not warrant that the use of any third-party–owned individual component or part contained in the work will not infringe on the rights of those third parties. The risk of claims resulting from such infringement rests solely with you. If you wish to re-use a component of the work, it is your responsibility to determine whether permission is needed for that re-use and to obtain permission from the copyright owner. Examples of components can include, but are not limited to, tables, figures, or images.

All queries on rights and licenses should be addressed to World Bank Publications, The World Bank Group, 1818 H Street NW, Washington, DC 20433, USA; e-mail: pubrights@worldbank.org.

ISBN (paper): 978-1-4648-1111-1
ISBN (electronic): 978-1-4648-1112-8
DOI: 10.1596/978-1-4648-1111-1

Cover photo: © Kay Fochtmann. Used with the permission of Kay Fochtmann; further permission required for reuse.
Cover design: Debra Naylor, Naylor Design, Inc.

Library of Congress Cataloging-in-Publication Data has been requested.

Contents

Boxes

Tables

Energy Pricing Policies for Inclusive Growth in Latin America and the Caribbean
http://dx.doi.org/10.1596/978-1-4648-1111-1

Energy Pricing Policies for Inclusive Growth in Latin America and the Caribbean
http://dx.doi.org/10.1596/978-1-4648-1111-1

Foreword

Energy helped fuel Latin America and the Caribbean's (LAC's) impressive growth over the past decade. The region created millions of jobs, halved extreme poverty, expanded the middle class, and put a dent in its historically high inequality. Since then, however, many countries have been hit by the shifting global economic winds that knocked down the prices for oil and other commodities produced in the region.

Today, as the LAC region emerges from six years of slow growth, including two years of recession, the energy sector can play a role in supporting the recovery while protecting the most vulnerable and the environment.

This report comes at a crucial time as the region is looking for ways to use limited fiscal resources as efficiently and effectively as possible. It reveals that improved energy pricing can be a key enabler to enhancing productivity and social equity and helping to boost growth.

Like many low- and middle-income regions of the world, the LAC region has fiscally onerous energy subsidies, a legacy of efforts to protect consumers from the sharp increases in oil prices in the 2000s. Although energy subsidies may help the poor, this report confirms that the lion's share is actually appropriated by the wealthy, wasting limited resources that could be better applied to priorities such as protecting the most vulnerable. Although politically popular, energy subsidies stress government budgets and crowd out social investments, such as education and health. These costs often go unseen, and so it is important that governments and citizens appreciate the trade-offs involved.

A key finding of this report is that the LAC region clearly has a window of opportunity for energy pricing reform because of the current relatively low international oil prices, which would minimize the potential impacts of reforms on the welfare of the poor and on competitiveness. This study makes it clear that through careful planning and the additional fiscal resources freed by the reforms themselves, it would be possible to offset the expected impacts. In addition, the report identifies more efficient and better-targeted alternatives to protect the poor from energy price hikes—for example, through direct cash transfer programs.

Meanwhile, the report offers practical, policy-oriented advice that draws extensively on successful experiences in the region. It contributes significantly to the practical knowledge of what works and what does not in energy pricing policies.

The good news is that the LAC region has already seen real progress in reforming energy subsidies. I am sure this report will further empower policy makers to take advantage of a unique opportunity for significant and long-lasting development impacts. This vision is consistent with the World Bank's commitment to helping the LAC region rekindle growth and continue on the path of social and economic development.

Jorge Familiar
Regional Vice President, Latin America and the Caribbean
The World Bank Group

Acknowledgments

This report was written by Guillermo Beylis, research economist, and Barbara Cunha, senior economist, at the World Bank. Substantive inputs were provided by Oscar Calvo, Daniel Lederman, and Ariel Yepez-Garcia. We thank David Coady, Gabriela Inchauste, Guido Porto, David Reinstein, David Rosenblatt, and Jon Strand, who kindly served as peer reviewers for our report. We also acknowledge the useful comments from Cecilia Briceño-Garmendia, Margaret Grosh, and Sameer Shukla. Emmanuel Chavez, Bianca Ravani Cecato, and Thiago Scot provided outstanding research assistance. Last, but not least, this report would not have been possible without the unfailing administrative support of Ruth Delgado and Jacqueline Larrabure.

About the Authors

Guillermo Beylis, an Argentine national, is a research economist in the World Bank's Office of the Chief Economist, Latin America and the Caribbean. He received a bachelor's degree and a master's degree in economics from Universidad Torcuato di Tella in Buenos Aires, and he holds a PhD in economics from the University of California, Los Angeles. His main areas of research are labor economics, inequality, and poverty. He joined the World Bank in 2013, and he has published reports on remittances and foreign direct investment (2014), inequality and growth (2014), and labor markets in Latin America and the Caribbean (2015).

Barbara Cunha is a senior economist in the Macroeconomics and Fiscal Management Global Practice at the World Bank. She received her master's degree in economics from the Fundação Getulio Vargas (Brazil) and her PhD in economics from the University of Chicago. Her work focuses on the links between economic policy and growth in low- and middle-income countries, including analysis of fiscal sustainability, efficiency of public spending, tax policy, and trade agreements.

Abbreviations

AMI	Advanced Metering Infrastructure
CIF	cost, insurance, and freight
DAC	Domésticas de Alto Consumo (Residential High Consumption)
DF	Distrito Federal, Mexico City
EA	East Asia
ECA	Europe and Central Asia
EPP	export parity price
EU	European Union
FEPC	Fondo de Estabilización de los Precios de los Combustibles (Fund for Fuel Price Stability)
FOB	freight on board
GDP	gross domestic product
HIC	high-income country
IBT	increasing block tariff
IEA	International Energy Agency
IMF	International Monetary Fund
IO	input-output
IPP	import parity price; independent power producer
LAC	Latin America and the Caribbean
LPG	liquefied petroleum gas
MENA	Middle East and North Africa
NIS	National Interconnected System
PI	producer income
SA	South Asia
SOE	state-owned enterprise
SSA	Sub-Saharan Africa
STMC	short-term marginal cost
VAT	value added tax
VDT	volume differentiated tariff

Units of Measure

GW	gigawatt
GWh	gigawatt-hour
kV	kilovolt
kW	kilowatt
kWh	kilowatt-hour
MPC	thousand cubic feet
MW	megawatt

Executive Summary

Introduction

A majority of countries in the Latin America and the Caribbean (LAC) region are currently facing tight fiscal conditions and low growth prospects. The end of the commodity supercycle has brought an end in turn to the economic bonanza that helped lift millions of families out of poverty, and it has left governments with challenging fiscal positions. Moreover, governments in the LAC region tend to be small as measured by their revenue as a percentage of their gross domestic product (GDP). As a result, their limited government resources have to be used wisely and need to better target the poor and vulnerable. One important area for improvement is energy pricing policies, where eliminating expensive subsidies and price controls can create fiscal space (now and in the future) and allow governments to better direct their resources toward protecting lower-income households. With international oil prices now relatively low, it is the ideal time to reform energy pricing policies.

Our Study

LAC governments intervene in energy markets to varying degrees, with a variety of stated objectives. Energy pricing policies, in particular, have been adopted as instruments for pursuing economic, social, and, in some cases, even environmental objectives. Common pricing policies include lowering prices through direct subsidies, increasing prices through taxation, granting tax exemptions, transferring funds to beneficiaries directly, assuming part of the risk of either the consumer or producer of energy, and imposing entry barriers on energy markets. Previous studies provided estimates of the fiscal costs of energy subsidies and, for some countries, estimated the distributional impacts by analyzing household consumption patterns. This report is the first to provide a comprehensive analysis of the fiscal, economic, and social impacts of energy pricing policies in the LAC region.

The overall objective of this study is to inform the policy debate on energy subsidy reforms by evaluating the costs and benefits of different energy pricing mechanisms. To this end, this study begins by characterizing the set of policies adopted by nine economies in the LAC region. The countries analyzed—Bolivia, Brazil, Colombia, the Dominican Republic, El Salvador, Haiti, Honduras, Mexico, and Peru—represent both oil producers and importers and big and small economies, and thus they provide a good overview of the region. Because oil prices are naturally volatile, we include six years of analysis (2008–13) to ensure that we measure the medium-term fiscal impacts of energy subsidies and identify the instances in which subsidy policies are consistent over time or just temporary measures. Within the scope of this study are all fossil fuel products (such as gasoline, diesel, kerosene, and natural gas) and electricity.

Our Findings

Crude oil prices are expected to hover at around US$55 a barrel in 2017 and slowly creep up to US$60 in 2018 and US$61.50 in 2019.[1] The findings in this report suggest that the aggregate price impacts and competitiveness effects of energy price increases are moderate to small and that they can be smoothed out through macropolicy responses. The removal of subsidies under such circumstances would likely mean smaller impacts on the welfare of poor and vulnerable populations.

Although some governments have made progress in reforming energy subsidies, the temptation and political pressures to reintroduce them will resurface if oil prices shoot up again. Thus it is important for governments to implement energy pricing processes that are protected from these pressures. For one thing, they could consider delegating pricing authority to independent agencies that are shielded from political pressure. They also could institute automatic price adjustment policies that cannot be easily modified, thereby further distancing themselves from pricing decisions. In exceptional circumstances of sharp price increases, governments could temporarily implement price-smoothing mechanisms that slowly pass price changes to end users. Households and industries would then have time to adapt to higher prices. But no matter what mechanisms are adopted, it is essential that they be clear and transparent so that all actors know the rules and that the rules are effectively enforced.

Although energy subsidies are an inefficient policy tool to protect the welfare of the poor, energy price increases can have a big impact on these households. This study finds that energy subsidies are highly regressive in an absolute sense—that is, the lion's share of every dollar spent on keeping energy prices low will benefit wealthier households. However, subsidies for fuels used widely for cooking and heating—liquefied petroleum gas (LPG), natural gas, and kerosene, as well as electricity—can be relatively neutral or progressive, implying that lower-income households capture benefits that are proportionate to their expenditures. In other words, although poorer households receive very little from every dollar spent on energy subsidies, that small amount can represent an important share of their expenditures.

It is important, then, that governments expand the coverage and depth of their social safety nets so they can provide relief for poor households if energy prices were to increase. Direct cash transfers or vouchers have been found to be a much more effective policy tool for protecting poor and vulnerable households from the negative welfare impacts of rising energy costs.

Organization of This Report

Chapter 1 describes the structure of the energy market in the LAC region. Although several countries have partially opened oil markets to private and foreign participation, most oil-producing countries continue to reserve a predominant role for state-owned enterprises, especially at the refining stage of the production chain. Meanwhile, independent and private generators are playing a bigger role in the generation segment of the electricity production chain, and new policy tools (such as auctions for long-term procurement of generation capacity) are lowering generating costs and incorporating renewable energy sources.

The pricing policies implemented by the LAC countries in both the oil and electricity markets are described in chapter 2. Government strategies for setting energy prices are not uniform across the LAC region. Rather, they span a full spectrum, ranging from discretionary price-fixing at one end to pure market-based approaches at the other. In between are a wide variety of other schemes such as price stabilization funds; regulation of commercial margins for refineries and wholesale and retail establishments; price smoothing using the levers of taxation; and direct price subsidies or vouchers that target specific groups. Moreover, price-setting processes may vary by fuel within countries. Governments can also influence final prices through tax policy. Excise and value added taxes, for example, are a significant form of fiscal intervention in the hydrocarbon sector.

Tightly related to the pricing strategy of choice is the magnitude of the distortion between domestic and international prices. These distortions are captured through the price-gap approach, which essentially measures the difference between a theoretical benchmark price and the actual prices paid by consumers. Multiplying the price gap by the total volume consumed yields an estimate of the fiscal cost to governments of subsidizing energy products. The main findings are summarized in chapter 3.

Energy is a key input for production in LAC countries. Although the region is less exposed than its international peers, it could still see its competitiveness affected by price increases resulting from energy pricing changes and subsidy reforms. The economic impacts of changes in pricing policies are examined in chapter 4 of this report.

Policy makers dealing with subsidy reform often cite concerns about its impact on poor and vulnerable households. Subsidy reform advocates counter with statistics that reflect how the biggest share of each dollar spent on fuel subsidies ends up in the pockets of wealthier households. And the reality is that attempts to reform or eliminate subsidies have been met with fierce resistance from citizens, often including the poor. At the heart of this discussion are concerns about the distributional impacts of eliminating subsidies. Chapter 5 presents a nuanced discussion on this topic, highlighting the many challenges that policy makers face.

Finally, chapter 6 presents a set of policy options that policy makers may want to consider for effective energy subsidy reform. Reform efforts are more likely to succeed if they are accompanied by wide-reaching educational and informational campaigns. According to the International Monetary Fund (IMF 2013), proactive public communication triples the likelihood of successful reform. Meanwhile, it is essential that governments and citizens alike recognize the true magnitude of fiscal support for lower oil prices as well as the distributional implications of the pricing policies. It is important as well to depoliticize the energy pricing mechanism. Energy pricing reforms will have a significant effect on the welfare of poor and vulnerable populations, and so implementing compensatory mechanisms can go far in ensuring the success of these reforms.

Note

1. Crude Oil Average Spot Price Forecast—World Bank Commodities Markets Outlook, January 24, 2017.

Reference

IMF (International Monetary Fund). 2013. *Case Studies on Energy Subsidy Reform: Lessons and Implications*. Washington, DC: IMF.

CHAPTER 1

Energy Markets in the Latin America and the Caribbean Region: Setting the Scene

Introduction

Energy pricing mechanisms are frequently used by governments to mitigate the impact of high and volatile oil prices on consumers. Interventions in the energy market have a long history in both high-income and low- and middle-income countries. Nevertheless, these policies come with high fiscal costs and introduce significant price distortions, which are generally associated with economic inefficiencies. Understanding and quantifying the costs and benefits of these policies are an important step toward designing and implementing more effective policy mechanisms.

Governments in the Latin America and the Caribbean (LAC) region intervene in energy markets to varying degrees, with a variety of stated objectives. Energy pricing policies, in particular, have been adopted as instruments to pursue economic, social, and, in some cases, even environmental objectives. Common pricing policies include lowering prices through direct subsidies, increasing prices through taxation, granting tax exemptions, transferring funds to beneficiaries directly, assuming part of the risk of either the consumer or producer of energy, and imposing entry barriers on energy markets. Previous studies provided estimates of the fiscal costs of energy subsidies and, for some countries, estimated the distributional impacts by analyzing household consumption patterns. This study will be the first to provide a comprehensive analysis of the fiscal, economic, and social impacts of energy pricing policies in the LAC region.

The overall objective of this study is to inform the policy debate on energy subsidy reforms by evaluating the costs and benefits of different energy pricing mechanisms. To this end, this study begins by characterizing the sets of energy pricing policies adopted in nine economies in the Latin America and the Caribbean region. The nine countries included in the analysis—Bolivia, Brazil, Colombia, the Dominican Republic, El Salvador, Haiti, Honduras, Mexico, and Peru—represent both oil producers and importers and big and small economies, and so they provide a good overview of the region. Because oil prices are naturally volatile, this study spans six years of analysis (2008–13) to ensure that it measures the medium-term fiscal impacts of energy subsidies and identifies the instances in which subsidy policies are consistent over time or just temporary measures. Within the scope of this study are all fossil fuel products (such as gasoline, diesel, kerosene, and natural gas) and electricity.

What follows is a brief overview of the structure of the energy market for hydrocarbons and electricity in the LAC region.

Structure of the Energy Market: Hydrocarbons and Electricity

The energy market in the LAC region is composed of two distinctive submarkets: hydrocarbon fuels—such as oil, natural gas, and liquefied petroleum gas (LPG)—and electricity. Each submarket is structured very differently. The main segments of the hydrocarbon market are exploration and production, refining, and distribution and retail. Importing is an additional segment for countries without proven reserves (and even for many of those that do). The electricity sector is made up of three main segments or subsectors as well: generation, transmission, and distribution and commercialization.

Most electricity markets in the LAC region are organized into a National Interconnected System (NIS), which covers the majority of regional demand. The rest of demand is met by isolated systems that serve rural and geographically isolated locations. In each LAC country, the NIS is operated by a government agency that dispatches electricity from power plants until demand is fully satisfied. The order of dispatch is based on the variable costs, which range from the least expensive to the most expensive units until total demand is covered. This feature, called economic or marginal cost dispatch, allows the system to operate at minimum cost.

Production Capacity

Hydrocarbons

The capacity of LAC countries to produce energy differs substantially. This is most obvious in the hydrocarbon sector, where some countries have substantially more natural reserves than others. For example, of the nine LAC countries analyzed in this report, Bolivia, Brazil, Colombia, and Mexico have large deposits of oil and gas. At the other end of the spectrum, countries such as the Dominican Republic, El Salvador, Haiti, and Honduras lack proven reserves and therefore depend heavily on fuel imports.

Just because a country produces and possibly even exports oil does not prevent it from importing refined fuel products. Mexico is a classic example. In addition to being a net exporter of crude oil (the 10th largest in the world in 2011), it is also a net importer of fuels (the fourth largest in 2011)—see IEA (2012). Brazil is another oil-producing nation that imports fuels. Because its supply is hampered by limits on its domestic refining capacity, Brazil imports fuels to keep up with the domestic demand for fuel products. Bolivia exports natural gas—mainly to Brazil and Argentina—but imports gasoline and diesel. Figure 1.1 highlights the net trade balance in both crude oil and fuels of the nine countries in the study sample.

Figure 1.1 Crude Oil and Fuels: Net Exports, Selected LAC Countries (Last Year Available)

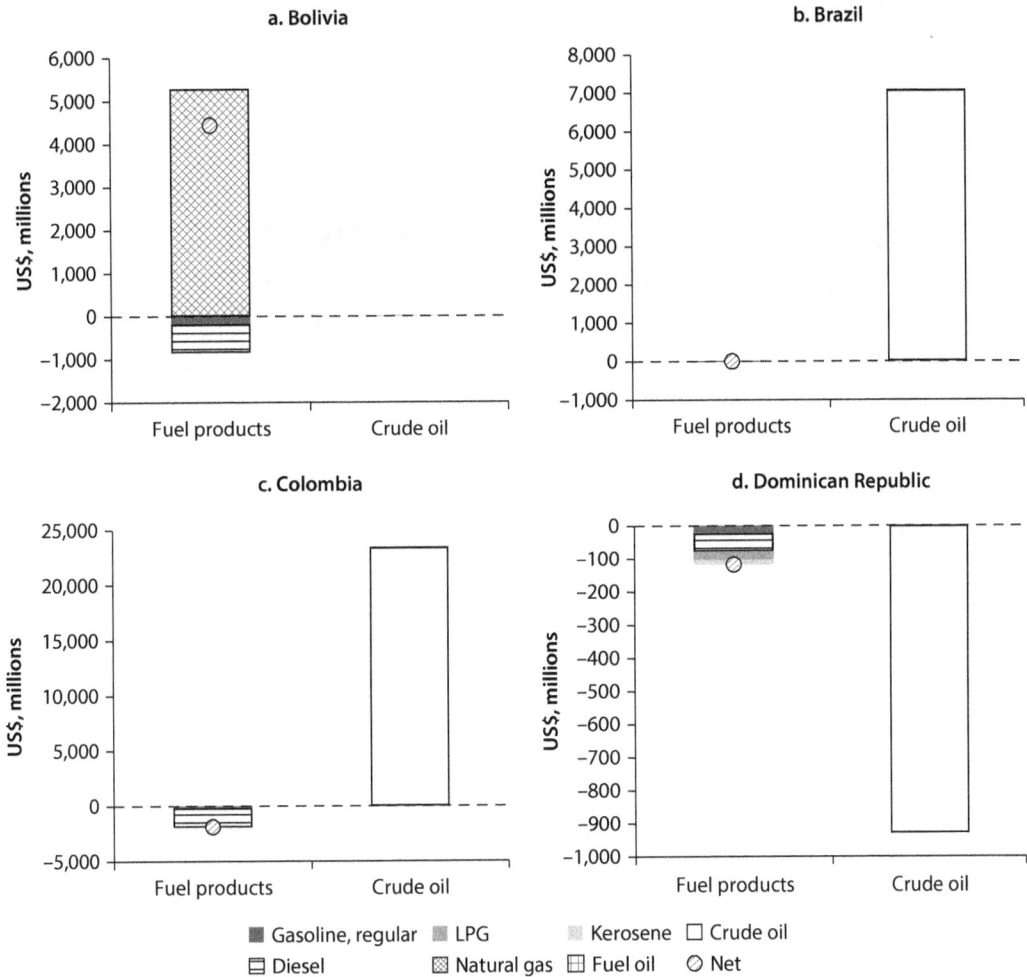

figure continues next page

Figure 1.1 Crude Oil and Fuels: Net Exports, Selected LAC Countries (Last Year Available) *(continued)*

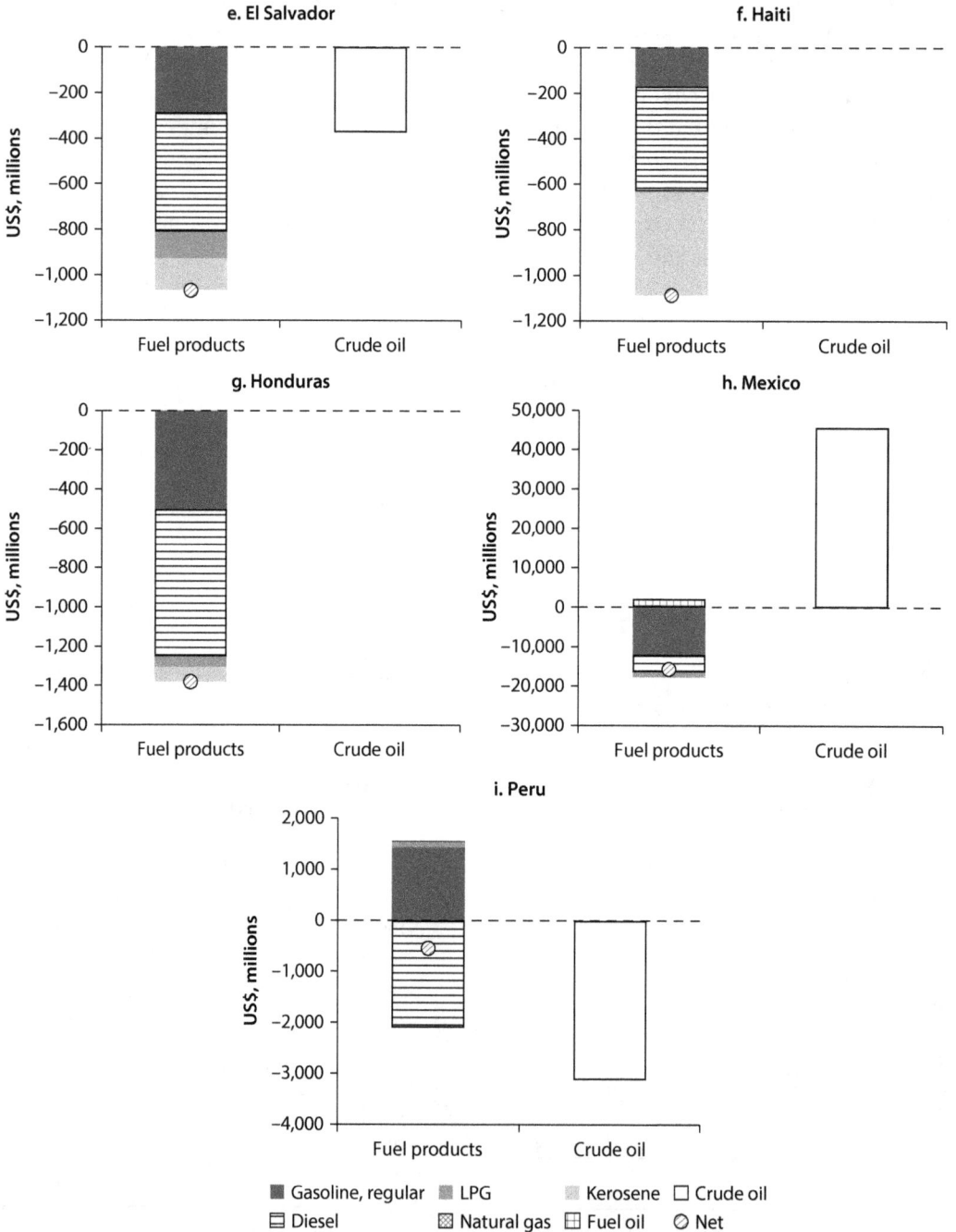

e. El Salvador

f. Haiti

g. Honduras

h. Mexico

i. Peru

Legend: ■ Gasoline, regular ▣ LPG ▨ Kerosene □ Crude oil
☰ Diesel ⊠ Natural gas ⊞ Fuel oil ⊘ Net

Sources: Imports and exports (liters): Central Intelligence Agency, *The World Factbook*; international spot (FOB) prices: International Energy Agency.
Note: Net exports of crude oil and fuels are shown for the following years: Bolivia—2012 for fuel products and 2013 for crude oil; Brazil—2013 for fuel products and 2012 for crude oil; Colombia—2012 for both fuel products and crude oil; Dominican Republic—2013 for fuel products and 2014 for crude oil; El Salvador—2013 for fuel products and 2012 for crude oil; Haiti—2013 for fuel products and 2012 for crude oil; Honduras—2013 for fuel products and 2012 for crude oil; Mexico—2013 for both fuel products and crude oil; Peru—2013 for fuel products and 2012 for crude oil. FOB = freight on board; LPG = liquefied petroleum gas.

Electricity

The electricity generation matrix within the sample of countries can be roughly divided into two groups. The first group comprises the countries in South America (Bolivia, Brazil, Colombia, Mexico, and Peru) that typically depend on natural gas and hydroelectric power plants to generate their electricity. The balance here depends on their natural assets. Bolivia, Colombia, Mexico, and Peru have generous natural gas deposits and therefore orient their electricity generation toward this technology. Brazil has generous hydrological resources, and thus it depends heavily on hydropower. The second group is made up of those countries in Central America and the Caribbean that are not especially blessed with either fossil fuels or hydrological resources. They tend to rely mostly on thermal generation powered by oil derivatives and coal (figure 1.2). A notable exception is El Salvador. In 2013 geothermal and hydroelectric power generation represented over 27 percent and 29 percent, respectively, of this Central American country's installed capacity.

Figure 1.2 Electricity Generation Matrix, Selected LAC Countries
GWh (percent)

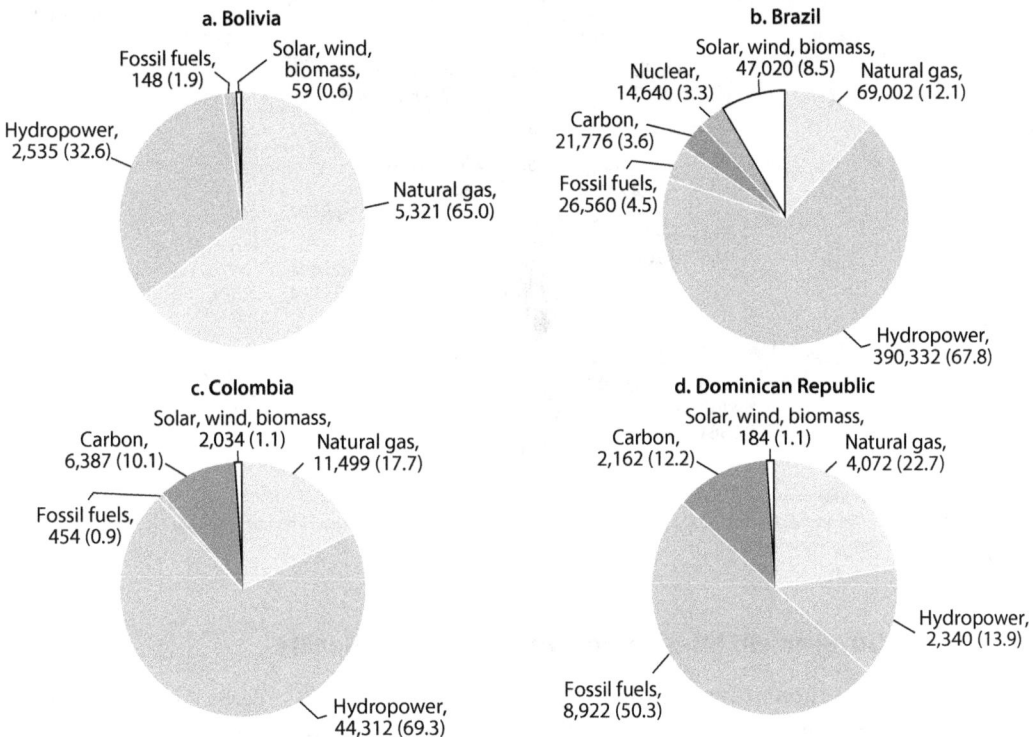

a. Bolivia

Fossil fuels, 148 (1.9)
Solar, wind, biomass, 59 (0.6)
Hydropower, 2,535 (32.6)
Natural gas, 5,321 (65.0)

b. Brazil

Solar, wind, biomass, 47,020 (8.5)
Natural gas, 69,002 (12.1)
Nuclear, 14,640 (3.3)
Carbon, 21,776 (3.6)
Fossil fuels, 26,560 (4.5)
Hydropower, 390,332 (67.8)

c. Colombia

Solar, wind, biomass, 2,034 (1.1)
Carbon, 6,387 (10.1)
Natural gas, 11,499 (17.7)
Fossil fuels, 454 (0.9)
Hydropower, 44,312 (69.3)

d. Dominican Republic

Solar, wind, biomass, 184 (1.1)
Carbon, 2,162 (12.2)
Natural gas, 4,072 (22.7)
Hydropower, 2,340 (13.9)
Fossil fuels, 8,922 (50.3)

figure continues next page

Figure 1.2 Electricity Generation Matrix, Selected LAC Countries *(continued)*
GWh (percent)

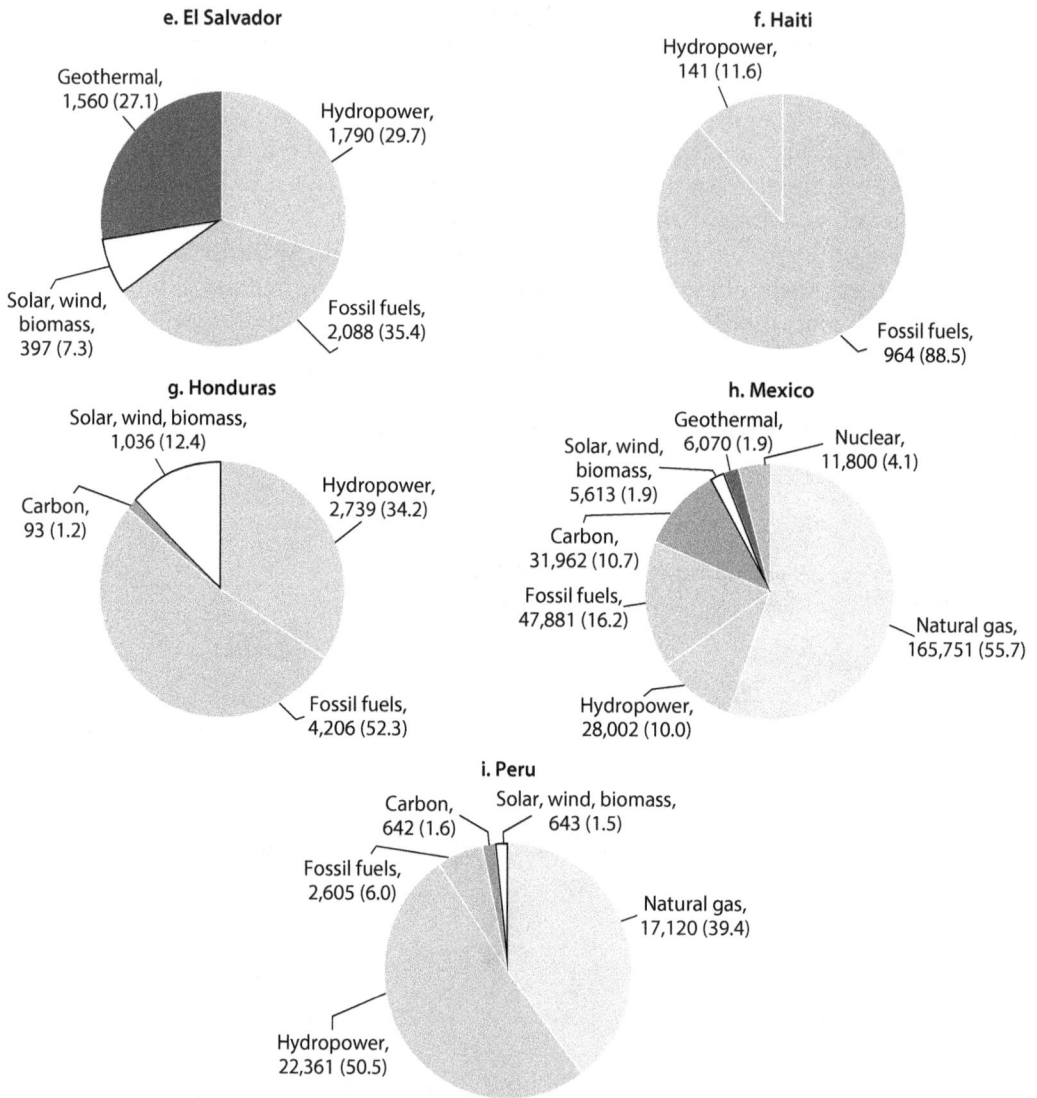

e. El Salvador

Geothermal, 1,560 (27.1)
Hydropower, 1,790 (29.7)
Solar, wind, biomass, 397 (7.3)
Fossil fuels, 2,088 (35.4)

f. Haiti

Hydropower, 141 (11.6)
Fossil fuels, 964 (88.5)

g. Honduras

Solar, wind, biomass, 1,036 (12.4)
Carbon, 93 (1.2)
Hydropower, 2,739 (34.2)
Fossil fuels, 4,206 (52.3)

h. Mexico

Geothermal, 6,070 (1.9)
Nuclear, 11,800 (4.1)
Solar, wind, biomass, 5,613 (1.9)
Carbon, 31,962 (10.7)
Fossil fuels, 47,881 (16.2)
Natural gas, 165,751 (55.7)
Hydropower, 28,002 (10.0)

i. Peru

Carbon, 642 (1.6)
Solar, wind, biomass, 643 (1.5)
Fossil fuels, 2,605 (6.0)
Natural gas, 17,120 (39.4)
Hydropower, 22,361 (50.5)

Source: Inter-American Development Bank (IDB) Energy Database.
Note: The figures show the electricity generation matrix by country in 2013 and the percentage of each source of energy used in the generation of electricity in 2013. GWh = gigawatt-hours.

Government Intervention: An Inescapable Reality

Government intervention is a fact of life in the LAC region's energy markets. State-owned enterprises (SOEs) continue to play an influential role in the production, transportation, and distribution of both hydrocarbon fuels and electricity, both of which are deemed "strategically important" sectors.

Governments primarily give three reasons for involving themselves in the energy sector: First, they maintain that long-term strategic planning is necessary in both the hydrocarbon and electricity markets if they are to function in a robust, sustainable manner over time. Second, the energy sector requires large upfront capital investments. Footing the bill for these investments may be beyond the capacity or the will of private sector investors. For the electricity sector, investments are needed not only in power plants, but also in transmission lines and distribution equipment. Similarly, the infrastructure needed for a well-functioning hydrocarbon sector—from production capacity to import and export facilities—is substantial. And, third, in the energy sector specifically government involvement serves as a defense against the natural vulnerability of the transmission and distribution subsectors to monopolies.

That said, the last two decades or so have seen a reduction in the influence of SOEs as private operators have moved into the market. This followed a wave of liberalization across LAC countries during the 1990s. Today, private companies and foreign investors are an integral part of the region's energy market. Even so, the architecture of the sector is still dominated by the state, and the spaces reserved for private sector involvement are heavily regulated. The nature of these regulatory restrictions varies. In every country, companies are understandably expected to meet mandatory safety and environmental standards. Other regulations can assume a more onerous guise, however. Some countries place restrictions on private sector involvement in energy production, for example, or place heavy controls on imports of oil and derivatives. In others, private sector distributers and retailers are required to operate within commercial margins set by regulators.

Although state control of the domestic energy sector varies from country to country, it is possible to identify some general patterns. In the hydrocarbon sector, for example, most oil-producing countries have opened up exploration to private participation.[1] By contrast, when it comes to refining and importing fuels many LAC countries grant monopolies to SOEs either explicitly or, more often than not, implicitly. Peru, for example, is the only oil-producing country in the study sample not to have a state monopoly governing the refining stage, whereas importing countries tend to allow for private and foreign participation in importing and refining. The only country in the study sample to break with this norm is Haiti. Table 1.1 details the openness of the main stages of the fuels market to private and foreign involvement in the nine countries studied.

In most countries, the state participates in the generation of electricity but has opened this segment to private participation under different types of arrangements. Meanwhile, a mix of state and private actors are present in the transmission and distribution segments of the electricity production chain. In most countries, SOEs are present in these segments, although some have instituted concession contracts with private agents (see table 1.2).

Energy Pricing Policies for Inclusive Growth in Latin America and the Caribbean
http://dx.doi.org/10.1596/978-1-4648-1111-1

Table 1.1 Fuels Market Structure, Selected LAC Countries

Country	Importing	Exploration/production	Refining	Distribution and retail
Bolivia	State monopoly (YPFB)	State (YPFB and Petrobras Bolivia) and private participation	State (YPFB and Petrobras Argentina) and private participation	State (YPFB and Empresa Tarijana de Gas SA) and private participation
Brazil	State (Petrobras) and private participation	State (Petrobras) and private participation	State (Petrobras) and private participation	State (Petrobras) and private participation
Colombia	State monopoly (Ecopetrol)	State (Ecopetrol) and private participation	State (Ecopetrol) and private participation	State (Grupo EPM) and private participation
Dominican Republic	State (Refinería Dominicana de Petróleo) and private participation	n.a.	State monopoly (Refinería Dominicana de Petróleo)	Private participation
El Salvador	Private participation	n.a.	Private participation	Private participation
Haiti	State monopoly	n.a.	n.a.	Private participation
Honduras	Private participation	n.a.	n.a.	Private participation
Mexico	State (PEMEX and Comisión Federal de Electricidad) and private participation	State monopoly (PEMEX)	State monopoly (PEMEX)	State (PEMEX) and private participation
Peru	State (Petroperu SA) and private participation	State (Petrobras) and private participation	State (Petroperu SA) and private participation	State (Petroperu SA) and private participation

Source: World Bank data.
Note: n.a. = not applicable; YPFB = Yacimientos Petrolíferos Fiscales Bolivianos.

Table 1.2 Electricity Market Structure, Selected LAC Countries

Country	Generation	Transmission	Distribution
Bolivia	State (ENDE) and private participation with IPP	State (ENDE) and private participation	State monopoly (Comité Nacional de Despacho de Carga)
Brazil	State (Eletrobras companies) and private participation via concession agreements	State (Eletrobras companies) and private participation via concession agreements	State (Eletrobras companies) and private participation via concession agreements
Colombia	Competitive/IPP State (EPM, ISAGEN) and competitive private participation with IPP	Competitive/IPP State (EPM, ISAGEN) and competitive private participation with IPP	Competitive/IPP State (EPM, ISAGEN) and competitive private participation with IPP
Dominican Republic	State (EGEHID) and private participation with IPP	State participation (ETED)	State participation with regional monopolies (EDEESTE, EDENORTE, EDESUR)
El Salvador	Competitive/IPP State (LaGeo) and competitive private participation with IPP	State monopoly (ETESAL)	Private participation
Haiti	State (EDH) and private participation with IPP	State monopoly (EDH)	State monopoly (EDH)
Honduras	State (ENEE) and private participation with power purchase agreements	State monopoly (ENEE)	State monopoly (ENEE)
Mexico	State (Comisión Federal de Electricidad) and private participation with IPP	State monopoly (Comisión Federal de Electricidad)	State monopoly (Comisión Federal de Electricidad)
Peru	State (Electroperú) and private participation with power purchase agreements	Private participation	State (Electronorte and others) and private participation with power concession agreements

Source: World Bank data.
Note: EDEESTE = Empresa Distribuidora de Electricidad del Este; EDH = Electricité d'Haiti; EGEHID = Empresa de Generación Hidroeléctrica Dominicana; ENDE = Empresa Nacional de Electricidad; ENEE = Empresa Nacional de Energía Eléctrica; EPM = Empresas Públicas de Medellín; ETED = Empresa de Transmisión Eléctrica Dominicana; IPP = independent power producer.

Note

1. In Mexico, the opening of the exploration segment of production was fairly recent. It came about with the important oil reform of 2014.

Reference

IEA (International Energy Agency). 2012. *Key World Energy Statistics 2012*. Paris: IEA.

CHAPTER 2

Energy Pricing Policies

Introduction

Government strategies for setting energy prices are not uniform across the Latin America and the Caribbean (LAC) region. Rather, they span the full spectrum, from discretionary price-fixing at one end to pure market-based approaches at the other. In between is a wide variety of other options, including price stabilization funds; regulation of commercial margins for refineries and wholesale and retail establishments; price smoothing through the levers of taxation; and direct price subsidies that target specific groups. Price-setting processes for individual fuels and electricity may vary within countries as well.

In addition to price-setting strategies, governments can influence final prices through tax policy such as the excise and value added taxes often levied in the hydrocarbon sector. Taxation policy on prices can be indirect or highly targeted. A good example is the tax breaks offered for specific fuel products such as liquefied petroleum gas (LPG) in Bolivia, the Dominican Republic, and Honduras; natural gas in Bolivia; and kerosene in Honduras. Another factor influencing final prices is the role of state-owned enterprises (SOEs) in the market. Where SOEs exercise monopoly control, their ability to influence prices is obviously very high.

Whatever a policy maker's particular tool of choice, one overriding factor governs domestic energy price policies in the LAC region: the international price of oil. Because of the volatility of the global oil price over recent years, governments across the LAC region have had to regularly adapt and refine domestic price-setting strategies, which requires a high level of policy flexibility.

Pricing Policy Toolkit: Hydrocarbons

Parity Prices

The parity pricing system seeks to mimic the outcomes of well-functioning competitive markets. This approach establishes formulas for pricing fuels based on one of two defined prices: the import parity price (IPP) or the export parity price (EPP). The IPP is the price at the border of a fuel that is imported, which includes international transport costs and tariffs. The EPP is the price that a producer can expect to get for its product if exported. This price is equal to the freight on board (FOB) price, minus the cost of moving the product to the border (also known as netback price).

Pricing formulas published by governments typically start with the IPP and then add on transport costs and distribution and retail margins, as well as the applicable custom, excise, and value added (VAT) taxes, to arrive at a final consumer price. In this way, pricing formulas based on the IPP seek to reflect international price fluctuations as a means of establishing compensation for producers and marketers. Pricing formulas based on the IPP or EPP can include subsidies. These can be made explicit by adding to the pricing formula a specific line with a negative number, thereby reducing the final price faced by consumers. Or, alternatively, governments can introduce subsidies implicitly by imposing or negotiating reductions in certain lines of the pricing formula, such as the commercial margins. This approach effectively forces agents in the production chain to absorb the subsidy. What follows are three examples of how LAC governments in the study sample have intervened in regulatory frameworks based on pricing formulas:

- *The Dominican Republic.* The Dominican Republic's pricing system is based on price formulas that start with the IPP and then add regulated margins and taxes. Thus fuel prices generally reflect fluctuations in the international market prices and the exchange rate. After substantial rises in global oil prices, the government temporarily suspended the pricing formula and froze fuel prices for final users. To absorb part of the fuel price increases, the government used a discretionary compensation fund. Neither measure is contemplated in any law nor in the country's regulatory framework. The revenue loss was made up either from the government's profit share in the state-run oil company REFIDOMSA, or from fuel tax revenues.

- *Haiti.* The government of Haiti introduced a pricing formula for fuels in the 1990s. The formula incorporated variations in international fuel prices, customs and excise duties and fees, as well as transport, distribution, and retail margins. The pricing rule established that final retail prices were adjusted only when international reference prices increased or decreased by more than 5 percent. For fluctuations below this figure, the government sought to absorb the changes through modifications to customs and excise taxes. However, the very sharp oil price increase in 2008 saw the government temporarily suspend the pricing formula. In March 2011, it went further and froze the prices of all liquid fuels.[1] More recently, the government reversed this policy.

- *Honduras.* The sharp increase in international oil prices from 2008 to 2013 proved particularly challenging for many governments using IPP-based price formulas. In Honduras, the government faced a public outcry in the wake of sharp hikes in the price of domestic fuels. In response, it began to regularly introduce a specific line in the pricing formula with a direct price subsidy. The net effect was an ipso facto reduction in tax revenue. At times, the subsidy exceeded the amount of tax collected for the same product, notably for kerosene and LPG. Also at times, the Honduran government renegotiated specific items in the pricing formula, thereby forcing importers, wholesalers, and retailers to accept a temporary reduction in their margins. Meanwhile, unilateral modifications of the formula allegedly forced importers to operate at a loss.[2] The government even went so far as to temporarily suspend the formula so that consumers would avoid the effects of sharp price increases.

Tax-Based Systems

Governments across the LAC region use a variety of fiscal strategies to influence fuel prices. In addition to the VAT, some governments levy excise taxes on domestic fuel prices, particularly of gasoline and diesel. Countries such as Bolivia, the Dominican Republic, and Honduras seek to make certain hydrocarbon products more affordable by lowering the sales taxes on, for example, LPG, kerosene, and natural gas, which are used widely for cooking and other domestic purposes. Meanwhile, Mexico uses a highly flexible excise tax on gasoline and diesel to smooth the fluctuations in domestic retail prices:

- *Mexico.* The Mexican government's pricing formula is based on the standard concept of the IPP. The government has to regulate the domestic price for gasoline and diesel because the state-run company PEMEX operates a legal monopoly over refining and distribution. The price received by PEMEX is based on the international oil price reference. Appropriate adjustments are then made to account for ancillary factors such as quality, transport, insurance, retail margins, and other miscellaneous costs. State duties, the VAT, and other taxes are significant considerations here as well. Where Mexico differs from the norm is its incorporation of an excise tax in its price-setting formula. Rather than pass international prices to the final consumers, the government uses the excise tax to buffer such movements based on monthly adjustments to keep prices on a less volatile trajectory. To establish the final retail price, the government sets a target based on the market prices of futures. It then updates the value of the excise tax monthly to achieve its target. However, should international reference prices increase faster than the target price, the excise tax becomes negative, making it a de facto subsidy. But if international prices fall faster than the target, the excise tax becomes positive. Consumers then end up paying higher prices than they would were prices set in a deregulated market. During the 2000s, the sharp increase in international oil prices resulted in the first of these two scenarios. As a result, the government subsidized the price of gasoline and diesel through the five years (2008–13) of this analysis.

Targeted Subsidies

Governments use targeted price subsidies to assist certain segments of the population such as the poor or vulnerable. Such subsidies can also be used to promote specific industries. This type of policy tool generally reduces the fiscal cost of energy subsidies and allows governments to direct their scarce resources toward the most affected or vulnerable populations. Case studies of Colombia and the Dominican Republic reveal how these two countries had different experiences with targeted subsidies:

- *Colombia.* In Colombia, both natural gas and LPG[3] are regulated as public domiciliary services. The government establishes a cross-subsidy price structure that directs subsidies to low-income households. The price formation reflects the efficient costs of gas supply, transport, and distribution. Gas supply prices are a result of the sale contracts between producers and marketers. Because of the natural monopolies that arise in the transport (through pipelines) and distribution of natural gas, the government regulates the prices. It does so by applying methodologies that pay investment as well as operation and maintenance costs.

  As part of Colombia's cross-subsidization price scheme, the government has designed a mechanism for social stratification based on the characteristics of individual households.[4] A household falls into one of six classes, with class 1 representing the lowest-income households and class 6 the highest. Households in classes 1 and 2 are assigned natural gas subsidies: class 1 households, 50 percent of the service cost per cubic meter, up to the quantity defined as the minimum consumption for subsistence (20 cubic meters per month); class 2 households, 40 percent. Classes 3 and 4 pay 100 percent of the cost, and classes 5 and 6 pay a 20 percent cross-contribution to help finance the subsidy. The commercial and industrial sectors pay an additional 8.9 percent contribution.[5] Consumer contributions do not cover all of the costs, so the government finances the difference with funds from the national budget. For LPG, subsidies are also assigned to class 1 and 2 households: subsidies of 50 and 40 percent of the service cost, respectively, up to 14.6 kilograms a month.

- *The Dominican Republic.* Since the 1990s, the government of the Dominican Republic had instituted a universal price subsidy for LPG. The original intention was to address two distinct problems associated with charcoal combustion in households: its impact on deforestation and its effect on people's health. By subsidizing LPG, the government hoped that households would switch from charcoal to this cleaner option. However, the price of subsidized LPG was so attractive relative to other fuel products that consumers began using it for both transport and domestic purposes such as cooking. LPG consumption consequently increased exponentially, as did the cost of the subsidy to the government. Between 2004 and 2008, the Dominican Republic's universal price subsidy for LPG averaged an estimated 0.5 percent of its gross domestic product (GDP) per year.

  At the end of 2008, the government applied a pricing formula for LPG that required most consumers to start paying market prices. It retained two targeted subsidy programs, however: one for low-income households (through the country's social safety program) and one for the public transport sector.

The program for low-income households, BONOGAS, assigns 6 gallons per household per month. The price is set by the program at US$6 per gallon. The program for public transport vehicles allocates them 90 gallons per vehicle per month. Approximately 19,300 vehicles rely on this initiative at a cost to the government of US$85 per vehicle per month. These measures have brought the overall subsidy cost down to 0.12 percent of GDP. They have not only freed up public spending, but also ensured that scarce government resources are now being better directed to the poorest segments of the population. LPG consumption has also dropped by 3 percentage points, to 34 percent of total fuel consumption (excluding liquefied natural gas), because consumers are returning to the use of gasoline and diesel for transport.

Price Stabilization Funds

Price stabilization funds are a means of insulating the domestic economy from volatility by maintaining smooth domestic fuel prices. These funds are generally used when sharp changes in the global oil price produce market volatility. Their format varies, but their basic structure is as follows. The government sets a target price (or a range of prices with a floor and a ceiling). Then, when world oil prices are low, it contributes the difference between the market-based price and the target price (or floor) to the fund (savings or over-recoveries). This contribution (savings) is then used to maintain the price at the targeted level (or ceiling) when international prices are higher than the targeted level or ceiling (dissaving or under-recoveries). As the following examples of Colombia and Peru reveal, LAC governments with price stabilization funds often find themselves having to finance a de facto fuel subsidy from the national budget:

- *Colombia.* In 2007 Colombia created the Fund for Fuel Price Stability (Fondo de Estabilización de los Precios de los Combustibles, FEPC) to reduce the volatility of gasoline and diesel prices and gradually eliminate subsidies. The fund applies only to the prices received by producers. It follows a pricing formula that calculates the price for producers, called producer income (PI). It also takes into account the margins for wholesale and retail distributors, along with transport charges and applicable taxes, as defined by the Ministry of Mines and Energy on a monthly basis. The price for producers is referenced according to the IPP or EPP. Parity prices, initially calculated daily, were set according to the reference of Gulf Coast commodities, with specific local adjustments.[6] But in September 2011, the government introduced an automatic adjustment scheme that defines the PI each month using prices based on parity price tendencies. The updated methodology takes into account the historical market price from the last 60 days, as well as the relationship between the current income of the producer and the parity price on day 60.

  The rules governing the PI seek to ensure that international parity price tendencies are followed, while guaranteeing that abrupt changes in international oil prices are not passed on to local consumers.[7] To keep local prices stable, the monthly price variation of PI is restricted.[8] For each producer and importer, the daily positive and negative differences between PI and parity prices are calculated on a quarterly basis, taking sales volumes into account.[9] When the balance shows a credit in favor of a producer or importer, the FEPC must recognize this value. If not, the agent must pay the FEPC. If for a certain quarterly balance settlement period the FEPC does not have enough resources to cover the balance, the government is authorized to grant the FEPC extraordinary loans that will be repaid in posterior savings periods. If this does not happen, the National General Budget will cover the deficit.[10] The FEPC was launched in 2007 with a credit balance of about US$277 million.[11] By the end of 2010, the balance had reached about US$1.6 million. This decrease stemmed from a series of factors, including an increase in diesel imports (which have a higher parity price compared with the import parity price), rising international fuel prices, and a reduction in global fuel price volatility. Since 2010, the FEPC has not been able to sustain itself. Between January 2010 and June 2013, the estimated diesel subsidy value was about US$1,625 million.

- *Peru.* In Peru, the government instituted a price stabilization fund (also called the FEPC) in 2004. This step followed a failed attempt to stabilize prices by modifying the excise tax rate. Peru has a deficit of crude oil and its by-products. Thus when the international oil price is high, domestic productivity is liable to slow. The effects of oil prices on inflation can also slow the national economy. The idea behind the stabilization fund was to steady prices without affecting the contractual freedom of fuel producers and importers.

  Like Colombia, Peru defines the reference price for producers and importers in relation to the import parity price for gasoline and diesel and the export parity price for LPG. This methodology incorporates a price banding system consisting of an upper and a lower limit. The limits establish the target prices for sale in the domestic market. When the reference price exceeds the upper limit, a compensation factor equivalent to the difference between the reference price and the upper limit is applied. In the opposite scenario, when the reference price drops below the lower limit, a contribution factor equivalent to the difference between the reference price and the lower limit is applied. When the reference price falls within the price band parameters, the price for producers is equal to the parity price, and so no contribution or compensation factor is necessary.

  In theory, Peru's FEPC should have been self-sustaining and should have delivered stable, predictable prices rather than subsidies. However, rising oil prices drained the fund's resources, forcing the government to enact emergency measures and permit the fund to pay the debt owed to producers and importers. That situation in turn forced the government to make regular contributions from the public treasury—US$2,231.3 million between 2006 and 2011. The fund was then essentially abandoned after this period.

Discretionary Pricing

Even though the regulatory regimes in LAC countries have become considerably more consistent and robust over recent decades, discretionary pricing remains relatively commonplace. This is particularly true in the light of sharp changes in the international oil price. The most extreme version of this tendency is price freezing—that is, governments set a fixed price for fuel products and thus totally delink the price-setting process from the international price. Recent years have seen Bolivia, the Dominican Republic, El Salvador, and Haiti pursue such an approach, even though it could cause major price distortions. On occasion, prices have even been set by parliamentary or presidential decree. Discretionary measures are generally motivated by the social and political tensions that arise from very sharp increases in domestic prices. Examples follow:

- *Bolivia and Haiti.* Since 2011, both Bolivia and Haiti have implemented discretionary price-freezing policies for most hydrocarbon products. These policies have created large distortions and have had large fiscal implications. In Bolivia, as a natural gas exporter and producer of fuels, most of the subsidy is absorbed through the SOE—YPFB (Yacimientos Petrolíferos Fiscales Bolivianos)—as foregone revenue, while in Haiti all of the subsidy is explicitly recorded as outlays of the state-controlled import company or the federal government.

- *Brazil.* Brazil is proof of how discretionary actions by government can occur even in de jure deregulated markets. Over the last 20 years, Brazil has liberalized and deregulated the price formation process for fuels. In 1997 the Petroleum Law[12] established a 36-month time frame for the conclusion of the price liberalization of all automotive fuels. After an extension of the deadline,[13] liberalization of the Brazilian fuel market ended on December 31, 2001, with the release of fuel prices at refineries that were still operating under price controls. Thereafter, the market prices of automotive fuels—in theory—became deregulated. Even so, the government continues to interfere with the price-setting process on a discretionary basis.

  The deregulation of fuel prices in Brazil is ineffective because state-owned Petrobras has an almost complete monopoly in the refining of most fuels. The federal government therefore exercises effective control over the sales prices of fuels produced and sold by Petrobras.[14] At times, the Brazilian government has decided to not increase fuel prices in spite of rising international prices, thereby effectively creating a subsidy financed by Petrobras stockholders. These stockholders are not only the federal government but also the general public and private investors. The costs of the government's interventions must also be covered by the sugarcane and ethanol industries because the price of hydrous ethanol is always bounded in 70 percent of the gasoline C price because a technical efficiency relationship exists between the two fuels.

- *Mexico.* In Mexico, refining is controlled by state-owned PEMEX. Since the 2000s, however, the downstream activities in LPG distribution have been open to private sector participation. The sales price of LPG sold by PEMEX takes the IPP as its reference. Despite liberalization of the nonrefining elements of the supply chain, consumer prices are still set on a monthly basis by presidential decree. To calculate the reference price received by PEMEX, this discretionary pricing formula starts with the final price and then subtracts the private distributors' margins. The result is a substantial misalignment with international prices.

Pricing Policy Toolkit: Electricity

In general, three components determine the final price of electricity: the energy generation costs, the transmission tolls, and the distributor's remuneration. Most tariffs are composed of a fixed charge, which covers remuneration for distributors and transmission network operators, and a variable charge, which depends on the amount of energy consumed and the costs of generation. Price intervention by governments has focused on this last component, thereby avoiding passing on hikes in generation costs to consumers. Large increases in oil prices can have significant impacts on generation costs, especially for countries that rely heavily on fossil fuel–based thermal generation.

One of the key responsibilities of national governments is to define a regulatory framework and price formation process that fairly remunerate all agents in the chain. It is also important that the final price incentivizes investment as well as promotes operational efficiency and quality compliance. The analysis here is split into two segments: upstream price formation, which refers to how prices are defined for generators, transmission network operators, and distribution companies, and downstream price formation, which refers to government pricing policies directed at consumers.

Upstream Price Formation
Generation
The primary calculation for governments when setting the price of electric power is the cost of generation. It is at this stage that governments typically intervene most. One advantage of doing so is the ability to protect consumers from excessive price increases when production costs rise sharply. Such increases typically occur when oil prices rise because the majority of power plants in LAC countries rely heavily on fossil fuel–based thermal generation.

The prices for electricity generators are defined in the wholesale market and retail market. The wholesale market is reserved for power generation plants and distribution companies overseen by regulators. The retail market serves unregulated users, with the participating agents at liberty to negotiate prices freely (box 2.1). The separation of the electricity generation market into retail and wholesale occurred in the 1990s, when many LAC countries undertook reforms of their power sectors. The reforms sought to introduce more competition at the generation stage by allowing independent power producers to compete evenly with integrated utilities in the provision of bulk energy. The emergence of a retail

---

Box 2.1 Brazil: Unregulated Consumers in the Electricity Market

Brazil is a positive example of where the introduction of a retail electricity market has led to relatively vibrant competition. The Brazilian government defines an unregulated consumer as a large company or institution that has a peak load of over 3 megawatts (or 500 kilowatts if the energy is from renewable sources). Many of those who qualify have exercised their option to be served by an alternative supplier. The retail segment of the market currently represents about 30 percent of the total energy consumed in Brazil. This figure could grow to 40 percent in the short term should the 3 megawatt threshold be reduced, as is being discussed.

At present, Brazil has about 70 independent marketers. They offer different types of products and services, depending on customer needs. The regulated tariff, which is normally taken into account in studies of industrial competitiveness, does not apply to the customers of these independent traders. Instead, "free" customers are able to obtain energy at more convenient prices and conditions. The exact terms and conditions in the retail market are not known because bilateral contracts are confidential. However, some estimates indicate that free customers benefit from energy that is 15–20 percent cheaper than that available to regulated consumers.

---

market allowed the introduction of so-called "unregulated" or "free" consumers. Unregulated consumers have the option of purchasing their own power through negotiated bilateral contracts with power plants or independent marketers (or traders). It is incumbent on distribution utilities to allow independent power producers or self-generating "free" consumers to use their existing grid services. All other consumers are regulated and have to abide by the tariff-setting procedures defined by regulators.

Transmission and Distribution

The power generation phase of the production cycle is followed by the transmission and distribution phase. The remuneration price for companies involved at this stage should, as a general rule, cover the company's operating costs, provide the company with a return on investment, give the company a margin large enough to allow it to invest in expanding the transmission network, and encourage the company to adopt operational efficiencies. These principles apply to both private companies and state-run utilities.

Most countries have instituted transmission and distribution fees (or tolls). They have two components: (1) a *fixed charge*, which should cover a healthy return to investments, as well as provide a margin for future investments and all other fixed costs related to the operation and maintenance of the network; and (2) a *variable component*, which should cover variable operation and maintenance costs, as well as transmission efficiencies, distribution losses, and the costs of energy purchases from generators.

The price paid by distributors to electricity generators has major implications for downstream prices. The upstream price is the result of two distinct

---

Box 2.2 Procuring Long-Term Electricity Contracts by Auction: A FAQ

What is an electricity auction? Simply defined, an auction is a selection process designed to procure (or allocate) goods and services competitively, often based solely on the lowest price offered for a set of qualified bidders.

How common are auctions? Dozens of auctions have been carried out to date across the Latin America and the Caribbean region, with noteworthy results particularly in terms of procuring new generating capacity.

What goods and services do the auctions cover? Auctions occur both "in the market" and "for the market." Some are for a specific technology (such as wind) or for a specific project (such as wind in location A). Occasionally, they are for all technologies (wind, solar, thermal, and so on) head to head.

What recent examples are there? Over the last few years, Brazil, Chile, Mexico, Panama, Peru, and Uruguay have conducted specific auctions for renewable energy capacity such as wind, solar, biomass, and small hydro. Argentina and Colombia are expected to follow suit presently.

How competitive are renewable resources? Large volumes of renewables have already been deployed, at prices reaching US$.04–$.05 per kilowatt-hour, making this form of energy extremely competitive even when compared with large-scale conventional energy. If these prices are locked into long-term contracts, then future prices for bulk generation should exhibit a declining pattern.

Source: Adapted from Maurer and Barroso (2011).

---

mechanisms: (1) the practice of long-term, freely negotiated contracts among distributors and generation plants and (2) the spot market, where prices are defined by auctions at regular intervals. Latin America has been leading the effort to introduce electricity auctions as a transparent instrument to promote competition in the procurement of long-term electricity contracts, including those for renewable sources of energy (see box 2.2).

Downstream Price Formation

Although electricity is a relatively homogeneous good, consumers pay a variety of prices rather than one universal price. This final price depends on a variety of factors.

One factor is the division of consumers into different categories, with a different price attached to each different category. The most common user categories in the LAC countries are residential, commercial, industrial, public lighting, agriculture, and government. Voltage level is another differentiating factor, with prices alternating according to whether the voltage received is high, medium, or low. At times, consumers are placed in different pricing categories according to their total level of consumption. A less common approach is to alter tariffs according to the season or to differentiate prices according to geographical location. Industrial consumers generally face tariffs that vary according to power levels and time of usage (peak versus nonpeak).

The most common consumer-based pricing mechanism in the LAC region is the *increasing block tariff (IBT)*. The IBT is also known as a progressive tariff, tiered-rate, or inverted block rate. The basic concept underpinning the IBT structure is that consumers who use low volumes of electricity pay lower rates, whereas high-volume users face higher prices. In this respect, IBTs generally have a cross-subsidy framework. The system works by defining a unit price on a per block basis. The precise volume of electricity (measured in kilowatt-hours) varies per block, but a set limit is established for each block. The first block, which is known as the base-line or lifeline, corresponds to essential or subsistence levels of consumption and has the lowest price. If consumers consume in excess of that limit, they will pay a higher unit price only on the excess kilowatt-hours consumed (see box 2.3 for an example). Each successive block has a higher price per kilowatt-hour.

Box 2.3 Increasing Block Tariffs, Volume Differentiated Tariffs, and Flat Pricing

The functioning of the different pricing mechanisms can be illustrated by a numerical example with fictitious prices and tariff block definitions.

Three customers have different levels of consumption. What is the average price per kilowatt-hour consumed as well as the total bill the final consumer would pay under the three different scenarios?

Consider two tariff schedules with identical block definitions: one increasing block tariff (IBT), one volume differentiated tariff (VDT). The first block is for consumption of less than 100 kilowatt-hours per month; the second block covers consumption from 101 to 300 kilowatt-hours; and the third block is for those consumers using more than 300 kilowatt-hours of electricity per month. For the sake of simplicity, assume that the prices per block for the IBT and VDT sched-ules are identical; 10 cents per kilowatt-hour in the first block, 15 cents for the second block, and 20 cents for the last block. The flat pricing schedule is set at 15 cents per kilowatt-hour.

Now, take three customers, A, B, and C, with different levels of consumption: 75, 150, and 400 kilowatt-hours, respectively. Table B2.3.1 shows the average per unit cost paid and their total bill under the three tariff schedules.

Table B2.3.1 Differences in Three Tariff Schedules, by Customer

Customer	Consumption (kWh/ month)	Flat pricing		IBT		VDT	
		Average price	Total bill	Average price	Total bill	Average price	Total bill
A	75	15 cents	$11.25 = 0.15 * 75	10 cents	$7.50 = 0.10 * 75	10 cents	$7.50 = 0.10 * 75
B	150	15 cents	$22.50 = 0.15 * 150	11.67 cents	$17.50 = 100 * 0.10 + 50 * 0.15	15 cents	$22.50 = 150 * 0.15
C	400	15 cents	$60.00 = 0.15 * 400	15 cents	$60.00 = 100 * 0.10 + 200 * 0.15 + 100 * 0.20	20 cents	$80.00 = 400 * 0.20

Note: IBT = increasing block tariff; kWh = kilowatt-hour; VDT = volume differentiated tariff.

The *volume differentiated tariff (VDT)* is based on consumers paying different prices per kilowatt-hour according to their total consumption. Consumption levels are divided into different blocks, with the per-unit price changing for each respective block. The higher the consumption block, the higher is the unit price for all kilowatt-hours consumed.

Some governments provide *targeted subsidies*—that is, explicit subsidies for specific sectors or user groups. These go beyond tax exemptions. An illustrative example is that of Bolivia, where the government provides a "Dignity Tariff"[15] (set at 25 percent of the electricity bill) for consumers using less than 70 kilowatt-hours per month. Meanwhile, consumers over 60 years of age receive a 20 percent discount for consumption levels below 100 kilowatt-hours per month.

The use of *flat pricing* is not widespread in the LAC region. Of countries in the study sample, only Brazil and Colombia use the approach. Under a flat price formula, all consumers pay the same unit price regardless of their total consumption.

Tariff Schedules in Practice

Consumers in the residential and commercial sectors generally have tariff schedules that depend on their consumption level (see table 2.1). The exceptions are Brazil and Colombia, which, as noted, have flat kilowatt-hour pricing strategies.

Increasing Block Tariffs

Most countries in the LAC region have implemented IBT schedules with different block definitions and levels of discount (relative to efficient costs) per block. In Bolivia and Honduras, the governments provide vulnerable consumers with additional subsidies. The governments of El Salvador and Haiti avoided or delayed the application of higher tariffs because generation costs increased between 2008 and 2013.

IBT with Cross-Subsidization

In Honduras, the government has implemented an IBT structure with an element of cross-subsidization. The government has defined four blocks and set the prices for consumers as a percentage of the service costs. The tariff structure, however, is quite generous in that only households with a consumption level of more than 1,450 kilowatt-hours per month pay the actual cost per kilowatt-hour. In addition to the highly subsidized tariff structure, some consumers receive direct subsidies. Until April 2010, the government contributed 80 lempiras per month toward the electricity bill of households that consumed less than 100 kilowatt-hours per month.[16] And between 2006 and April 2010, the government also provided between 53 and 105 lempiras per month for customers with consumption of between 150 and 300 kilowatt-hours, and between 72 and 120 lempiras per month for those consuming 300–500 kilowatt-hours per month. After May 2010, all subsidies were unified. This reform resulted in the government covering all the electricity expenses for households consuming less

Table 2.1 Electricity Tariff Schedules, Selected LAC Countries

Country	Consumer categories	Tariff components
Bolivia	Residential, general, mining, industrial, public lighting, others	Consumer charge, maximum power charge, power charge, energy charge
Brazil	Groups A (>2.3 kV) and B (<2.3 kV): commercial, self-consumption, public lighting, industrial, public, residential, rural, and public service	– Group A: binomial rate: energy consumed + max demand power – Group B: binomial rate: energy consumed
Colombia	Industrial, commercial, and residential low demand (regulated consumers)	Cost of energy, cost of usage, cost of distribution, commercialization margins, transmission constraints, cost of purchase and transport
Dominican Republic	BTS-1 (residential low voltage) and BTS-2 (nonresidential low voltage)	– Energy charge (distribution companies): BTS-1: <200 kWh, >200 and <300 kWh, >300 and <700 kWh, >700 kWh BTS-2: <200 kWh, >200 and <300 kWh, >300 and <700 kWh, >700 and <1,000 kWh, >1,000 kWh – Power charge: installed capacity
El Salvador	– Small users (<10 kW): residential, general use, street lighting – Medium users (>10 kW and <50 kW): low and medium voltage – Large users (>50 kW): low and medium voltage with hour meter	– Fixed charge (two parts): (1) small and medium; (2) large demands – Distribution charge (two parts): (1) small and medium; (2) medium demands
Haiti	Residential, commercial, public organizations	– Fixed charge: same for all categories – Capacity charge: for industrial users – Energy charge (by consumption level): <30 kWh, >30 kWh and <200 kWh, >200 kWh
Honduras	Residential, residential with peak demand, high voltage, public service, water pumping, industrial	Fixed price and peak demand charge (for demands >250 kWh)
Mexico	Residential, public services, agricultural, commercial; low (commercial), medium (industrial), and high voltage (industrial)	Varies according to voltage level within each consumer category
Peru	– Free users (determined by generation price) – Regulated users: rural and urban areas with high/low density	– Busbar price: power payments + energy payments – Transmission toll – VAD (distribution value added): changes based on rural or urban area with high/low density

Source: World Bank data.

Note: kV = kilovolt; kW = kilowatt; kWh = kilowatt-hour.

than 150 kilowatt-hours per month. The most recent modification, in March 2014, instituted a contribution of 120 lempiras per month toward the electricity bill for customers who consumed less than 75 kilowatt-hours per month and who were categorized as low income.

In El Salvador, the government designed IBT schedules with cross-subsidies. This tariff arrangement has an adjustment formula that incorporates the cost of fuels used for generation. However, because of the large increase in generation costs, the government temporarily suspended the pricing formula in 2011.

The original tariff instituted a price subsidy for consumers in the first block (<100 kilowatt-hours), setting the price at 10.5 percent of the indexed tariff, plus US$.0671 per kilowatt-hour. Since 2011, the government has fixed the tariff for consumers in the range of 100–200 kilowatt-hours per month to the tariff prevailing in 2011. In this way, it avoided the higher generation costs in the years that followed.

Haiti has a tariff structure in which residential customers with consumption levels of less than 200 kilowatt-hours per month are heavily subsidized; all other end-user consumers pay above efficient cost prices. As in El Salvador, because of the increase in fuel prices during the period of this study, the tariff structure did not cover supply costs, let alone energy losses. In 2009 the Haitian government doubled tariffs for all consumers except those with consumption of less than 200 kilowatt-hours

IBT with Volume Differentiated Tariffs

The Dominican Republic and Mexico have tariff schedules that combine the IBT with the VDT for very high levels of consumption. As noted earlier, a VDT scheme is one in which the unit price increases with the level of consumption for *all* units consumed.[17]

The IBT/VDT adopted in 2001 by the Dominican Republic initially started from a negotiated base. It was indexed by the price of Fuel Oil No. 2, the consumer price index of the Dominican Republic and the United States, as well as the exchange rate. However, the government has impeded application of the automatic adjustments implied by the formula. Thus the applied tariff significantly subsidizes the price of electricity for consumers at all levels and has diverged significantly from the indexed tariff.[18] The current formula considers consumer type along with voltage level.

The tariff is designed to facilitate cross-subsidies from high-tranche residential and commercial consumers, as well as from industrial users. In this way, the cost of consumption by low-use households (defined as those using less than 300 kilowatt-hours per month) is supported. The tariff schedule has four blocks, the first three of which (representing consumption of 700 kilowatt-hours per month and less) operate under the IBT structure. As a result, the lowest-consuming block pays the cheapest price, with a higher price set for each successive block. The fourth block requires high-use residential and commercial customers in this block to pay a higher rate surcharge for consumption levels above their 700 kilowatt-hour monthly threshold. The surcharge was set at US$.27 per kilowatt-hour in 2013. The system has recently been impeded by government changes to the adjustment criteria.

Mexico has a complex system of residential tariffs that vary according to geographical location and season. Different tariff schedules (1, 1A, 1B, 1C, 1D, 1E, 1F) apply to households based on the average temperature in the municipality of residence. Each tariff schedule is divided into blocks (each tariff has different thresholds), with different prices per block. In addition, the tariff is different during the summer months. The VDT element applies to the highest-consuming block. All but this last block pay electricity according to the IBT structure. Those in the higher-consuming block are charged an above-production price for all the electricity they consume. Despite the inflated per kilowatt-hour price for this last block, the cross-subsidy system still does not fully cover the cost of consumption by blocks 1, 2, and 3. The shortfall is made up by the SOE that controls distribution.

IBT with Subsidized Inputs

Even if a country has an IBT-based formula that achieves self-financing through cross-subsidy mechanisms across consumers, policy makers may still choose to subsidize the inputs for power generation beyond mere tax exemptions. For example, both Peru and Bolivia have elected to provide natural gas for electricity generation at prices below the international market prices. Prompting this move is the fact that both countries have considerable domestic gas reserves.

With prices set at below-market levels, the marginal cost for generating companies in Peru is significantly reduced. As a consequence, it can offer electricity at lower prices.[19] In light of the extraordinary peaks in the marginal cost in 2008, the government changed the calculation of generation prices from actual marginal costs to idealized marginal costs. Under the revised methodology, the spot price is calculated as though there are no restrictions on gas pipelines and transmission lines are not congested.[20]

Like Peru, Bolivia set the natural gas price for use in power generation below the international market rates, thereby reducing the cost of generation and enabling lower electricity prices for consumers. In addition, the Bolivian government implements the Dignity Tariff. The Bolivian government has also funded a program that replaces old light bulbs with low-consumption fluorescent versions at no cost to consumers.[21] The first stage of the program ran from March 2008 to April 2009. According to the Ministry of Hydrocarbons and Energy, the initiative produced savings of 72 megawatts, or US$25 million. The project represented an investment of US$11.6 million.[22] The second stage of the program, implemented at the end of 2011, provided 10 million free light bulbs. The US$11 million initiative produced savings of 89 megawatts, or US$65 million.[23]

Flat Pricing

In general, a flat pricing strategy is better aligned with the actual cost of producing and delivering each unit of electricity. It does not exclude the possibility of subsidizing certain consumers, as exemplified by the case of Brazil.

The overall pricing framework in Brazil is characterized by a flat pricing scheme for all low-tension consumers. Built into the system, however, is a subsidy for rural areas (set at a 10 percent reduction in the final tariff) and for public services (set at a 15 percent discount). Meanwhile, low-income residential consumers enrolled in the Social Programs Registry of the federal government benefit from a pro-poor IBT mechanism. This provides them with a 65 percent discount for the first 30 kilowatt-hours of electricity consumed. The reduction falls to 40 percent for consumption of between 31 and 100 kilowatt-hours, and then to 10 percent for the 101–200 kilowatt-hour range. The Brazilian government was obliged to intervene in 2012 and 2013 to avoid passing along a spike in prices to consumers. The spike resulted from a combination of high global oil prices and domestic drought conditions. The government slowed down the rate increases and made contributions through the treasury and loans. As a consequence, adjustments came in less robust doses over a longer than usual period.

Notes

1. Gasoline prices were fixed at 200 gourdes, diesel at 162 gourdes, and kerosene at 161 gourdes.
2. Executive order PCM-02-2007, January 13, 2007, which lasted to December 2009.
3. The subsidy for LPG was established as a pilot program in 2013. In 2014 the program was formalized, and the subsidy conditions and regional coverage were delegated to the Ministry of Mines and Energy.
4. Variables considered include the area of the house, the presence of a pedestrian walk, the size of the front yard, whether there is parking and what type, the material used for the roof, and the front of the house.
5. As of 2013, the contribution for the industrial sector was eliminated.
6. Adjustments are then made in light of three main factors: (1) the local production or supply quality (when it is regarding export parity); (2) freight and land and sea transport insurance; (3) and local transport to domestic supply reference centers (for example, Cartagena or Barrancabermeja in the center of the country), depending on whether it is an import or export.
7. The following rules govern the changes in producer income: (1) *raising the producer income:* if there is an upward trend on parity prices and the producer income is below the parity price of day 60 of the historical series; (2) *reducing the producer income:* if there is a downward trend on parity prices and the producer income is above the parity price of day 60 of the historical series; (3) *maintaining the producer income without modifying it:* if there is an upward trend on parity prices and the producer income is above the parity price of day 60 of the historical series; if there is a downward trend on parity prices and the producer income is below the parity price of day 60 of the historical series.
8. The maximum change rate for gasoline is ±3 percent and for diesel ±2.8 percent.
9. Decree 4839 (2008) regulated the Fuel Price Stabilization Fund.
10. Decree 1067 (June 2014) establishes that the resources the FEPC needs to function will come from (1) returns of the fund's resources; (2) resources from the extraordinary loans of the National Treasury; and (3) resources from the General National Budget earmarked for this purpose.
11. The source of these funds was Ecopetrol savings in the Oil Savings and Stabilization Fund (FAEP), which was created in 1995 to achieve fiscal savings and macroeconomic stabilization. FAEP is a management system of foreign accounts without a legal personality.
12. Law 9,478 (August 6, 1997).
13. Law 9,990 (July 21, 2000).

14. The federal government controls the governance system of Petrobras, as well as its assets. The minister of finance, for example, is the president of Petrobras's management board. All other board members are appointed by the federal government.

15. The Tarifa Dignidad was implemented on March 21, 2006.

16. This program was called Bono 80. At times, the direct transfer amount varied between 80 and 100 lempiras per month, depending on government directives.

17. As opposed to IBT pricing in which only the units consumed within a block are given a different (higher) price.

18. Even with the applied tariff, high-tranche consumers pay electricity prices below those implied by the indexed tariff. The government provides a direct subsidy for low-income households registered in the federal government assistance program, BONOLUZ. This program subsidizes the total value of the first 100 kilowatt-hours consumed, with beneficiaries paying the (subsidized) tariff for any consumption beyond the threshold. The government picks up the tab for these consumers but remunerates the distribution companies at the applied tariff rate, not the indexed rate.

19. This was achieved by the Natural Gas Industry Development Promotion Act, Law 27133 (1999).

20. The modification was motivated by three main factors: (1) congestion of the natural gas transportation pipeline, which was unable to supply the total natural gas demanded by power plants; (2) congestion of the National Interconnected System transmission lines; and (3) production restrictions in some power plants.

21. Programa de Eficiencia Energetica.

22. http://www.hidrocarburosbolivia.com/noticias-archivadas/156-contenidogeneral -archivado/contenidogeneral-01-01-2008-01-07-2008/3984-focos-ahorradores -permitieron-al-estado-economizar-us-25-millones-en-inversi.html.

23. http://www.hidrocarburosbolivia.com/noticias-archivadas/387-energia-archivado /energia-01-01-2012-01-07-2012/49139-focos-ahorradores-generaron-un-ahorro-de -8904-megavatios-en-2011.html?start=16.

Reference

Maurer, Luiz T. A., and Luiz A. Barroso. 2011. *Electricity Auctions: An Overview of Efficient Practices*. Washington, DC: World Bank. http://documents.worldbank.org/curated /en/114141468265789259/Electricity-auctions-an-overview-of-efficient-practices.

CHAPTER 3

Fiscal Cost of Energy Pricing Policies

Introduction

As described in chapter 1, countries in the Latin America and the Caribbean (LAC) region have adopted a range of measures to keep energy prices low for final consumers. Any interventions that result in a difference in the domestic price of energy paid by final consumers from the actual market price of that energy must, by definition, incur a cost. This fiscal burden is generally picked up by the government, but the expenditure is not always immediately obvious, nor is it necessarily recorded in the budget.

Fiscal costs can be explicit, such as government subsidies, or implicit, such as stabilization funds. Sometimes these costs are very high, especially during sustained periods of high international oil prices. This was the experience of many LAC governments during our study. Fiscal costs are not just influenced by external factors, however. Our research finds that the type of price formation process chosen by national governments can have a profound effect as well.

Government interventions in energy pricing take on different forms in different markets. For fuels, subsidies for gasoline and diesel are generally delivered through interventions in the price-setting process. By contrast, subsidies for liquefied petroleum gas (LPG) and kerosene are typically delivered through a combination of government intervention in the price formation strategy and differentiation in taxation rates. Most of the subsidies for electricity are directed to the residential sector. These typically carry a much lower fiscal cost than interventions in the fuel sector. That said, a unique problem that hits the revenues of electricity utilities hard is technical and nontechnical losses. Such losses can destabilize utility finances and effectively curtail improvements in investment infrastructure.

Measuring Fiscal Costs: The "Price-Gap" Approach

To measure the fiscal burden, we follow the price-gap approach (Koplow 2009). On a conceptual level, the approach is relatively straightforward. It is based on taking the theoretical benchmark price and the actual price paid by consumers and then calculating the difference between the two. That difference is what we call the price gap. The essential formula governing the approach is

$$\text{price gap} = \text{benchmark price} - \text{consumer (producer) price}.$$

The price-gap methodology is popular among international agencies (IEA 2010; IMF 2013), not least because of its relative simplicity in quantifying price distortions. Among its other advantages are its ability to capture expenditures that are not explicitly recorded in government budgets, as well as its requirement for relatively few data points. On a more pragmatic level, the approach is ideally suited to cross-country analysis, and thus it fits neatly in the sample-based framework of this study. The price-gap method also provides a basic metric for tracking comparable price distortions across products and time, as well as across countries.

The approach is not without its challenges, however. Its primary deficiency is that it fails to account for interventions and transfers that do not affect final prices.[1] According to the International Energy Agency (IEA), "the price-gap approach establishes a lower bound for the impacts of [subsidies] on economic efficiency and trade" (IEA 1999). For example, large transfers or tax expenditures that distress government budgets but do not affect market clearing prices[2] would be missed by the price-gap approach. Also missed would be the full effects of subsidies on marginal investment decisions. A further weakness of the price-gap methodology is its failure to capture how subsidies can create competitive barriers to new technologies.[3]

Challenges to Establishing the Price Gap

In establishing the price gap, the benchmark price is the first reference point needed. In the fuels market, this price differs, depending on whether a country is a net importer or a net exporter of hydrocarbon products. For a net importer, the benchmark price is defined as the one at which the same hydrocarbon product can be brought in from the border. For a net exporter, the benchmark refers to the netback price (equivalent to the final sales price minus transportation costs) that producers would obtain from delivering the product to the border. The final price to consumers is a matter of record. The difference between the final price and the benchmark price is the price gap. To arrive at the fiscal burden to the government for a specific fuel product, we multiply the price gap by the total consumption of that product.

The price-gap approach requires a number of preliminary calculations. The first of these is identifying the reference price of products such as crude oil, gasoline, or diesel that are internationally traded. We do this by pinpointing the relevant international price, adjusted for quality differences. For the study sample of countries, with few exceptions, the international reference prices are those of the U.S. Gulf Coast for gasoline, diesel, and kerosene; Mont Belvieu for LPG (propane); and Henry Hub for natural gas (see table 3.1). The case of natural gas in Bolivia is an exception because Bolivia is a landlocked country. Here, the weighted average export price (to Argentina and Brazil) is the most appropriate reference price because of the opportunity cost that producers forgo for every unit they do not export.

An additional calculation is the estimated costs of transporting and distributing fossil fuels. Inevitably, some assumptions and simplifications are needed here. When the objective is to quantify all price distortions regardless of their nature, the costs of transportation and distribution[4] should assume perfect market competition with an efficient transportation infrastructure. Several approaches have been used in the literature to proxy for these efficient or idealized costs.[5] In this study, we focus only on price distortions resulting from explicit government pricing policies and not those arising from imperfect competition or inefficiencies. As such, we construct the benchmark price using the actual transportation and distribution cost structure currently in place. In other words, we compare the final consumer prices with a benchmark price that assumes the current infrastructure and market conditions.

Table 3.1 Reference Fuel Prices, Selected LAC Countries

Country	Gasoline	Diesel	Kerosene	LPG	Natural gas
Bolivia	U.S. Gulf Coast Conventional Gasoline	U.S. Gulf Coast Ultra-Low Sulfur No. 2 Diesel	n.a.	Mont Belvieu Propane	Export price[a]
Brazil	U.S. Gulf Coast Conventional Gasoline	U.S. Gulf Coast Ultra-Low Sulfur No. 2 Diesel	U.S. Gulf Coast Kerosene-Type Jet Fuel	Mont Belvieu Propane	n.a.
Colombia	U.S. Gulf Coast Conventional Gasoline	U.S. Gulf Coast Ultra-Low Sulfur No. 2 Diesel	n.a.	n.a.	n.a.
Dominican Republic	U.S. Gulf Coast Conventional Gasoline	U.S. Gulf Coast Ultra-Low Sulfur No. 2 Diesel	n.a.	Mont Belvieu Propane	n.a.
El Salvador	U.S. Gulf Coast Conventional Gasoline	U.S. Gulf Coast Ultra-Low Sulfur No. 2 Diesel	n.a.	Mont Belvieu Propane	n.a.
Haiti	New York Harbor Conventional Gasoline	New York Harbor Ultra-Low Sulfur No. 2 Diesel	U.S. Gulf Coast Kerosene-Type Jet Fuel	n.a.	n.a.
Honduras	U.S. Gulf Coast UNL87	U.S. Gulf Coast Ultra-Low Sulfur No. 2 Diesel	U.S. Gulf Coast Avjet 54	Mont Belvieu Propane	n.a.
Mexico	U.S. Gulf Coast Conventional Gasoline	U.S. Gulf Coast Ultra-Low Sulfur No. 2 Diesel	n.a.	Mont Belvieu Propane	n.a.
Peru	U.S. Gulf Coast Conventional Gasoline	U.S. Gulf Coast Ultra-Low Sulfur No. 2 Diesel	n.a.	Mont Belvieu Propane	U.S. Gulf Coast Henry Hub

Source: World Bank data.
Note: LPG = liquefied petroleum gas; n.a. = not applicable.
a. Weighted average export price to Argentina and Brazil.

In the electricity market, establishing the benchmark price presents a different set of challenges. Unlike hydrocarbon products, electricity is not traded, and so it typically does not have an international reference price.[6] Usually, the benchmark price is calculated as the efficient cost recovery price for generation, transmission, and distribution. In practice, the efficient cost recovery price differs from the actual cost recovery price of utilities that operate inefficiently because of elevated technical and nontechnical losses.[7] In this study, we do not consider these inefficiencies as part of consumer subsidies. As such, the price gap that we calculate using efficient costs may not be sufficient to cover the total costs of inefficient utilities.[8]

Factoring in Taxes: Pretax and Post-Tax Price Gaps

How to treat taxes is an important methodological issue not yet fully resolved.[9] Several studies by the World Bank simply bypass the issue of taxes altogether, subtracting all taxes from final consumer prices. This approach implicitly assumes that the same tax structure will be applied to the benchmark price, thereby leaving the absolute size of the price gap equal. To draw comparisons with previous studies and to abstract from the tax issue altogether, we introduce here the pretax price gap. Because some governments use tax policy to affect final end-user prices, we also introduce post-tax price gaps or counterfactual tax price gaps. This exercise requires defining an appropriate benchmark tax rate, which is not straightforward. To do so, we follow the approach adopted by the IEA (1999, 81) in setting the benchmark tax rate at the standard value added tax (VAT). Our rationale is that virtually all products and services in an economy have some level of taxation that can be considered "normal." Any deviation from this "normal" implies special treatment by the government, either by encouraging consumption via lower tax rates or by discouraging it by applying higher tax rates. This approach allows us to shed light on government pricing policies carried out through tax expenditures.

However, at least three important caveats should be noted. First, our measures of counterfactual tax price gaps and tax expenditures are not comparable across countries. Both measures are relative to the benchmark tax structure defined for each country. Thus the fact that one particular country has higher than average tax expenditures does not necessarily imply that it provides a higher level of support or lower taxation levels. Second, by fixing the benchmark tax rate at the standard VAT rate, we abstract from optimal taxation considerations. Optimal taxation theory says that products whose demand is relatively inelastic should have higher tax rates. Oil products are a case in point.[10] Finally, this approach also abstracts from tax considerations meant to address externalities. In economic theory, excise taxes are implemented so that agents internalize the total social costs of certain products (that is, Pigouvian taxes). An example is the call for fossil fuels to have taxation levels well beyond the standard VAT to mitigate the negative externalities they cause such as pollution, health problems, and climate change (see box 3.1).

Box 3.1 Taxing for Negative Externalities

In *Getting Energy Prices Right: From Principle to Practice,* published by the International Monetary Fund (IMF), Heine et al. (2014) contend that countries are setting most of their energy prices incorrectly. According to the report, current pricing levels fail to reflect environmental damage—notably, global warming, air pollution, and the various side effects of motor vehicle use such as congestion, accidental deaths, and injuries. The worst culprit is coal (which is generally underpriced) not only because of the carbon emissions it causes, but also because of the health costs of local air pollution. Air pollution from natural gas is modest relative to that of coal, but natural gas is still underpriced relative to its carbon emissions.

The solution proposed by Heine et al. is to increase the taxes on fossil fuels. Assuming these fiscal instruments are appropriately fixed, the fossil fuel prices paid by firms and consumers would then reflect the full costs to society. Corrective taxes would not only substantially reduce pollution-related deaths and significantly reduce CO_2 emissions, but yield large revenue gains as well. The main conceit underpinning this conclusion is that taxation (or tax-like instruments) is a highly effective mechanism for influencing consumer behavior and discouraging overuse of environmentally harmful energy sources.

According to Heine et al., a full-cost energy tax system would best be structured as follows. First, levy a charge on fossil fuels in proportion to their CO_2 emissions multiplied by the global damage from those emissions. Second, levy additional charges on the fuels used in power generation, heating, and other stationary sources in proportion to the local air pollution net emissions from these fuels. And, third, consider whether additional charges for local air pollution, congestion, accidents, and pavement damage attributable to motor vehicles are warranted. Ideally, some of these charges would be levied according to distance driven or time of use, such as during peak periods.

The authors' argument in favor of taxing the negative externalities of fossil fuels does not imply that the overall tax burden on a country's economy has to increase. Higher fuel taxes, for example, could partially replace broader taxes on income or consumption. If new revenue sources are needed, corrective energy taxes may be an attractive option because "unlike most other options, they improve economic efficiency by addressing a market failure" (Heine et al. 2014, 3).

Price-Gap Results

This section presents the fiscal costs to governments considering the pretax price-gap measures for fuels and natural gas and puts these subsidy expenditures in context by comparing them with GDP and with total government revenues and expenditures. These results are followed by an analysis of tax expenditures using our counterfactual tax price gap.

For electricity, this section presents the results using the pretax price-gap measures by end user (residential, commercial, and so forth). The results by consumption block are also included when this information is available, allowing for an analysis of the allocation of subsidies in terms of beneficiaries. These findings are followed by the results of our counterfactual tax price-gap measures in the electricity sector. The section then moves to a discussion of the significant costs imposed on electricity systems by technical and nontechnical losses. Some of the countries analyzed have worrisome levels of losses that need to be addressed urgently because they affect not only the viability of utility companies, but also the prices and quality of service for consumers.

Pretax Price Gap in Hydrocarbons

This section considers the pretax price gap measures for the five primary hydro-carbon fuels used in the LAC region: gasoline, diesel, LPG, natural gas, and kero-sene. As noted earlier, the pretax price gap compares the benchmark price with end-user prices without taking taxes into account. This measure allows us to compare distortions that arise from government intervention in the direct sales price of fuels. Figure 3.1 presents the estimated yearly average fiscal cost of the

Figure 3.1 Average Yearly Fiscal Support for Fuels: Selected LAC Countries, 2008–13

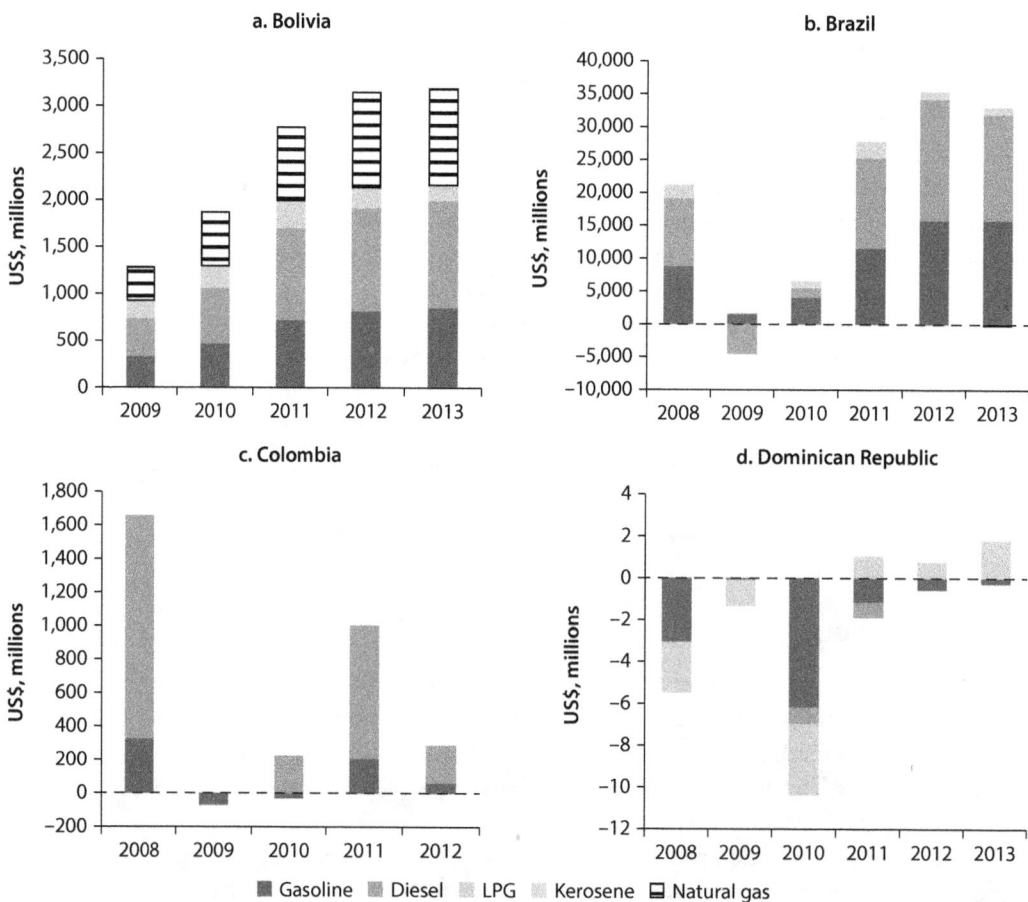

figure continues next page

Figure 3.1   Average Yearly Fiscal Support for Fuels: Selected LAC Countries, 2008–13 *(continued)*

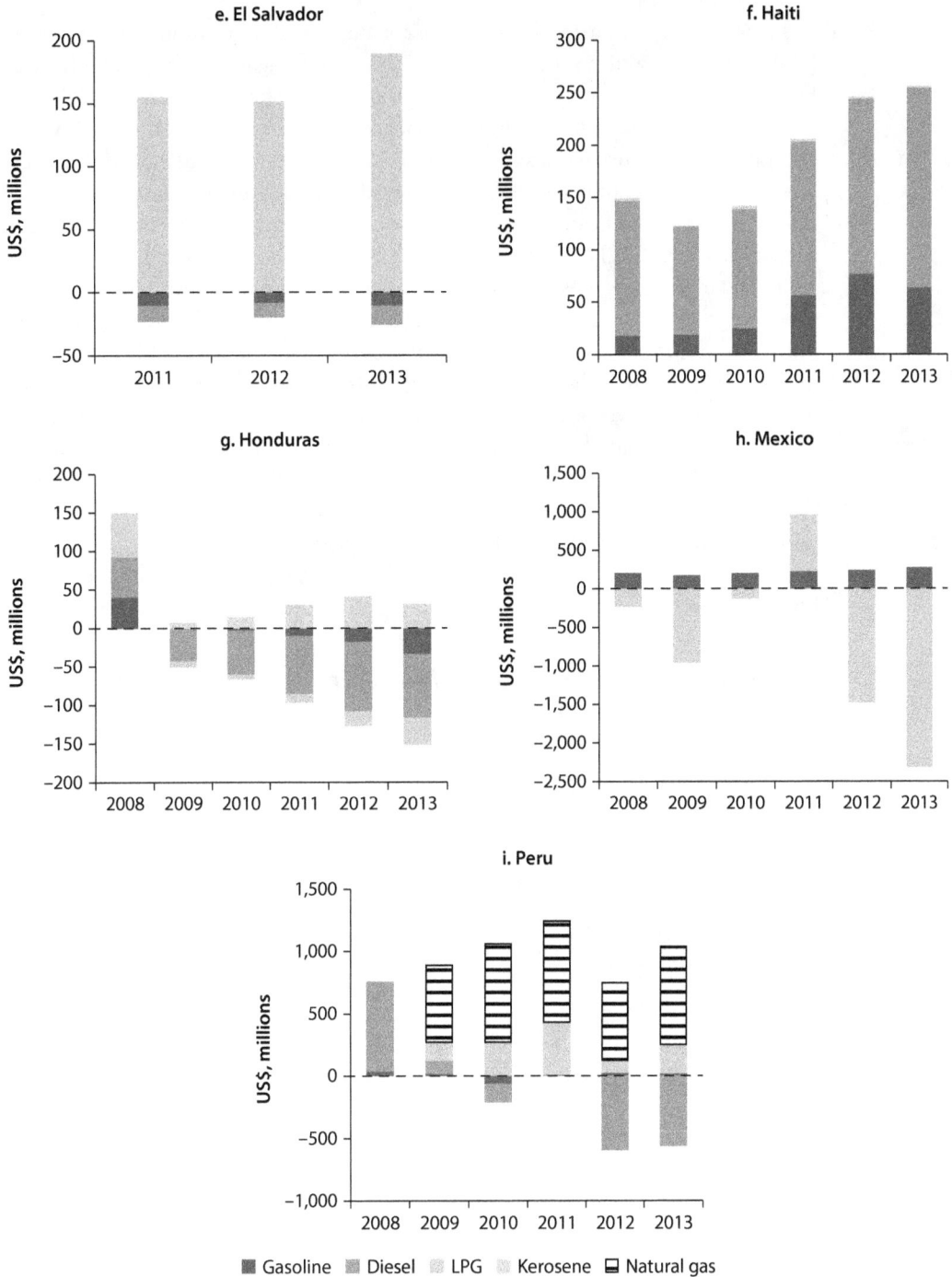

e. El Salvador

f. Haiti

g. Honduras

h. Mexico

i. Peru

■ Gasoline   ▨ Diesel   ▧ LPG   ▨ Kerosene   ▤ Natural gas

Source: World Bank data.
Note: The figures show yearly (net) fiscal costs of subsidies for each fuel by country. Results are calculated using the pretax yearly average price gap multiplied by the total consumption of each fuel within the year. LPG = liquefied petroleum gas.

Energy Pricing Policies for Inclusive Growth in Latin America and the Caribbean
http://dx.doi.org/10.1596/978-1-4648-1111-1

support for fuels during the 2008–13 period. For more details on the calculations, see appendix A.

In figure 3.1, countries with pricing formulas based on import parity prices (IPPs)—the Dominican Republic, El Salvador (for gasoline and diesel), and Honduras—generally present small or negative price gaps. By construction, the pricing formulas are linked to international prices, and so local retail prices (without taxes) track fluctuations closely and do not present significant distortions. Thus, with the exception of LPG in El Salvador and kerosene in Honduras, there is generally no significant fiscal support of fuels. However, the universal price subsidy for LPG in El Salvador did incur a cost of US$150–200 million per year.

In Mexico, which uses an excise tax to smooth the domestic retail prices of gasoline and diesel, the producer prices received by PEMEX are also directly linked to international prices via an IPP-based pricing formula. When the issue of taxes is abstracted, there are few or no distortions in the retail prices of gasoline and diesel. However, as described earlier, LPG prices are set in a discretionary manner by presidential decree. In 2011 the government kept LPG prices artificially low, at a cost of over US$500 million. Over the sample period (2008–13), the fiscal cost of this policy represented on average 0.6 percent of Mexico's gross domestic product (GDP)—see figure 3.2.

Colombia and Peru, which use price stabilization funds, yield some interesting results. Colombia did not have an automatic adjustment mechanism for its

Figure 3.2 Average Fiscal Support for Fuels as a Percentage of GDP: Selected LAC Countries, 2008–13

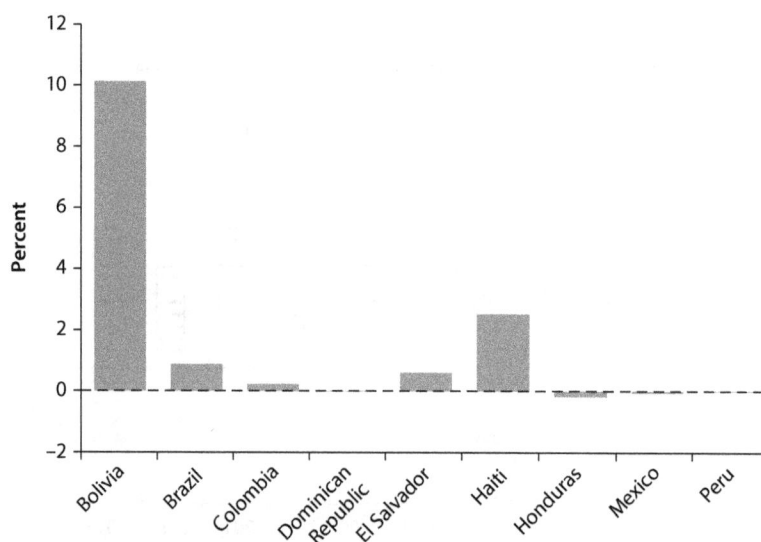

Source: World Bank data.
Note: The figure shows the average for the years 2008–13 of the fiscal costs (net) of subsidies for each fuel by country, divided by the gross domestic product (GDP). Results are calculated using the pretax yearly average price gap multiplied by the total consumption of each fuel within the year. This result is divided by the GDP of the year and then averaged across the years available for each country.

producer income (PI) target price until 2009. In 2008 Colombia had subsidies in place that reached US$1.7 billion, mostly toward underpricing diesel. After introduction of the automatic adjustment formula, this fiscal support diminished significantly. The only exception was in 2011 in response to a rapid rise in global oil prices. By contrast, in 2010 Peru abandoned its price stabilization fund for gasoline and diesel. The fund continued to cover LPG, however. Our results mirror this situation, showing how the fiscal cost disappeared for gasoline and diesel but stayed in place for LPG. The Peruvian government continued to support low prices for natural gas as a way of promoting its use in electricity generation, as well as for consumption in the residential, industrial, and automotive sectors. The cost of doing so was significant, amounting to between 0.3 and 0.5 percent of GDP over the 2008–13 period.

Finally, Bolivia, Brazil, and Haiti show large levels of fiscal support for hydrocarbons. As shown in figure 3.1, the total magnitude of subsidies moves closely with international prices. Subsidies fell in 2009 when international prices fell and then consistently grew as oil prices recovered. In figure 3.2, on average the fiscal support for fuels and natural gas in Bolivia represented 10 percent of GDP. In Haiti, the average was closer to 2.5 percent of GDP because of the government's policy of freezing prices. The average in Brazil is by far the lowest of the three, at 0.87 percent of GDP. However, between 2011 and 2013 fiscal support reached almost 1.3 percent of GDP, with a peak of 1.43 percent in 2012.

Counterfactual Tax Price Gap in Hydrocarbons

This section analyzes the fiscal expenditures to support hydrocarbons. Anyone interpreting the figures should keep in mind that tax expenditures are measures of support only relative to the benchmark tax structure of the country in question. They are not comparable across countries.

As explained earlier, we define the benchmark tax rate as the standard VAT applied to all other goods and services. When energy products are taxed at less than the standard VAT rate, we consider it to be a positive tax expenditure. In other words, the government is actively encouraging consumption of the particular energy product in question. If a product has a total tax rate (VAT plus excise taxes) that is higher than the standard VAT rate, we consider it to be a negative tax expenditure (or revenue), implying that the government is discouraging consumption of that good relative to the average good or service in the economy.

The results show a variety of tax policies across the LAC region. For the most part, a majority of countries tax gasoline and diesel more than the standard level for all other goods and services (see figure 3.3). The only exception is Mexico, which uses an excise tax to smooth the price path of these products. During 2008–13, Mexico's excise tax was negative, thereby pushing the total taxation rate of these products to below the standard VAT rate, resulting in large tax expenditures. These tax expenditures exceeded US$15 billion in 2008, 2011, and 2012, representing almost 2 percent of GDP.

Countries such as Bolivia, the Dominican Republic, and Honduras provide support through lower taxation rates on LPG, kerosene, and natural gas,

Figure 3.3 Fuels Fiscal Expenditures: Selected LAC Countries, 2008–13

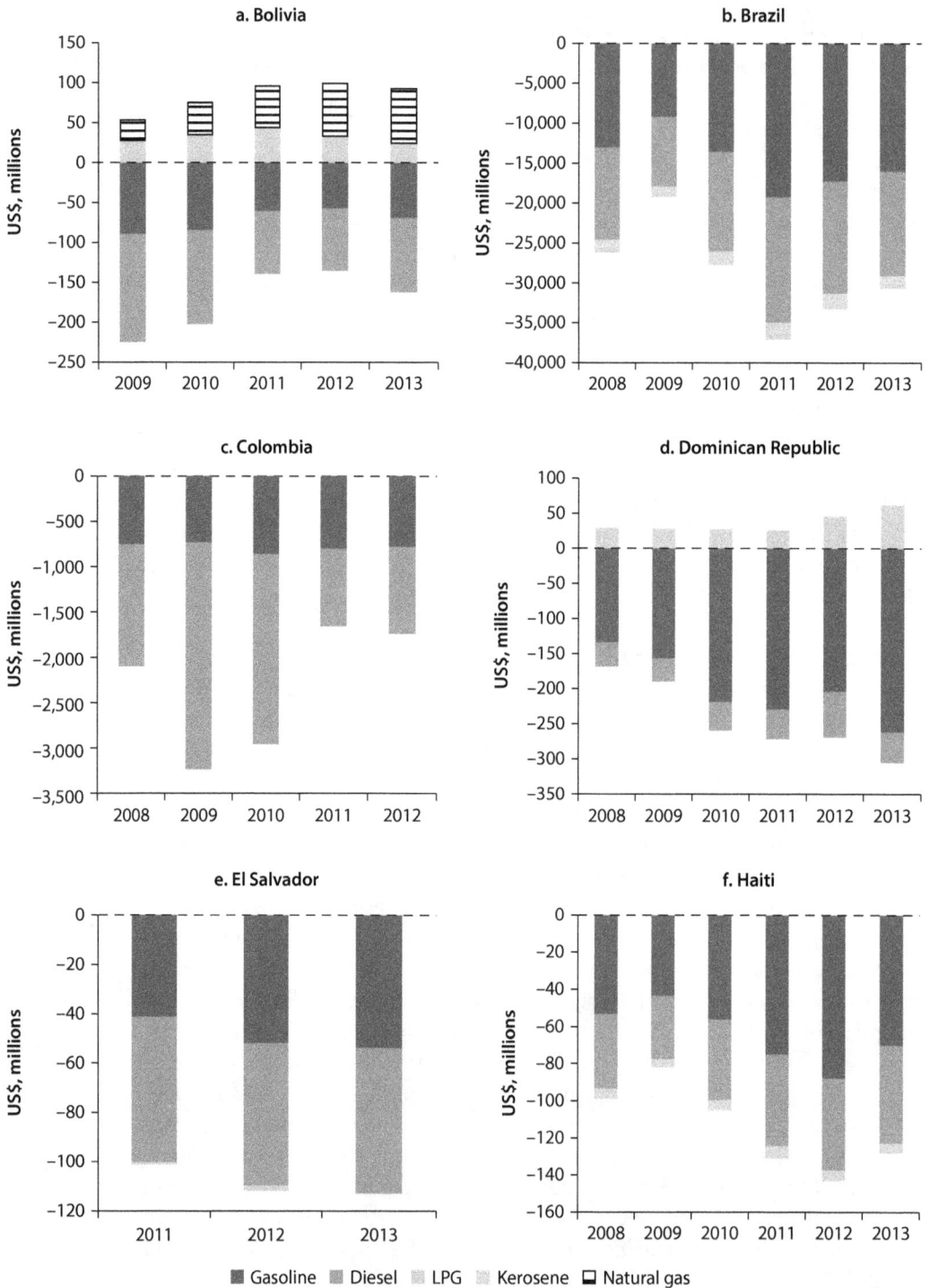

Gasoline Diesel LPG Kerosene Natural gas

figure continues next page

Figure 3.3 Fuels Fiscal Expenditures: Selected LAC Countries, 2008–13 *(continued)*

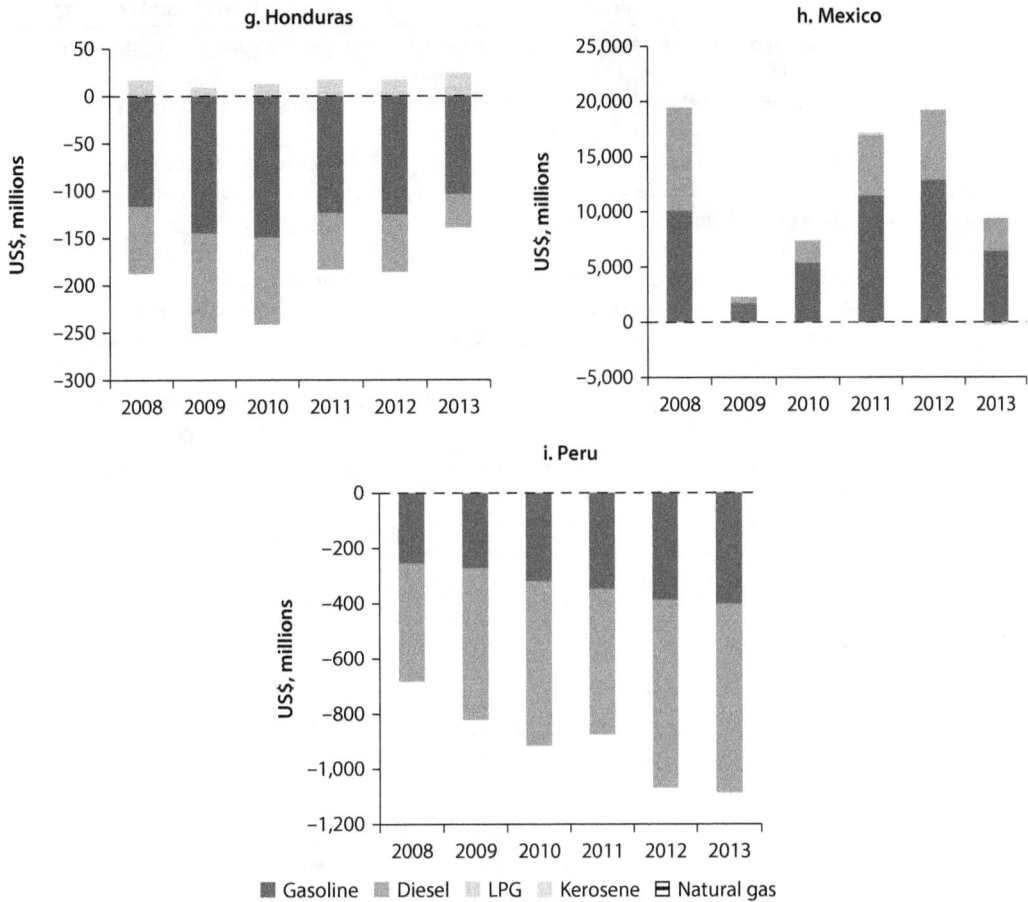

g. Honduras

h. Mexico

i. Peru

■ Gasoline   ▨ Diesel   ▨ LPG   ▨ Kerosene   ⊟ Natural gas

Source: World Bank data.

Note: The figures show the yearly (net) tax expenditures for each fuel by country. Results are calculated using the difference between the price gap with the counterfactual tax rate for the value added tax and the yearly average price gap. This difference is multiplied by the total consumption of each fuel within the year. LPG = liquefied petroleum gas.

all of which are widely used by the population, principally for cooking. The fiscal expenditures are generally small, amounting to US$50–100 million in Bolivia and less than US$50 million in the Dominican Republic and Honduras.

Pretax Price Gap in the Electricity Sector

This section turns to analysis of the fiscal cost of pricing policies in the electricity sector as measured by the pretax price gap. To clarify again, whenever possible we have used the efficient cost structure rather than the actual cost structure because the latter may reflect the large inefficiencies observed in some of the LAC electricity systems in our sample. We judge it unfair to characterize these inefficiencies as part of consumer subsidies. Although complete data are not

Table 3.2 Electricity Price Formation, Selected LAC Countries

Country	Supply cost	Transmission and distribution costs
Bolivia	Yearly average of the tariff of each distributor of natural gas, weighted by total sales	Included in natural gas tariff per company
Brazil	Balance liquidation value (PDL): weighted average of weekly prices by region and by transmission level	Weighted average by company
Colombia	Average cost of electricity	Tariff publications by Codensa
Dominican Republic	Indexed tariff: weighted average of world reference prices of fuel, natural gas, and coal	Included in the supply cost
El Salvador	Short-term marginal cost (STMC) of electricity: energy price (variable) plus capacity charge	Based on efficient tariff schedules defined by regulatory entity
Haiti	System average generation cost, including power purchase agreements of independent power producers with Electricité d'Haiti (EDH)	Assumes transmission and distribution costs of US$.05 per kilowatt-hour
Honduras	System average generation cost: total amount billed by all private power generation companies	System average cost of transmission and distribution
Mexico	Average cost of electricity, divided between consumer segment and voltage level within that segment	Included in average cost of electricity
Peru	Natural gas prices at international spot prices, system-wide averages (actual prices), and Santa Rosa Busbar prices (idealized spot price)	Transmission: system average cost of transmission toll. Distribution: costs of aggregate value of low and medium tension levels

Source: World Bank data.

available for all LAC countries in our sample, table 3.2 provides details of the data that were available. (For more details on the calculations, see appendix B.)

Several facts are of interest here. First, every country in the sample subsidized its electricity sector at some point during the period of analysis. Most countries subsidized electricity rates consistently over the sample period (see figure 3.4). The exception is Brazil, which subsidized electricity rates in 2008 and between 2012 and 2013 because of the sharp jump in generation costs resulting from a combination of low rainfall and high oil prices. In general, however, the magnitude of fiscal support for electricity is moderate compared with that for fuels. This is not to imply that the costs involved are insignificant (see figure 3.5). In Bolivia and Honduras, for example, fiscal support for electricity averages around 1.2 percent and 1 percent of GDP, respectively. For Brazil, the average over the sample period appears to show no fiscal support, and yet this masks the cost of maintaining low electricity prices during 2012–13, which averaged over 0.4 percent of GDP.

Second, the residential sector corners the lion's share of subsidies. In some countries such as Colombia and Haiti, domestic households are the sole recipients of subsidies. To sustain such a system, all other sectors must pay either above-efficient costs in a cross-subsidy framework or efficient costs. In the study sample, governments generally do not subsidize electricity consumption by the industrial sector, although some countries do provide industry with limited tax exemptions. For example, Peru subsidizes the inputs for electricity generation (through lower-priced natural gas). This subsidy indirectly benefits industries even though there is no direct subsidy through the tariff structure.

Figure 3.4 Average Yearly Fiscal Support for Electricity: Selected LAC Countries, 2008–13

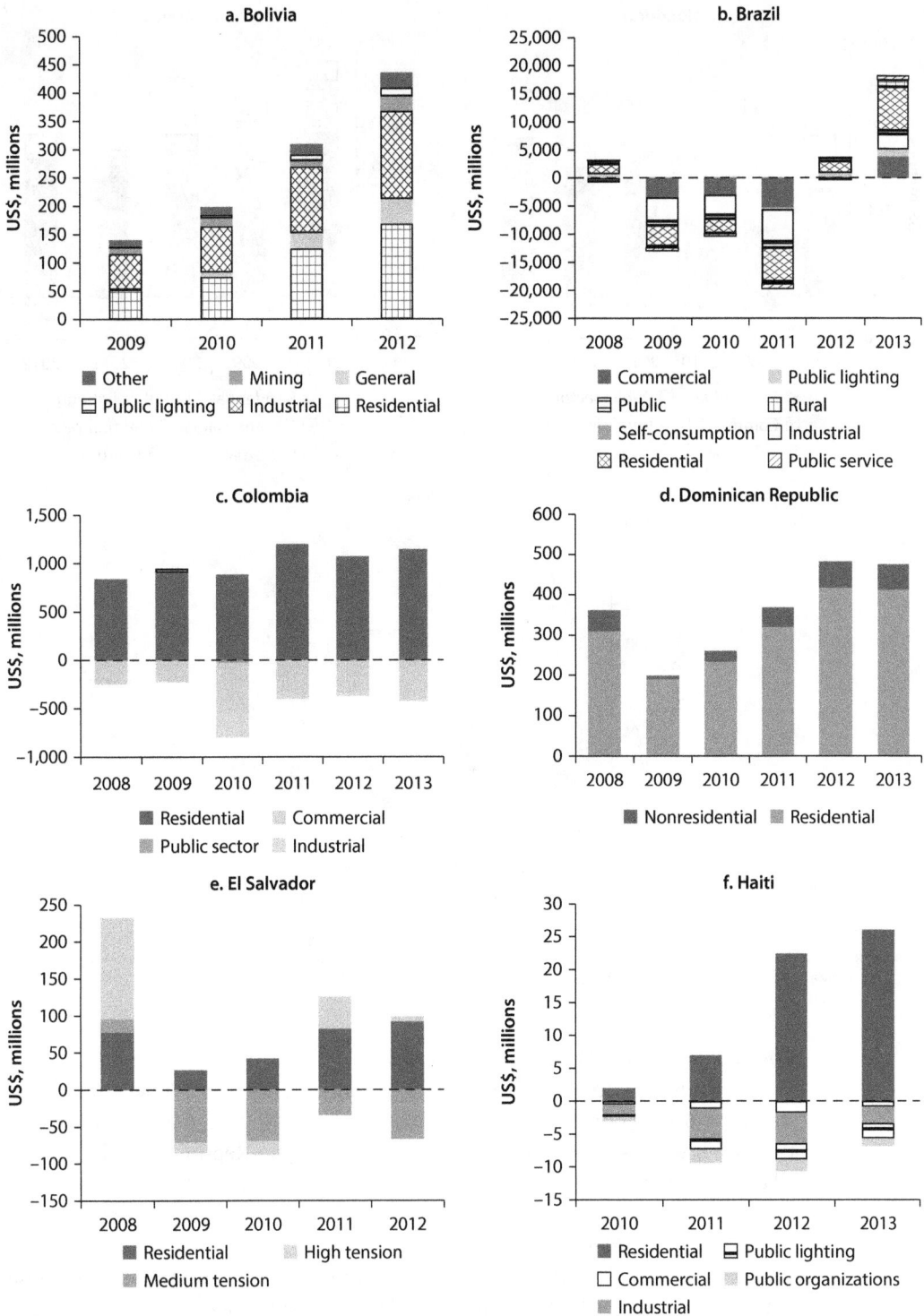

a. Bolivia

Legend: Other, Mining, General, Public lighting, Industrial, Residential

b. Brazil

Legend: Commercial, Public lighting, Public, Rural, Self-consumption, Industrial, Residential, Public service

c. Colombia

Legend: Residential, Commercial, Public sector, Industrial

d. Dominican Republic

Legend: Nonresidential, Residential

e. El Salvador

Legend: Residential, High tension, Medium tension

f. Haiti

Legend: Residential, Public lighting, Commercial, Public organizations, Industrial

figure continues next page

Figure 3.4 Average Yearly Fiscal Support for Electricity: Selected LAC Countries, 2008–13 *(continued)*

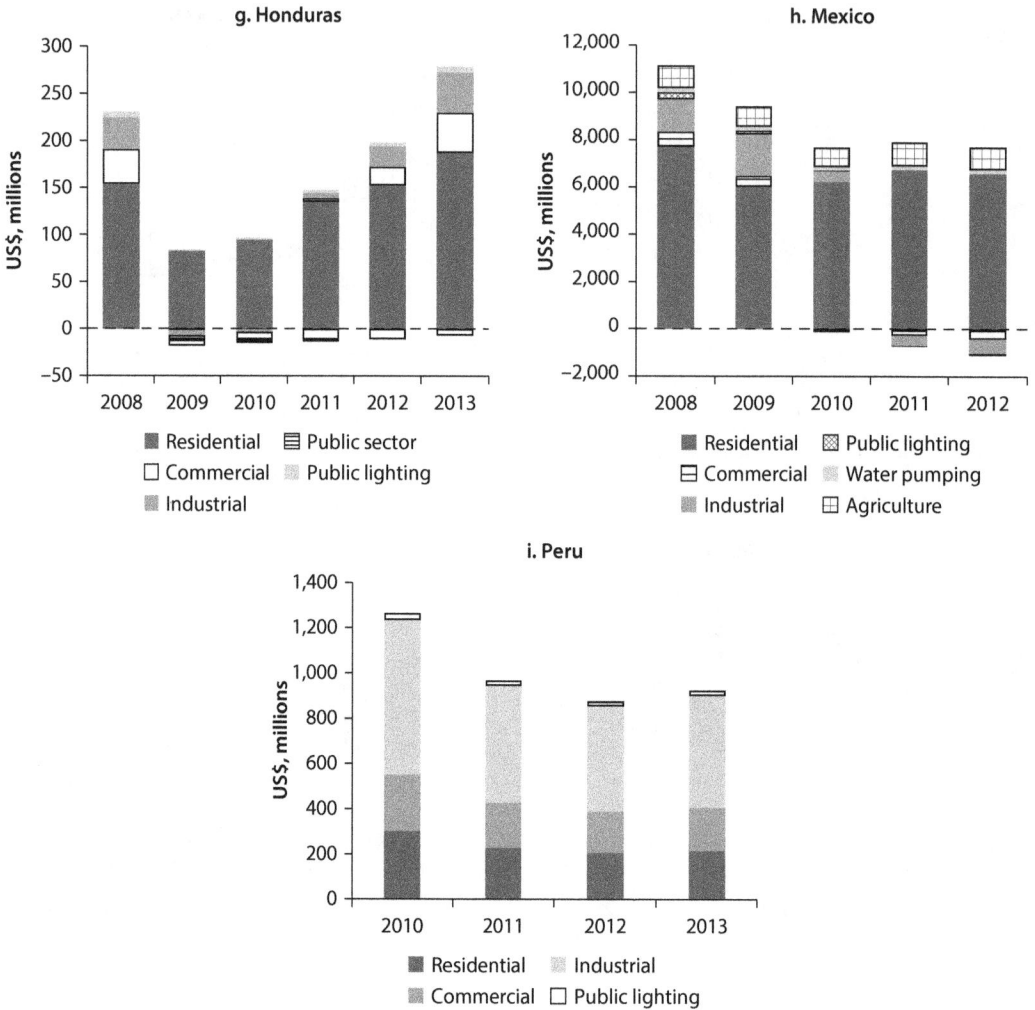

g. Honduras

h. Mexico

i. Peru

Source: World Bank data.

Note: The figures show the yearly (net) fiscal costs of subsidies for each electricity consumption category by country. Results are calculated using the pretax yearly average price gap multiplied by the total consumption of each segment within the year.

Counterfactual Tax Price Gap in Electricity

Tax expenditures are measures of support only relative to the benchmark tax structure of the country in question. Thus they are not comparable across countries. With the exception of Colombia, the Dominican Republic, and Honduras, all countries in the study sample tax electricity consumption at the standard VAT rate (the benchmark tax rate for the purposes of this analysis). Therefore, no tax expenditures generally exist.

In Honduras, all residential consumers are exempt from taxes except those with consumption of more than 750 kilowatt-hours per month. This high-consuming group is required to pay the VAT rate, which was set at 12 percent in

Figure 3.5 Average Yearly Fiscal Support for Electricity as a Percentage of GDP: Selected LAC Countries, 2008–13

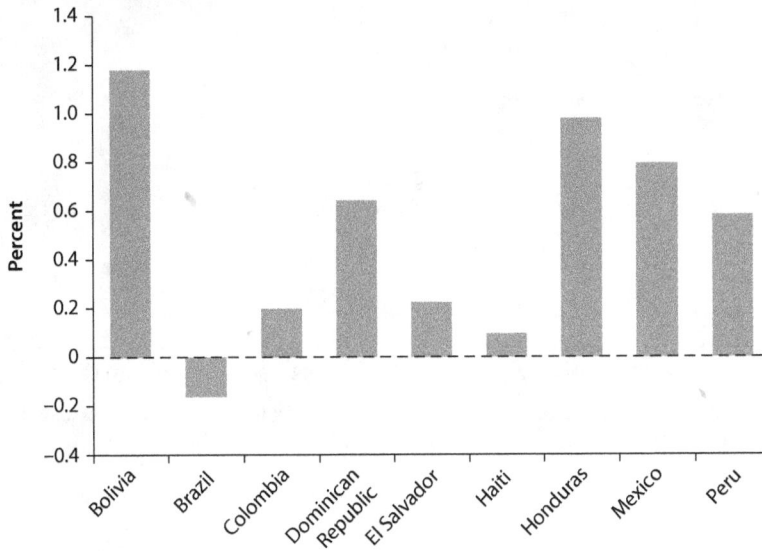

Source: World Bank data.

Note: The figure shows the average across the years of fiscal costs (net) of subsidies for electricity by country, over the gross domestic product (GDP). Results are calculated using the pretax yearly average price gap multiplied by the total consumption of electricity within the year for each segment. The total for all segments within the year are divided by the GDP of the year and then averaged across the years available for each country.

2010 and raised to 15 percent in 2014. Tax expenditures in Honduras increased along with electricity costs during the study period, exceeding US$130 million in 2013 (see figure 3.6). Colombia exempts by law all public domiciliary services—including electricity—from taxation. The implied tax expenditure for electricity alone amounted to around US$1.2 billion in 2013. Similarly, in the Dominican Republic, the whole electricity chain from production to distribution is exempted from consumption and import taxes. The tax expenditures in this case have been growing over time, rising to over US$130 million in 2013.

Peru is a special case because it has positive tax expenditures, albeit decreasing over time. The rise in tax expenditures stems from the subsidized inputs for generation rather than a differential taxation regime for electricity relative to all other goods and services as might reasonably be assumed.[11] Our measure of fiscal expenditure captures the lost revenue from the Peruvian government attributable to lower generation costs (and thus a lower tax base) than would be implied if natural gas prices were set at international levels.

System Losses in Electricity

Losses in the electricity systems are very high in some LAC countries. These losses pose significant costs to their governments and can even threaten the financial viability of utility companies. Electricity losses are categorized as technical and nontechnical. Technical losses are caused by actions internal to the power system

Energy Pricing Policies for Inclusive Growth in Latin America and the Caribbean
http://dx.doi.org/10.1596/978-1-4648-1111-1

Figure 3.6 Electricity Fiscal Expenditures: Selected LAC Countries, 2008–13

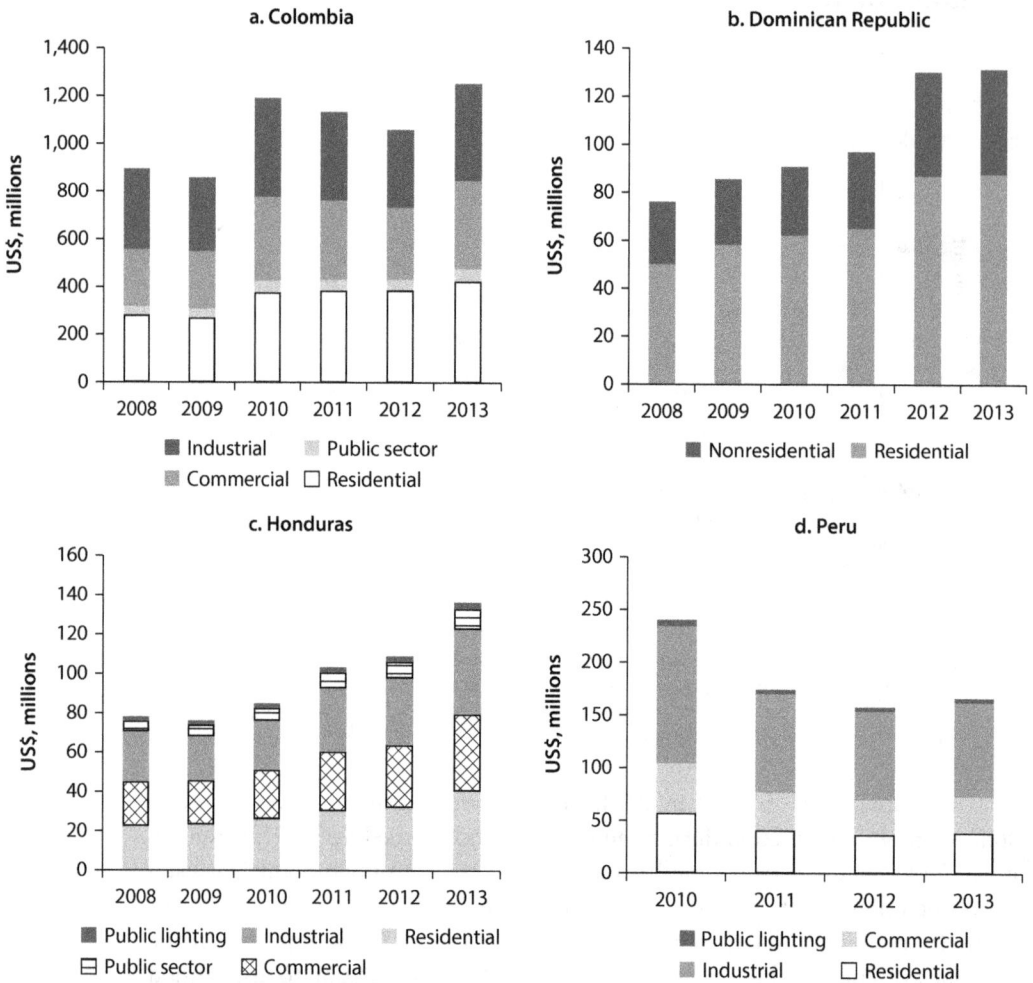

a. Colombia

b. Dominican Republic

c. Honduras

d. Peru

Source: World Bank data.
Note: The figures show the yearly (net) tax expenditures of subsidies for each electricity consumption category by country. Results are calculated using the difference between the price gap with counterfactual tax rate for the value added tax (VAT) and the yearly average price gap. This difference is multiplied by the total consumption of electricity in each segment within the year.

and consist mainly of power dissipation in electrical system components such as transmission lines, power transformers, and measurement systems. Nontechnical losses stem from actions external to the power system. Common causes include electricity theft, nonpayment by customers, and errors in metering or accounting.

In several countries, the nontechnical losses are acute. In Honduras, for example, the total loss during 2013 was more than 31 percent of the power generated, or more than double the total admissible loss set by law (15 percent). The reduction in billing from the energy loss amounted to US$206 million. Although electricity theft is a major problem, it is not the sole cause of losses. It is estimated that at least 100,000 users in Honduras do not have energy meters.

In addition, arrears of more than 120 days (which can be considered unrecoverable) contributed a further US$177 million in financial losses. As a result, the country's electricity system ended 2013 with a deficit of US$340 million, plus accumulated debts with power generation companies of over US$430 million. All these costs have to be covered by the government.

Mexico faces a similar scenario, with electricity losses accounting for about 18 percent of the total electricity produced. Two-thirds of these losses are due to nontechnical factors. Although electricity theft, metering errors, and failure to bill consumption properly occur across the country, the problem is especially severe in the metropolitan area of Mexico City. In 2012 nontechnical losses amounted to slightly over US$320 million. Theft of electricity accounted for 43 percent of these losses, metering for 31 percent, and billing errors for 25 percent. These losses are not factored into our electricity subsidy calculations, but they are covered by the state nonetheless and therefore form part of the fiscal burden deriving from inefficient electricity systems.

When technical losses are combined with nontechnical losses, the impact can be particularly egregious for the financial viability of utilities. In the Dominican Republic, the rate of electricity losses is about 33 percent because of the prevalence of theft and a high number of nonmetered clients, on the one hand, and aging infrastructure with high levels of power dissipation, on the other. The overall deficit of the country's distribution and state-owned generation companies amounted to 2.2 percent of GDP in 2013. Haiti finds itself in an even more extreme situation, with distribution losses a staggering 61 percent. Haiti's electricity distribution company, EDH, is in such a financially precarious state that its recovered funds cover a mere 27 percent of the fuel costs for generating electricity. Because of the scale of these losses, the utilities in these countries are not able to invest in new equipment and infrastructure, thereby exacerbating the problem of technical and nontechnical losses.

Notes

1. For more detail on policies often missed by the price-gap approach, see Steenblik and Coroyannakis (1995) and Steenblik and Wigley (1990).

2. For example, accelerated depreciation allowances for capital, investment credits, additional deductions for exploration and production, and preferential capital gains treatment. In the United States, a policy known as "excess of percentage over cost depletion option" for the natural resource sector is estimated to have provided a subsidy of US$1.190 billion in fiscal 2011 (OMB 2012). In Canada, the accelerated capital cost allowance for oil production in the oil sands was estimated at Can$300 million a year (0.02 percent of the gross domestic product) for the 2007–11 period.

3. For a full discussion of the advantages and disadvantages of the price-gap approach, see Koplow (2009).

4. These costs include wholesale and retail margins.

5. One approach uses the delivered price of a fuel in an efficient market as the benchmark (for example, for oil, see Rajkumar 1996). Another approach examines the differences between import and export values in countries with efficient transportation

networks (for example, for coal, see Koplow 1998). For multicountry studies, the IMF (2010, 2013) has used a single value for all countries based on the cost per liter in the United States.

6. There are some instances of bilateral international trade such as between Argentina and Uruguay, but that is the exception, not the rule.

7. Actual or accounting cost recovery prices include implicit subsidies to producers that operate inefficiently.

8. In cases in which losses in electricity systems are beyond reasonable benchmarks, any reform of pricing policies should be accompanied by reforms that address these inefficiencies.

9. See OECD (2013, chap. 1) for a thorough discussion of the issues surrounding the measurement of tax expenditures.

10. This is mainly because viable substitutes are lacking.

11. In fact, electricity is taxed at the standard VAT rate of 18 percent.

References

Heine, Dirk, Eliza Lis, Ian W. H. Parry, and Shantung Li. 2014. *Getting Energy Prices Right: From Principle to Practice.* Washington, DC: International Monetary Fund.

IEA (International Energy Agency). 1999. *Looking at Energy Subsidies: Getting the Prices Right.* World Energy Outlook 1999. Paris: IEA.

———. 2010. *World Energy Outlook.* Paris: IEA.

IMF (International Monetary Fund). 2010. *Petroleum Product Subsidies: Costly, Inequitable, and Rising.* Washington, DC: IMF.

———. 2013. *Case Studies on Energy Subsidy Reform: Lessons and Implications.* Washington, DC: IMF.

Koplow, Doug. 1998. "Quantifying Impediments to Fossil Fuel Trade: An Overview of Major Producing and Consuming Nations." Unpublished analysis prepared for the Organisation for Economic Co-operation and Development Trade Directorate, OECD, Paris.

———. 2009. "Measuring Energy Subsidies Using the Price-Gap Approach: What Does It Leave Out?" International Institute for Sustainable Development (IISD), Ottawa, Canada.

OECD (Organisation for Economic Co-operation and Development). 2013. *Inventory of Estimated Budgetary Support and Tax Expenditures for Fossil Fuels.* Paris: OECD.

OMB (U.S. Office of Management and Budget). 2012. *Analytical Perspectives: Budget of the US Government, Fiscal Year 2013.* http://www.whitehouse.gov/sites/default/files /omb/budget/fy2013/assets/spec.pdf.

Rajkumar, Andrew. 1996. *A Study of Energy Subsidies.* Washington, DC: World Bank.

Steenblik, Ronald, and Panos Coroyannakis. 1995. "Reform of Coal Policies in Western and Central Europe: Implications for the Environment." *Energy Policy* 23 (6): 537–53.

Steenblik, Ronald, and Kenneth Wigley. 1990. "Coal Policies and Trade Barriers." *Energy Policy* 18 (5): 351–67.

CHAPTER 4

Economic Impacts of Energy Pricing and Subsidy Reforms

Introduction

The decision by governments to introduce or maintain energy subsidies and price control policies is generally motivated by several objectives: to provide welfare, to create price stability, and to promote competitiveness.

All these objectives are well intentioned. For welfare, policy makers believe subsidies and price control policies can help make energy products more affordable for the population as a whole and for low-income individuals in particular, thereby increasing their standard of living. Seeking to avoid excessive changes in energy prices is also advantageous to firms and consumers because it protects them from fluctuations in production costs. Finally, these policies also seek to promote competitiveness, especially among energy-intensive production sectors.

This chapter focuses on the last two of these objectives—price stability and competitiveness—and analyzes how changes in pricing policies can potentially lead to increases in energy prices. This analysis focuses in particular on the impacts of pricing policies on aggregate prices, overall competitiveness, and specific sectors of production for selected countries in the Latin America and the Caribbean (LAC) region.

As noble as a government's objectives may be, energy subsidies and pricing policies can create unintended costs and distortions that could undermine their desired impacts. Most notably, subsidy expenditures tend to aggravate fiscal imbalances, which are financed through distortive taxes or public debt, crowding out priority public spending and private investment. Other possible negative knock-on effects of making energy costs artificially low include encouraging excessive consumption and creating reduced incentives for investments in renewable energy sources or energy-saving technologies. Furthermore, subsidies exert pressure on the balance of payments of net energy importers and encourage smuggling to neighbors with higher domestic prices. Finally, by promoting higher energy usage, subsidies increase greenhouse gas emissions and aggravate global warming, thereby affecting the welfare of future generations (Parry et al. 2014).

Energy is an important input in the production of many intermediary and final products, and so energy prices have implications for the overall cost of production. Although economists do not often model energy as a primary input, most sectors of production depend directly or indirectly on energy products. Some of these direct links are straightforward. Most transport vehicles, to cite an obvious example, require fuel to move. Other links are less intuitive although no less strong. For example, cleaning and pumping water for distribution might require large amounts of electricity, depending on geographical conditions.

As for the indirect effects of energy prices, consider food manufacturing. Companies manufacturing processed food and other food products typically rely on agricultural goods, often produced with hydrocarbon-based fertilizers and produced on farms located at a distance from their operations. Thus the goods have to be transported to the factory, which entails fuel consumption. The same is true for any other kind of manufacturer that needs to transport raw materials or finished goods. The more a particular sector relies on energy usage (whether directly or indirectly), the more affected it will be by shifts in energy prices.

Energy prices also can shape the structure of production, the adoption of particular technologies, and the overall level of energy consumption. In practice, in their decision making firms take energy costs into account in much the same way they do for capital rents and labor costs. In the short term, this decision making is reflected in the quantity of goods a firm produces and how much energy it uses versus other inputs. In the medium to long term, such decision making is reflected in the production technologies a company chooses to adopt or whether it elects to produce a particular product, perhaps for export. The sum of these effects determines the energy consumption in the economy. The shale gas revolution, which significantly increased natural gas production in the United States over the last decade, is illustrative. Arezki and Fetzer (2016) note how gas prices declined as a consequence of the spike in the gas supply. The lower prices led in turn to a significant rise in the construction and expansion of manufacturing plants in energy-intensive industries. The authors also find implications for competitiveness, with the shale gas revolution being credited with a rise in U.S. manufacturing exports of about 6 percent, mainly driven by energy-intensive segments.

As described earlier in this report, the LAC region is home to a diverse range of pricing policies and energy market structures, resulting in most countries in a below-cost market price. Even so, the final post-tax prices of energy products remain relatively high across most LAC countries when compared with those of their international peers (figure 4.1). This difference is especially true for electricity, with industrial consumers being particularly hard-hit. The average prices in LAC countries are 50 percent higher than the averages for East Asia and Europe, for example, and twice the average for the United States. Fuel prices provide a more nuanced picture. Bolivia, Mexico, and El Salvador are on the low side, for example, whereas Uruguay, Paraguay, and Costa Rica have some of the highest prices

Figure 4.1 International Comparison, Retail Prices of Energy, 2012 Products

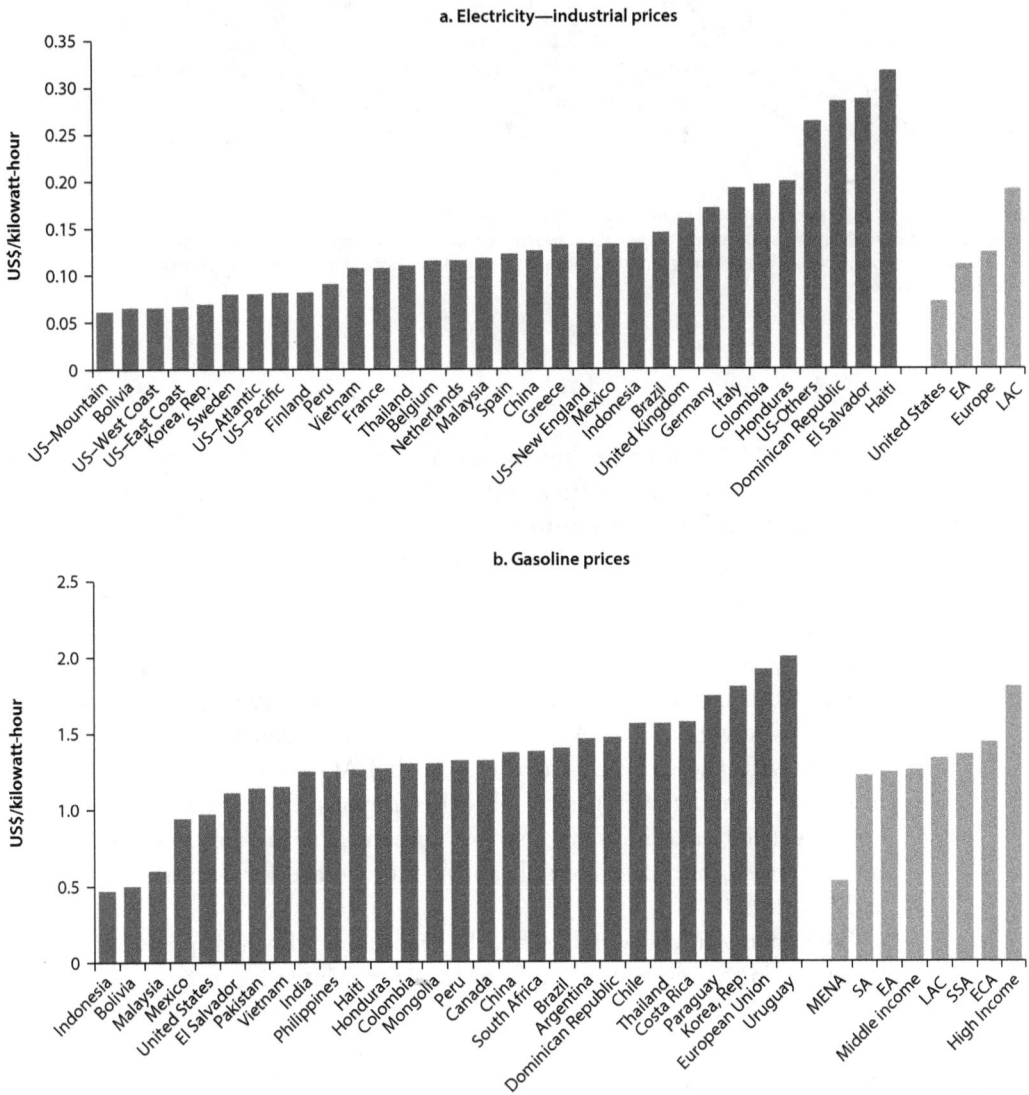

a. Electricity—industrial prices

b. Gasoline prices

figure continues next page

Figure 4.1 International Comparison, Retail Prices of Energy, 2012 Products *(continued)*

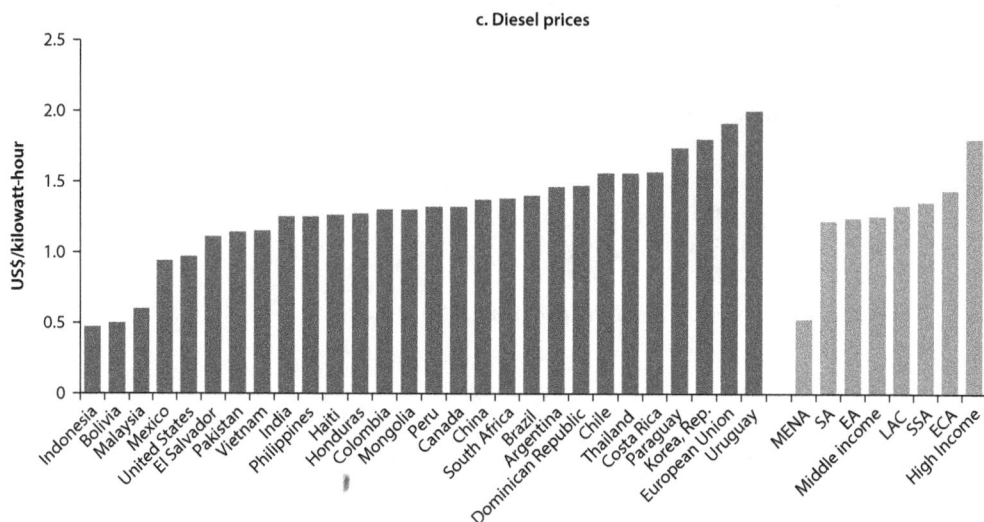

c. Diesel prices

Sources: Global Petrol Prices, Eurostat, and World Bank, World Development Indicators (database).
Note: The figures show electricity and fuel average prices for 2012 by country, region, and country income groups. EA = East Asia;
ECA = Europe and Central Asia; LAC = Latin America and the Caribbean; MENA = Middle East and North Africa; SA = South Asia;
SSA = Sub-Saharan Africa.

in the world. On average, however, LAC prices are well above those in the Middle East and North Africa and South and East Asia, but below those in Sub-Saharan Africa and Europe.

LAC countries tend to be less energy-intensive than other countries with similar income levels, suggesting these economies have somehow adapted to relatively high prices. The region consumes relatively low amounts of both electricity and fuel energy products (figure 4.2). Energy intensity, as measured by the amount of energy consumed per unit of gross domestic product (GDP), is lower in the LAC region than in any other region in the world. In fact, with the exception of Bolivia, all LAC countries are less energy-intensive than the averages for all other regions. Per capita energy use for both electricity and fuels are also low compared with countries with similar income levels, but they are still above the average for Sub-Saharan Africa and South Asia.

Low energy intensity is likely to be associated with high prices. Economic structures in the region seem to favor less energy-intensive sectors, especially in countries with higher prices. For example, the top three energy-intensive sectors account for 21 percent of GDP in Bolivia and about 10 percent in Mexico—the two countries with among the lowest fuel prices in the region. By contrast, these sectors account for about 5 percent in Colombia and 3 percent in the Dominican Republic, where prices are higher.

Figure 4.2 International Comparison, Energy Intensity and GDP per Capita

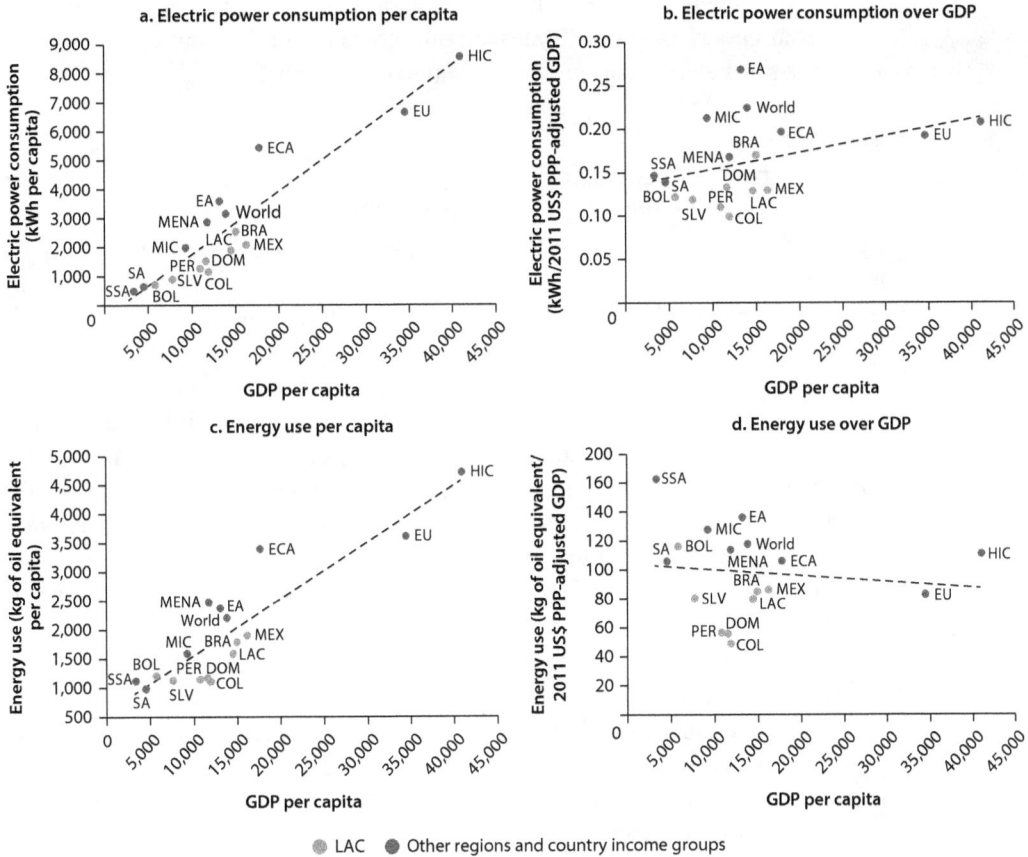

a. Electric power consumption per capita

b. Electric power consumption over GDP

c. Energy use per capita

d. Energy use over GDP

● LAC   ● Other regions and country income groups

Source: World Bank, World Development Indicators (database).

Note: Panels a and b plot electricity consumption by country and region (per capita and per unit of GDP, respectively), against GDP per capita (in constant 2011 US$, PPP-adjusted) in 2012. Panels c and d plot primary energy use by country and region (per capita and per unit of GDP, respectively), against GDP per capita (in constant 2011 US$, PPP-adjusted) in 2012. GDP = gross domestic product; kg = kilogram; kWh = kilowatt-hour; PPP = purchasing power parity. BOL = Bolivia; BRA = Brazil; COL = Colombia; DOM = Dominican Republic; EA = East Asia; ECA = Europe and Central Asia; EU = European Union; HIC = high-income countries; LAC = Latin America and the Caribbean; MENA = Middle East and North Africa; MEX = Mexico; MIC = middle-income countries; PER = Peru; SA = South Asia; SLV = El Salvador; SSA = Sub-Saharan Africa.

Analysis

Although they are less exposed than their international peers, LAC countries could still be affected by increases in prices resulting from energy reforms. It is therefore important to assess the potential impacts of these reforms. The fact that energy remains a key input for production in the LAC countries compounds this importance. For this reason, an assessment of economic and sectoral impacts merits particular attention. To undertake such an assessment, we simulate the effects of changes in energy prices and estimate the impacts on economic sectors and overall production prices for our sample of LAC countries.

As a preliminary contextual step, it is necessary to understand the production structure of an economy and how each sector of production depends on energy (that is, whether a particular sector uses energy as an input or uses it indirectly through the energy embedded in another production input). In this context, changes in energy prices will translate into changing production costs for most sectors, with the energy-intensive ones affected the most.

Methodology

We use a standard methodology to analyze the distributional impacts of subsidy reform. Our exploration of economic and sectoral implications, however, brings a new lens to the analysis. Our methodology has been used in a number of studies as an intermediary step in estimating the impacts of energy price changes and subsidy reforms on households (see Coady and Newhouse 2006; Coady et al. 2010; Parry et al. 2014). It requires a price-shifting model that makes it possible to identify how higher oil and electricity costs are shifted to other sectors of the economy. Analysts typically use the Leontief input-output (IO) price model here. Under this model, the total price of one unit of output is equal to the total cost of its production, including intermediate purchases and primary inputs. By inverting the IO matrix (that is, creating a Leontief inverse), one can express production costs as a function of primary inputs only.[1] In turn, one is also able to assess how changes in the cost of primary inputs—in this case, energy—affect the cost of different final products.

This approach has many factors in its favor. Most obvious is the transparency and simplicity of the methodology. In addition to being well established, the IO approach allows an intuitive analysis of industry interdependency. Furthermore, it provides a satisfactory approximation of simplified general equilibrium models through its focus on short-term analysis of one-shot policy shocks. In using the IO approach for our analysis, we make certain basic assumptions and considerations. First, we assume that the technology of production in the economy is fully described by the IO matrix. In addition, we consider the full transmission of price shocks but do not take into account potential lags in adjustment of prices. Industries are unable to adjust their input structure in reaction to prices because technical coefficients are fixed (short-term approximation). This implies that the quantities of intermediary goods and primary inputs demanded may remain constant. For the purposes of our analysis, we treat the sector/input initially shocked as a controlled price sector. Thus it does not experience feedback effects from changes in the prices of other intermediary goods. All other sectors are assumed to be "cost-push" sectors—that is, they fully push higher input costs into output prices.

This methodology is subject to a few important caveats as well. First, the accuracy of the findings depends on the quality of the data and harmonization assumptions. Our analysis focuses on countries that have a relatively recent IO matrix (less than 8 years old). It is therefore likely that the structure of these economies has changed, at least moderately, since the data were collected because even low- and middle-income economies are frequently exposed to new products and technologies. In addition, the period of analysis includes the highs and lows of international commodity prices to which economic agents are gradually adjusting. Another important caveat mentioned earlier is that industries are unable to respond to price changes by adjusting technological processes. The full transmission of price shocks to costs happens only if no compensatory measures have taken place. In practice, firms might adjust by substituting inputs or changing the quantity they produce, particularly in tradable sectors where they have to compete with international prices. In this respect, the results presented here can be seen as an upper bound of the actual price/cost impacts on the economy. Finally, a harmonization exercise was carried out to ensure comparability across countries. We aggregated each country's IO matrixes into 27 sectors and industries. This process may imply loss of precision.

Price Shock Simulation

The central exercise in the analysis is simulating the effects of relative large price shocks on electricity and fuel prices for selected LAC countries. The exercise provides a conservative estimate of the potential effects of changing existing pricing policies and subsidies. It covers seven of the nine sample countries in the LAC region: Bolivia, Brazil, Colombia, the Dominican Republic, El Salvador, Mexico, and Peru—countries for which the data required for the analysis are available. The sample represents the regional diversity in size, geography, and net export status (net energy exporter versus net energy importer). It also takes into account the availability and quality of country data.

Each country individually faces one of the following shocks: a US$.25 per liter increase in different oil products (gasoline and diesel, liquefied petroleum gas [LPG]), or a US$.05 per kilowatt-hour increase in electricity prices. These are relatively large shocks (figure 4.3). Depending on the country, they result in an increase in electricity prices of from 17 percent to 77 percent, an increase in gasoline and diesel prices of from 17 percent to 59 percent, and an increase in LPG prices of from 32 percent to 184 percent. Indeed, the shocks are sufficiently

Figure 4.3 Price Shocks Relative to Energy Prices and Price Gap, Selected LAC Countries

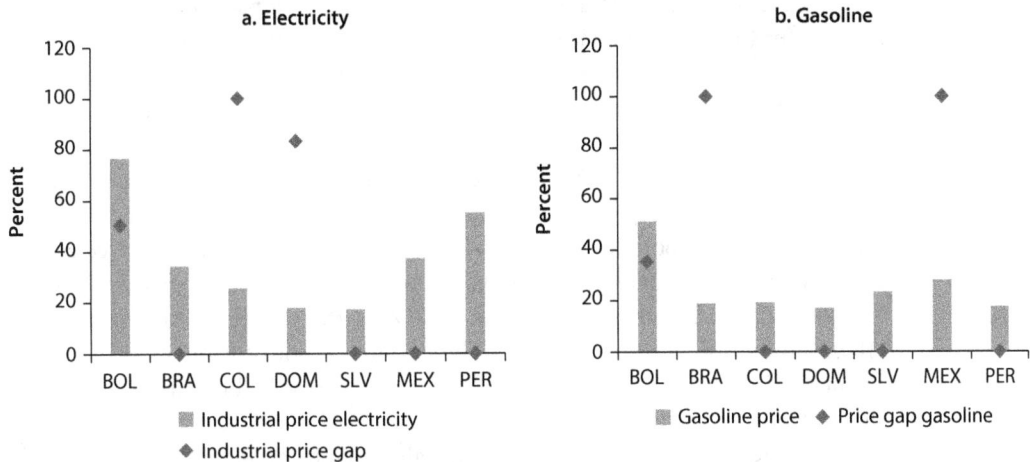

figure continues next page

Figure 4.3 Price Shocks Relative to Energy Prices and Price Gap, Selected LAC Countries *(continued)*

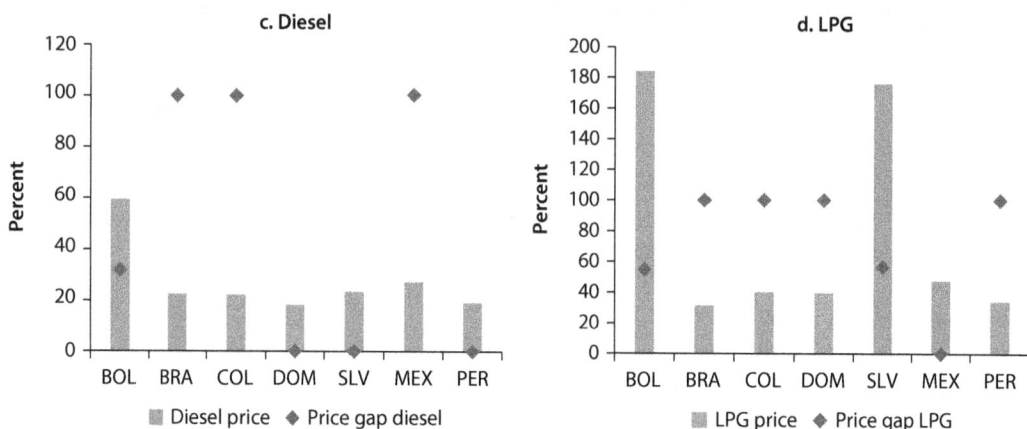

c. Diesel

d. LPG

■ Diesel price ◆ Price gap diesel

■ LPG price ◆ Price gap LPG

Source: World Bank calculations.

Note: In the figures, the orange bars plot the percentage change in electricity and fuel prices of a US$.05 per kilowatt-hour increase in electricity prices and a US$.25 per liter increase in fuel prices, respectively. The blue diamonds plot the shock as a share of the price gap, which has been capped at 100 percent. Where there is a 0 or negative price gap initially, the shock as a share of the price gap appears as 0. LPG = liquefied petroleum gas. For country codes, see note, figure 4.2.

large to close the existing price gap for each product in most countries (with Bolivia a notable exception).

One important caveat of this exercise is that it assumes subsidies are simply cut out of the economy and not reallocated elsewhere. This inevitably leads to negative outcomes, which manifest themselves in this case as higher prices. This issue is discussed in more detail later in the chapter.

Main Findings

Aggregate and Average Impacts

For most countries, the aggregate impacts of a shock on oil prices are relatively modest. Figure 4.4, panel a, shows the overall price impacts of a shock on oil prices after full transmission to all sectors of the economy. The figure reports the average cost of impacts per sector and the aggregate cost calculated as a weighted average of sectoral impacts, taking into account the sector's contribution to total value added.

For the sample countries, the aggregate and average impacts are relatively small, amounting to approximately half of a country's annual inflation rate. These impacts can be smoothed through relatively minor policy changes in countries that have room for policy adjustments. In only two countries in our sample is this not realistic. In El Salvador, inflation is so low that relatively small aggregate effects could double inflation rates. The country's room for policy responses is further reduced by its status as a dollarized economy. In Bolivia, the proposed shock represents a very large price increase. In this country, the impacts are almost as large as the annual inflation rate, although its monetary policy is slightly more flexible than that of El Salvador.

The aggregate cost impacts vary substantially across fuel products, with larger impacts arising from changes in transport fuels. Figure 4.4, panel b, compares the aggregate impacts of a shock on LPG prices to the impacts of a shock on transport fuel prices—gasoline and diesel. For all countries, the aggregate cost impacts of changes in transport fuel prices are significantly larger than those of changes in LPG prices, ranging from 0.5 percent for Mexico to 4 percent for Bolivia. This result reflects the importance of transport as an intermediary service in many sectors of the economy. Although the shock on LPG prices is larger than the shock on transport fuels in relative terms, the aggregate impacts are negligible, reflecting LPG's weak links with other sectors.

The aggregate impacts of a shock on electricity prices are even smaller and easier to accommodate. Figure 4.4, panel c, shows the overall price impacts of a

Figure 4.4 Overall Price Impacts of a Shock on Fuel and Electricity Prices, Selected LAC Countries

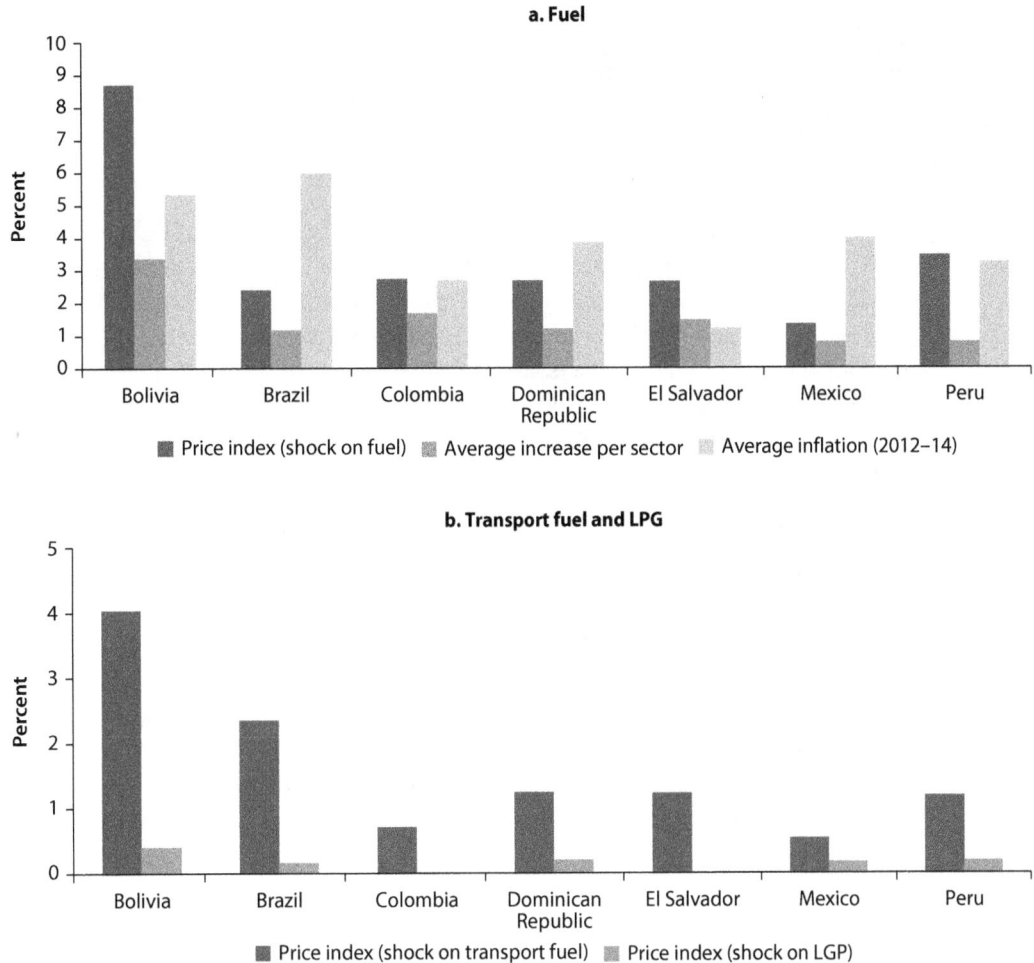

a. Fuel

Price index (shock on fuel) ■ Average increase per sector ■ Average inflation (2012–14)

b. Transport fuel and LPG

■ Price index (shock on transport fuel) ■ Price index (shock on LGP)

figure continues next page

Figure 4.4 Overall Price Impacts of a Shock on Fuel and Electricity Prices, Selected LAC Countries *(continued)*

c. Electricity

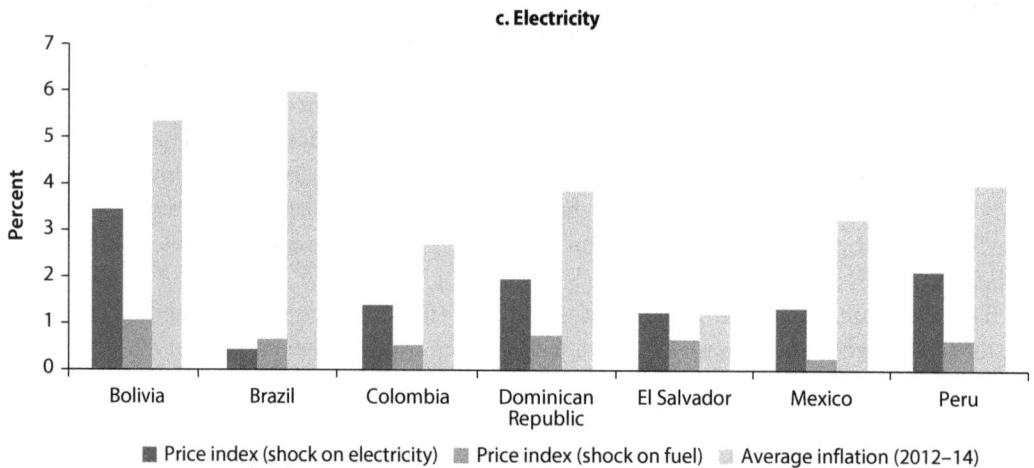

■ Price index (shock on electricity)   ■ Price index (shock on fuel)   ▨ Average inflation (2012–14)

Source: World Bank calculations.

Note: Panel a plots the percentage change in the price index and the average price increase per sector of a US$.25 per liter price shock on gasoline and diesel, as well as the average annual inflation rate from 2012 to 2014. Panel b plots the percentage change on the price index of a US$.25 per liter price shock on transport fuels (gasoline and diesel) and on LPG. Panel c plots the percentage change on the price index of a US$.05 per kilowatt-hour price shock on electricity and a US$.25 per liter price shock on fuels, as well as the average annual inflation rate from 2012 to 2014. LPG = liquefied petroleum gas.

shock on electricity prices. Across the sample, the aggregate and average impacts are small, amounting to approximately one-third of a country's annual inflation rate (one-half for El Salvador). As noted previously, these impacts can be accommodated through relatively small policy changes. Even for countries with little policy room, aggregate impacts are unlikely to cause major disruptions by themselves. Nevertheless, these results could hide large heterogeneities across sectors. If exporting or tradable sectors are disproportionally affected, for example, countries might experience a loss in competitiveness. Moreover, impacts might be concentrated in few activities that are disproportionally burdened by the changes.

Tradable and Nontradable Sectors

To explore the implications of energy shocks to competiveness, the results are broken down into tradable and nontradable sectors. Tradable sectors are agriculture, extractives, manufacturing, and utilities.[2] Construction and services are considered nontradable sectors. This categorization is overly simplistic and ignores nuances such as tradable segments within services, but because the services sector represents only a minor share of trade for most countries in the sample, the approximation seems reasonable.

The total impact in the nontradable sector is calculated as the weighted average of the impact in each industry, using the shares in total (nontradable) value added as weights. For the tradable sector, the total impact is calculated as the weighted average of the impacts in each tradable sector. Weights in this case are based on a tradability measure that captures the relative importance of imports in the consumption and of exports in the production of each sector. Sectors that are more exposed to international competition have a higher weight than inward sectors.

The tradability measure used is defined as

$$\text{Tradability}_{i,j} = \frac{\text{Exports}_{i,j}}{\text{Production}_{i,j}} + \frac{\text{Imports}_{i,j}}{\text{Consumption}_{i,j}} \qquad (4.1)$$

where i is a tradable sector in country j. The consumption of each sector is calculated as production plus imports minus exports.

As figure 4.5 reveals, the cost impacts of a shock on oil prices are larger on tradable segments than on nontradable segments. The same is true of a shock on electricity, with the exception of Mexico, where the impact on nontradables is higher. The disproportionate impacts on the tradable sectors are more acute for the oil shock than for the electricity shock. In Colombia, for example, the oil shock on tradables is 2.82 percentage points higher than the oil shock on nontradables, whereas this gap drops to 0.1 percentage points when electricity prices increase.

The relatively small impacts identified for shocks on electricity prices in this exercise can be easily accommodated by macroadjustments and exchange rate fluctuations. The issue of competiveness, however, may emerge as a legitimate concern in countries in which these policy tools are not available. This is especially true when facing a shock on oil prices, which has a significant impact on tradable sectors. The overall export cost impacts shown in figure 4.5 seem to be generally manageable by macropolicy tools. Even in Bolivia, the 7.3 percent export cost impact is less concerning when compared with the recent fluctuations in the nominal exchange rates in countries with floating regimes. Between 2012 and 2014, the average annual exchange rates depreciated almost 19 percent in Brazil, 12 percent in Colombia, 10 percent in the Dominican Republic, and about 7 percent in Mexico and Peru. However, even small impacts can be harsh on economies that are dollarized or pegged to the dollar because firms in the tradable sectors will likely absorb most of the cost impacts. Currently, this would be true for Bolivia and El Salvador in the study sample, although there are other examples in the region such as Ecuador, Nicaragua, and Panama.

Cost impacts on individual sectors can be significant. Transport is among the sectors most vulnerable to oil price shocks in all countries in the LAC region, and manufacturing of metal products is among those most affected by electricity price shocks. Naturally, the cost impacts of price shocks on specific sectors vary with the structure of the economy and the relative size of the initial shock.

Notwithstanding, a few common patterns emerge among almost all LAC countries. Figure 4.6 reports the cost impacts on the three most affected sectors in each country for both oil and electricity price shocks. The cost impacts of an oil price shock on the most affected sectors of the economy are three to four times larger than the aggregate cost effects, which range from 3.7 percent in Mexico to 13 percent in Bolivia. Transport is one of the two most affected sectors

Figure 4.5 Price Impacts of Price Shocks on Tradable Sectors and Export Baskets, Selected LAC Countries

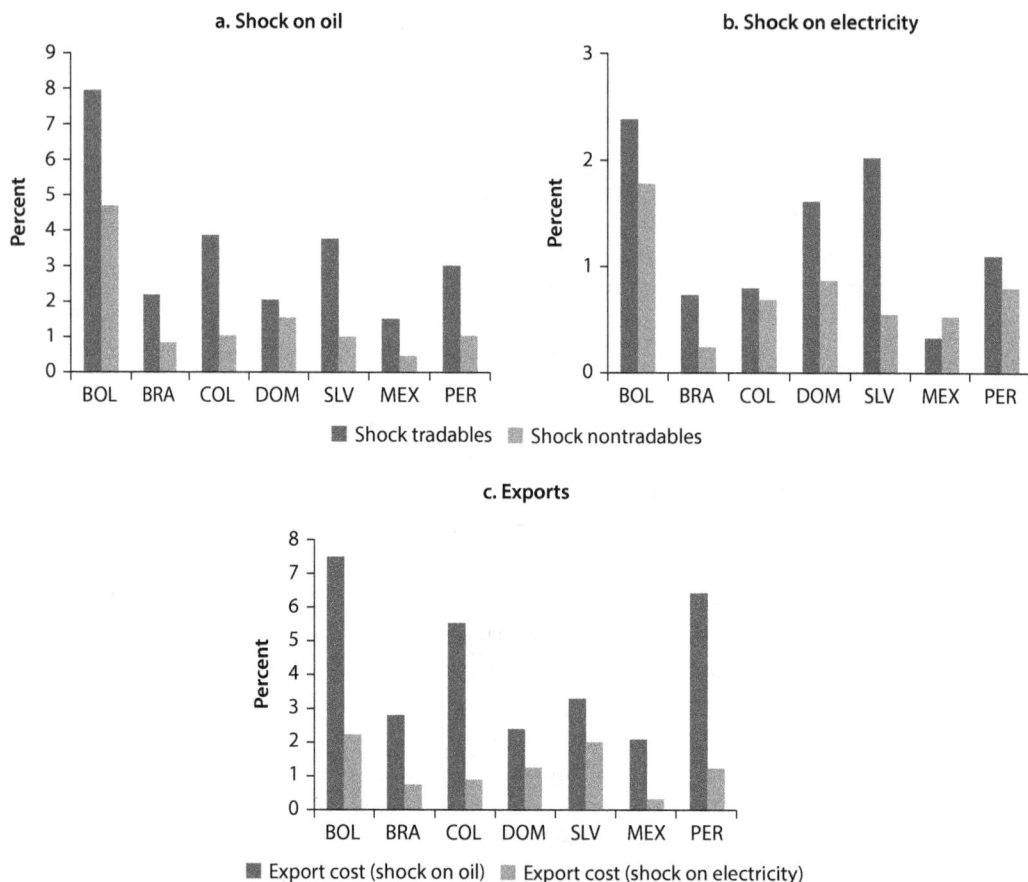

a. Shock on oil

b. Shock on electricity

■ Shock tradables   ■ Shock nontradables

c. Exports

■ Export cost (shock on oil)   ■ Export cost (shock on electricity)

Sources: Gross domestic product (GDP) data: World Bank, World Development Indicators (database); trade data: World Integrated Trade Solution (WITS).

Note: Panels a and b plot the price impacts on the tradable and nontradable sectors of a US$.25 per liter price shock on gasoline and diesel and a US$.05 per kilowatt-hour price shock on electricity, respectively. The shocks for the tradable sectors are calculated as a weighted sum of the price impacts in each tradable sector, weighted by a tradability index specific to the sector. The shocks for the nontradable sectors are calculated as a weighted average of the price impacts in each nontradable sector, weighted by the share of each nontradable sector in the total value added. Panel c plots the percentage change in the export price index of a US$.25 per liter price shock on gasoline and diesel and a US$.05 per kilowatt-hour price shock on electricity. It is the weighted sum of the sector's specific shocks, weighted by the share of each sector in total exports. For country codes, see figure 4.2.

across all countries in the sample. Electricity is significantly affected for those countries that rely on oil for power generation. Heavy manufacturing and mining are also affected in some countries.

In an electricity price shock, the sectoral impacts are also approximately three to four times larger than the aggregate impacts but relatively smaller in size. Two important exceptions to this pattern are water services in El Salvador and mining activities in the Dominican Republic. These activities are atypically reliant on electricity as a primary input, possibly reflecting country-specific factors such as geography. Although the vulnerable sectors vary from country to country, manufacturing of metal products appears among the top three most affected sectors in all countries.

The sectoral impacts of price shocks can be better assessed when compared with sector profit margins. If a specific sector has a large profit margin, it is more likely to accommodate larger changes in total cost. Similarly, if a sector has a small profit margin, even small changes in costs can undermine a firm's profitability. This factor helps explain the resistance of some sectors to subsidy reforms. Table 4.1 compares the estimated cost impacts of price shocks to operating margins in four economic sectors. The table reports profit margins in the United States as a proxy for sectoral margins in the region. This is a significant shortcut, but it is an unavoidable one because of data constraints. Although the cost effects of price shocks in low-impact countries are relatively small compared with sectoral margins, the cost effects in high-impact countries such

Figure 4.6 Top Three Sectors Most Affected by Subsidy Reforms, Selected LAC Countries

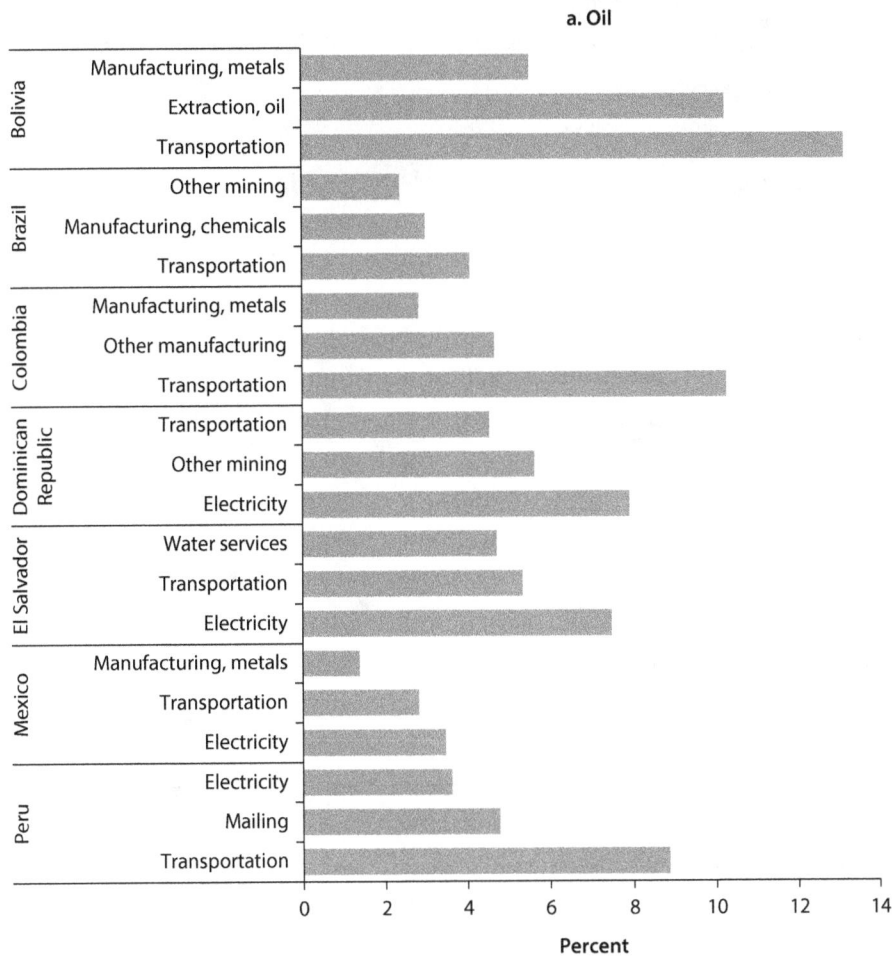

a. Oil

figure continues next page

Energy Pricing Policies for Inclusive Growth in Latin America and the Caribbean
http://dx.doi.org/10.1596/978-1-4648-1111-1

Figure 4.6 Top Three Sectors Most Affected by Subsidy Reforms, Selected LAC Countries *(continued)*

b. Electricity

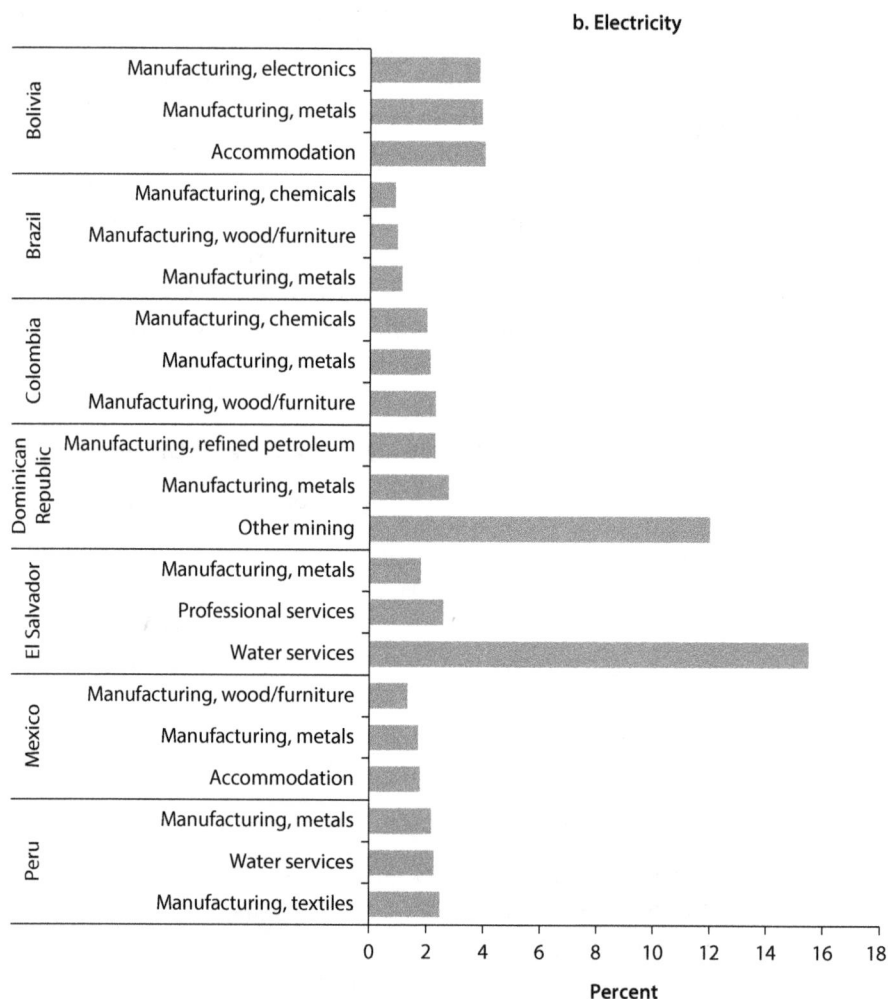

Source: World Bank calculations.
Note: The figures plot the percentage change in prices of the top three most affected sectors by country of a US$.25 per liter price shock on gasoline and diesel and a US$.05 per kilowatt-hour price shock on electricity.

Table 4.1 Cost Impacts versus Profit Margins in Selected Sectors of LAC Countries
percent

Sector	Cost impacts (oil shock)	Cost impacts (electricity shock)	Average operating profit margins
Agriculture	0.9–2.0	0.3–0.7	5–15
Accommodation	0.6–1.8	0.5–4.0	16–18
Manufacturing, textile	0.6–3.3	0.7–2.6	5
Manufacturing, electronics	0.6–5.1	0.8–3.8	8

Sources: World Bank calculations and IBISWorld sector reports.
Note: Average operating profit margins are calculated for the United States. Agriculture profit margins refer to soy, vegetables, and cattle. Accommodation profit margins refer to large hotels and casinos and small hotels and bed and breakfasts.

as Bolivia can be significant because firms' profitability is undermined compared with other economic activities. Without compensatory policies or adjustments, price shocks could swallow up to 40 percent of agriculture profits and about two-thirds of manufacturing profit margins.

These results help shed light on the political economy aspects of reforms. When one looks at the aggregate results, it seems puzzling that subsidy reforms are so unpopular. And yet when one considers specific sectoral impacts, it makes more sense why certain groups emerge as vocal opponents of subsidy reforms and dedicate significant resources to lobbying against them.

Final Considerations

This chapter analyzes some of the economic impacts—price, competitiveness, and sectoral—associated with energy price increases. The analysis uses the well-established Leontief input-output approach to simulate the sectoral impacts associated with the full transmission of oil and electricity price shocks to the economies of seven LAC countries. The exercise considers relatively large price shocks and provides upper-bound estimates of the potential impacts in each country.

The results of our analysis suggest that aggregate price impacts and competitiveness effects are moderate and can be smoothed through macropolicies in countries in which there is room for policy responses. However, impacts on specific sectors can be large and undermine profitability. The aggregate price effects associated with both oil and electricity price shocks are relatively small compared with each country's annual inflation rates and can be smoothed through monetary policy adjustments. The only exception is Bolivia, where the initial price shocks and overall impacts are much larger than in other countries.

Overall competiveness effects, assessed via changes in the cost of tradable products and the export basket, are similar to overall price effects. Even the largest competitiveness impacts could be potentially mitigated by exchange rate fluctuations. However, a few countries in the region have fixed exchange regimes or are dollarized economies. As a result, they lack the room for such policy responses. Finally, the cost impacts in specific sectors can be significant. Transport and electricity generation are among the sectors most affected by oil price shocks. Manufacturing of metal products is particularly vulnerable to changes in electricity prices.

The analysis presented in this chapter focuses only on the implications of cutting energy subsidies. The emphasis in this analysis on the costs of extracting resources from the economy inevitably leads to negative outcomes. In practice, however, the net impacts of subsidy reforms are likely to be positive because the fiscal resources saved by subsidy reform revert back to the economy through lower taxes or public spending and investments. Moreover, subsidy reforms should be thought of as a package that should include alternative uses for saved resources, as well as complementary reforms that can help increase efficiency and compensate for the additional cost burden in vulnerable sectors.

Energy Pricing Policies for Inclusive Growth in Latin America and the Caribbean
http://dx.doi.org/10.1596/978-1-4648-1111-1

Our findings can help inform decisions about where public resources could be allocated to mitigate large cost impacts. They can also be used to assess which structural reforms may prove the most effective at increasing efficiency and compensating for a larger cost burden. For example, the transport sector is heavily affected by changes in oil price changes. It is also plagued by inadequate infrastructure and low connectivity in many countries in the LAC region (Calderón and Servén 2010). Thus well-thought-out investments in road infrastructure may help reduce transport costs. Regulatory changes could also play a role. For example, Pachón, Araya, and Saslavsky (2013) suggest that regulations preventing trucks from picking up cargo in another country are one of the main causes of high transport costs in Central America. Changing this regulation could fully compensate for the estimated impacts of the subsidy reform.

Notes

1. $\Delta p' = \Delta v'(I - A)_{-1} = \Delta v'L$, where p' is a row vector of prices of goods produced in each sector; v denotes a row vector of primary inputs; I is the identity matrix; A is the matrix of input-output coefficients; and $(I - A)_{-1}$ is the Leontief inverse L or total requirements matrix.
2. The utilities sectors in Bolivia and the Dominican Republic are considered to be nontradable, according to our tradability measure.

References

Arezki, Rabah, and Thiemo Fetzer. 2016. "On the Comparative Advantage of U.S. Manufacturing: Evidence from the Shale Gas Revolution." CEP Discussion Paper dp1399, Centre for Economic Performance, London School of Economics and Political Science.

Calderón, César, and Luis Servén. 2010. "Infrastructure in Latin America." Policy Research Working Paper 5317, World Bank, Washington, DC.

Coady, David, Robert Gillingham, Rolando Ossowski, John Piotrowski, Shamsuddin Tareq, and Justin Tyson. 2010. "Petroleum Product Subsidies: Costly, Inequitable, and Rising." IMF Staff Position Note 10/05, International Monetary Fund, Washington, DC.

Coady, David, and David Newhouse. 2006. "Ghana: Evaluating the Fiscal and Social Costs of Increases in Domestic Fuel Prices." In *Poverty and Social Analysis of Reforms: Lessons and Examples from Implementation*, edited by A. Coudouel, A. Dani, and S. Paternostro. Washington, DC: World Bank.

Pachón, María Claudia, Gonzalo Araya, and Daniel Saslavsky. 2013. *Road Freight in Central America: Five Explanations for High Costs of Service Provision*. Report 75100. Washington, DC: World Bank.

Parry, Ian W. H., Dirk Heine, Eliza Lis, and Shanjun Li. 2014. *Getting Energy Prices Right: From Principle to Practice*. Washington, DC: International Monetary Fund.

CHAPTER 5

Distributional Impacts of Energy Pricing

Introduction

Policy makers dealing with subsidy reform often cite concerns about the impacts of such reform on poor and vulnerable households. Subsidy reform advocates counter with statistics that reflect how the lion's share of each dollar spent on fuel subsidies ends up in the pockets of wealthier households. The reality is that attempts to reform or eliminate subsidies have often been met with fierce resistance from citizens, especially the poorer ones. At the heart of this discussion is a preoccupation with the distributional impacts of eliminating subsidies.

To understand such an impact, it is necessary to consider two important effects: direct and indirect. On the one hand, households will be affected because of their *direct* consumption of energy products—that is, if subsidies for gasoline were removed, customers would have to pay more at the pump. Likewise, if electricity subsidies were reformed, consumers could expect heftier electricity bills. This, then, is the direct effect of a price increase in energy. However, households also consume energy *indirectly* because energy is an input in the production of many consumer goods. Thus increases in energy prices will translate into changes in production costs. To the extent that they are passed on to consumers, they will likely imply higher prices for consumer goods as well. This, then, is the indirect effect of a price increase in energy products.

Another important issue often overlooked in distributional studies is the proportion of households—particularly, lower-income households—that actually consume energy or have access to energy products. Consequently, we present two measures of the direct welfare impact by income decile of the expenditure distribution: the conditional average and the unconditional average. The unconditional average reflects the average effect for *all* households within a decile—that is, it includes households that purchase an energy product as well as those that do not. The conditional average reflects the average effect *only* for those households that actually purchase a particular energy product. Presenting both measures is important for at least two reasons. First, it highlights the fact that the regressive nature of some energy subsidies derives substantially from poorer households not consuming specific energy products (sometimes because they simply do not have access to them). Second, it demonstrates a tricky policy dilemma: poor households that consume energy products are hard-hit by higher energy prices (conditional average). This impact can be hidden because the impact on lower-income households as a whole may remain small (unconditional average).

Our findings reveal that the conditional average welfare effect can be substantially higher than the unconditional average welfare effect. For example, the conditional average effect of a price increase in liquefied petroleum gas (LPG) can be three times larger than its unconditional equivalent. In addition, focusing on conditional average effects can significantly alter the distributional profile of subsidy benefits. Even for gasoline subsidies—which are generally the most regressive of all subsidies—the conditional average effect is more important for poorer households than for richer ones. Thus policy makers will have to balance the relatively low impact on poorer households as a whole with the significant impact on the relatively few lower-income households that do consume energy products.

Consistent with the previous literature, we find that indirect effects significantly increase welfare losses and markedly alter the distributional profile of these losses. For gasoline and diesel subsidies, the indirect effect is actually the largest contributor to welfare loss. Moreover, the indirect channel generally affects the poorest income deciles the most, whereas its importance diminishes for the richer deciles. Thus subsidy reform strategies should include compensatory measures that account for both the direct and indirect welfare losses that households would face.

It is hoped that our analysis clearly reveals that any discussion of the regressivity or progressivity of energy subsidies is more nuanced than generally presented. In this respect, it is important to clarify two distinct distributional concepts: that of *relative progressivity* and that of *absolute progressivity*. In the context here, a subsidy scheme is said to be relatively progressive (regressive) if after a price increase—or subsidy reduction—the percentage increase in expenditures is higher (lower) for low-income households. This statement will prove true if low-income households' share of the subsidy benefits is higher than their share of the total expenditures. A subsidy is absolutely progressive (regressive) if after a price increase—or subsidy reduction—the increase in expenditures of low-income households is higher (lower) than the population share of these households. This statement will prove true if, for example, the increase in the expenditures of the bottom 20 percent of the population is more than 20 percent of the total increase in expenditures.

There is no doubt that energy subsidies are regressive in the absolute sense. Even accounting for indirect effects (which disproportionately affect the poor), between 40 percent and 60 percent of gasoline and diesel subsidies end up favoring the richest 20 percent of households, and only 10–20 percent of the benefit goes to the bottom 40 percent. Even for LPG, which enjoys one of the most progressive price subsidies, 34 percent of the benefits are enjoyed by the wealthiest 20 percent on average, whereas only 23 percent is received by the bottom 40 percent. In other words, of every dollar spent on LPG subsidies, 34 cents go to the 20 percent richest households and only 23 cents end up in the pockets of the 20 percent poorest households. However, this kind of analysis misses a very important point: those 23 cents can represent a lot to poorer households.

Our findings suggest that some energy subsidies can be relatively progressive or neutral. Although gasoline and diesel subsidies are regressive in both the absolute and relative senses, those for LPG, as noted, are relatively progressive (although absolutely regressive)—that is, poorer households are capturing a higher share of benefits relative to their expenditures from LPG subsidies. These findings help explain why subsidy reform efforts are often met with strong opposition from consumers.

The sections that follow first briefly describe the methodology used in our analysis and its implementation, followed by the results for households. In that discussion, energy products are divided into three broad groups: (1) petroleum-based fuels—gasoline and diesel—that are mainly used by households for private transport; (2) LPG, a fuel mainly used for cooking; and (3) electricity.

Methodology and Implementation

Any assessment of the distributional impact of removing subsidies and reforming energy pricing mechanisms must rely on an understanding of the consumption patterns of households—not only the *direct* consumption of energy products (such as buying gasoline for one's car or paying an electricity bill), but also the *indirect* consumption of energy products, which may be even more important. As highlighted in chapter 4, many of the final goods consumed by households have energy embedded in them. Therefore, changes in energy prices will translate into changes in production costs, with industries that are energy-intensive affected the most. As a result, to correctly identify the distributional impact of changing energy prices, one has to assess two distinct effects: the distributional effects of the direct consumption of energy products and the indirect effects of changes in the prices of all other consumer goods that have energy embedded in them.

For the purposes of our analysis, we adopt a methodology that has become standard for assessing the distributional impact of subsidy reform (see, among others, Coady and Newhouse 2006; Coady et al. 2006; Kpodar 2006). This methodology relies on estimating a monetary measure of compensation that would leave households indifferent to facing low (before the price change) or high (after the price change) energy prices. In short, we estimate the amount of money that if offered to households would leave them indifferent between the two levels of price.

The proposed methodology allows us to clearly disentangle the direct and indirect effects. Moreover, it is relatively straightforward to implement and easy to understand. For the analysis, there are essentially two data requirements, an input-output (IO) matrix of the economy and a household expenditure survey.

The IO matrix depicts the connection between sectors of the economy, showing how any one sector is both a supplier of inputs and a buyer of outputs from other industries. The IO matrix is used with two purposes: (1) to identify how increases in costs resulting from increases in energy prices will propagate throughout the economy; and (2) to ascertain how much the prices of final consumer goods will rise. We use household expenditure surveys to describe the consumption baskets across the expenditure distribution, allowing for the distinction between direct expenditures on energy products and expenditures on all other consumer goods.

The analysis is developed in three stages. The first stage assesses the magnitude of the direct effect, which is estimated by multiplying the share of a household's budget allocated to a particular energy product by the percentage price increase of that energy product. The second stage calculates in two steps the magnitude of the indirect effect: (1) the change in prices of all goods and services (excluding energy products) consumed by households is estimated using IO matrixes (as described in the previous chapter); and (2) the indirect effect is obtained by multiplying the shares of a household's budget devoted to the various goods and services by the estimated percentage price increase in each of these goods and services. Finally, the total welfare effect on a household is calculated as the sum of both effects. To assess the total distributional impacts, we look at the magnitude of the total welfare effect across the distribution of households' per capita expenditure.

The main advantage of this approach is that it delivers sensible estimates of the impact of changes in the price of energy products on households based on a relatively simple methodology and just a few data requirements that are met by many countries in the region. However, there are some limitations. The proposed approach delivers estimates of welfare impacts that ignore any substitution effects at the household level.[1] In other words, the estimates do not account for the fact that households shift their consumption baskets toward relatively cheaper goods as products become more expensive. Moreover, the methodology assumes that cost increases for industries will be fully passed on to consumer prices. Both assumptions bias the results toward overestimating the magnitude of the welfare effects. However, as noted by Deaton (1989), if the price elasticity of households does not depend on the level of income, then the relative distributional effects will be measured accurately, even if the absolute welfare impact is overestimated. In addition, because the methodology relies on the use of matrixes to trace the impact of changing energy prices on the prices of all other goods and services, it is subject to all the assumptions and limitations of the IO model described previously.

Some data constraints are present as well. Because both household expenditure surveys and IO matrix calculations are carried out infrequently, the reference year of the household survey will generally not match the reference year of the IO matrix. Although it is standard to assume that the proportion of inputs does not change significantly from year to year, this discrepancy is an unavoidable feature of the data. Our goal in this chapter is not only to identify the distributional impacts within countries, but also to compare and contrast the effects across countries.

In this regard, some data constraints do limit cross-country comparisons. First, as noted, the reference year of the household survey and the IO matrix will typically be different across countries. Second, because we use expenditure surveys designed by national statistical agencies rather than standardized surveys, there will be cross-country differences in the order, number, and level of detail of questions. Finally, as noted earlier, the exact composition of industries in each broad sector definition used in the IO matrixes may differ across countries.

Another limitation of this approach is that it does not consider general equilibrium effects through, for example, the labor market. Changes in domestic energy prices will affect input costs and have significant effects on the relative prices of other goods and services. As the relative prices of both change, some sectors will expand while others will contract. Depending on the relative factor intensities used in different sectors, this will cause changes in relative factor demand. Consequently, wages would likely adjust, and, in particular, wages for different types of labor (skilled, semiskilled, and unskilled) would likely adjust in different ways.

In Latin American countries, labor income is generally the main component of total household income. Therefore, ignoring the general equilibrium effects on wages could substantially bias the estimated welfare effects on households. Moreover, because households differ in the type of labor they supply, labor income effects could be heterogeneous across the expenditure distribution. This heterogeneity implies that the relative distributional effects of energy pricing reforms could be biased as well. Because these effects are the result of complex general equilibrium interactions, it is hard to assess a priori the direction of the bias.

We adapted the methodology developed in Porto (2006) to analyze the welfare impacts of changes in domestic energy prices on households. This approach takes into account the effects on households both as consumers *and* as income earners. Moreover, the analysis considers the general equilibrium effects on prices of nontraded goods as well as the general equilibrium effects on labor income as factor demand and wages adjust. Although our results suggest no statistically significant impact on wages, further research is needed to confirm the results.[2] Meanwhile, policy makers should remain attentive to changes in employment and wages, particularly in industries that rely heavily on oil and electricity as inputs.

In contrast to other energy products, electricity typically does not have a unique price for all residential consumers. Most utilities have tariff schedules that discriminate among consumers. This price discrimination is generally based on total consumption, but it sometimes relates to specific regions, proxies of household wealth, or some combination of these. The implication is that the distributional analysis of changes in electricity prices requires the additional step of matching a household to the correct tariff category. Thus we develop country-specific algorithms that use household expenditure data on electricity and the electricity tariff schedule to match households to the correct tariff category.

In what follows, we present the welfare effects of price increases in energy products on households. The three broad groups of energy products—gasoline and diesel (primary transport fuels), LPG (used mostly for cooking), and electricity—not only mirror the distinct purpose of each energy product, but also reflect the differences in magnitude and composition of the welfare effects. As noted earlier, the results presented simulate the impact of an absolute price increase in each energy product category. For gasoline and diesel and for LPG, we estimate the welfare impact of a US$.25 per liter price increase. Because LPG is cheaper than gasoline or diesel, this absolute price implies a much larger percentage increase. For electricity, we measure the welfare effects of a US$.05 per kilowatt-hour increase in all consumption block tariffs.

Distributional Impacts of a Price Increase in Gasoline and Diesel

Because countries in the Latin America and the Caribbean (LAC) region have widely different prices for gasoline and diesel, an absolute price increase is a very different percentage increase in each country (table 5.1).

The results of our analysis are presented in three steps: (1) direct effects, (2) indirect effects, and (3) total welfare impacts across the expenditure distribution. The results focus on the seven countries in our sample (Bolivia, Brazil, Colombia, the Dominican Republic, El Salvador, Mexico, and Peru) for which the data required for the analysis were available.

Direct Effects

The direct benefits of keeping gasoline and diesel prices low are disproportionately captured by the richest deciles of the expenditure distribution. As mentioned earlier, the direct effect relates to how much households actually spend on gasoline and diesel. These expenditures in turn are inextricably linked to the ownership of vehicles (motorcycles or cars) by a household. In fact, the correlation between households that report positive consumption of transport fuels and those that own vehicles is above 0.99 for all countries in our sample. Therefore, because the distribution of vehicle ownership is highly skewed toward the higher-income deciles (see figure 5.1), it is the richest households that retain the highest direct benefit from subsidized gasoline and diesel prices.

Table 5.1 Percentage Increase in Transport Fuel Prices from a US$.25 per Liter Price Increase, Selected LAC Countries

percent

Country (reference year)	Gasoline	Diesel
Bolivia (2012)	51.2	59.2
Brazil (2008)	18.3	22.5
Colombia (2009)	30.4	37.5
Dominican Republic (2008)	19.9	21.9
El Salvador (2011)	26.7	25.8
Mexico (2012)	32.3	31.3
Peru (2011)	33.3	30.7

Source: World Bank staff survey.

Analysis of the unconditional average effects per income decile reveals the regressive nature of transport fuel subsidies. In all households, regardless of whether they actually purchase these fuels, the share of total expenditure devoted to gasoline or diesel increases with the overall level of expenditure. In other words, higher-income deciles spend a higher share of their income on transport fuels. This result is not unexpected because the richest deciles report a higher proportion of households with positive expenditures on fuels, and thus we are including fewer households that report zero expenditures. The orange bars in figure 5.2 confirm, then, that an increase of US$.25 per liter affects the richest segment of the distribution the most. Conversely, the benefits of subsidies and pricing policies that keep transport fuels inexpensive are concentrated at the top of the distribution.

However, when considering only the households that consume gasoline and diesel (the conditional average), the message is less straightforward. For countries such as Bolivia, Brazil, Colombia, and Mexico, the direct welfare impact of a US$.25 per liter increase is actually relatively more important for the poorer

Figure 5.1 Vehicle Ownership by Income Decile, Selected LAC Countries

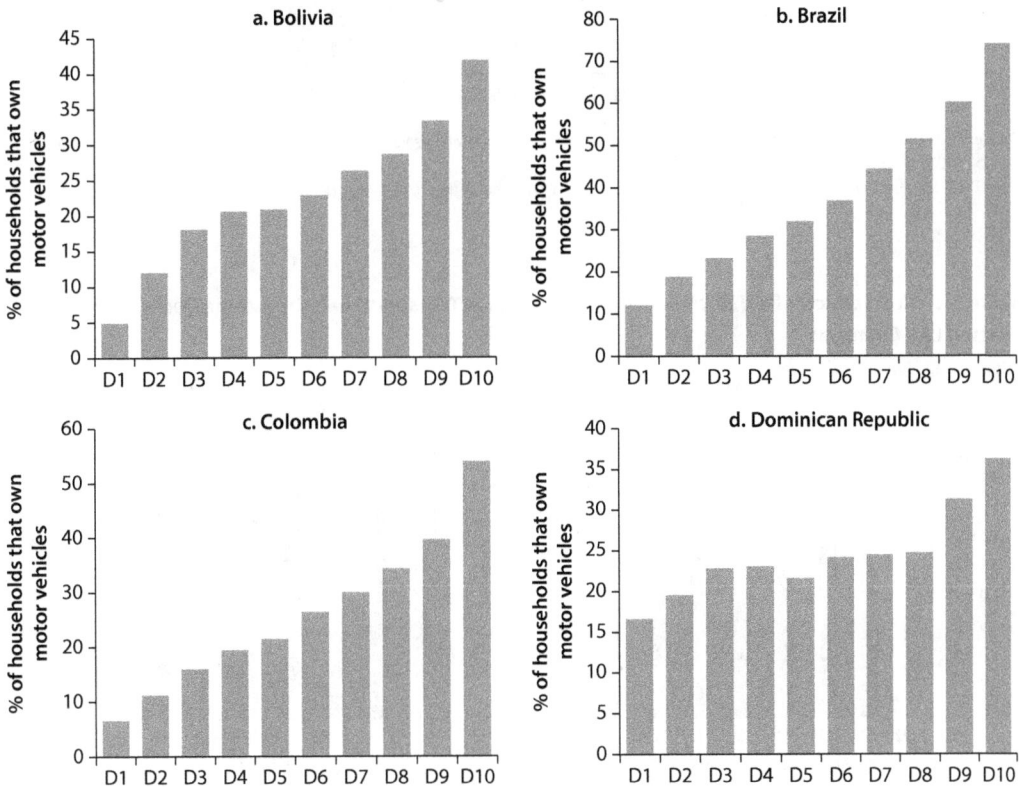

figure continues next page

Figure 5.1 Vehicle Ownership by Income Decile, Selected LAC Countries *(continued)*

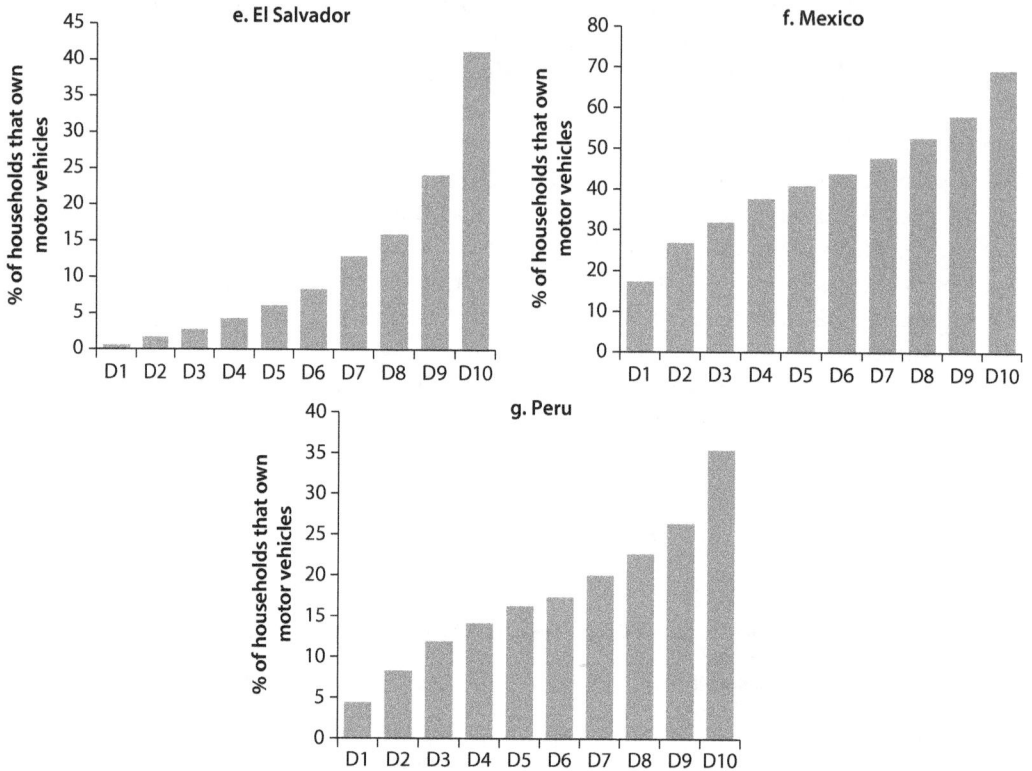

Source: Calculations using national household expenditure surveys for the following years: Bolivia, 2012; Brazil, 2008; Colombia, 2009; Dominican Republic, 2008; El Salvador, 2011; Mexico, 2012; Peru, 2011.

Note: The figures show by income decile the average percentage of households that own at least one vehicle.

Figure 5.2 Direct Effects of a US$.25 per Liter Price Shock in Transport Fuels, by Income Decile, Selected LAC Countries

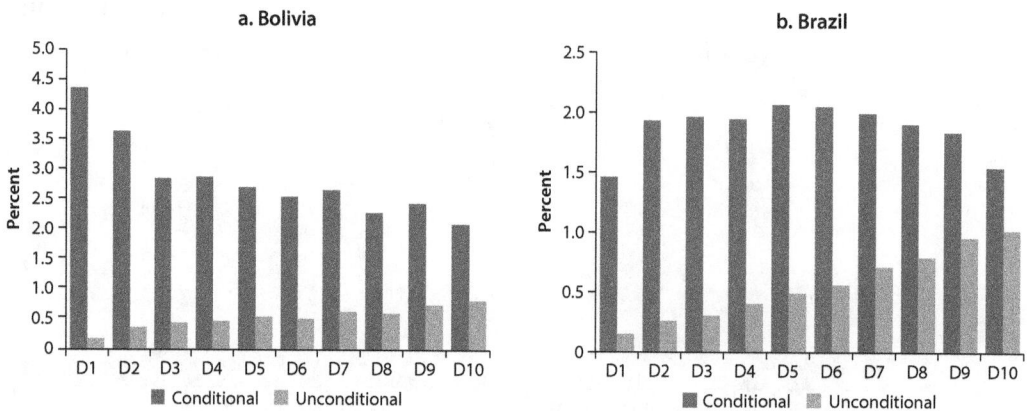

figure continues next page

Figure 5.2 Direct Effects of a US$.25 per Liter Price Shock in Transport Fuels, by Income Decile, Selected LAC Countries *(continued)*

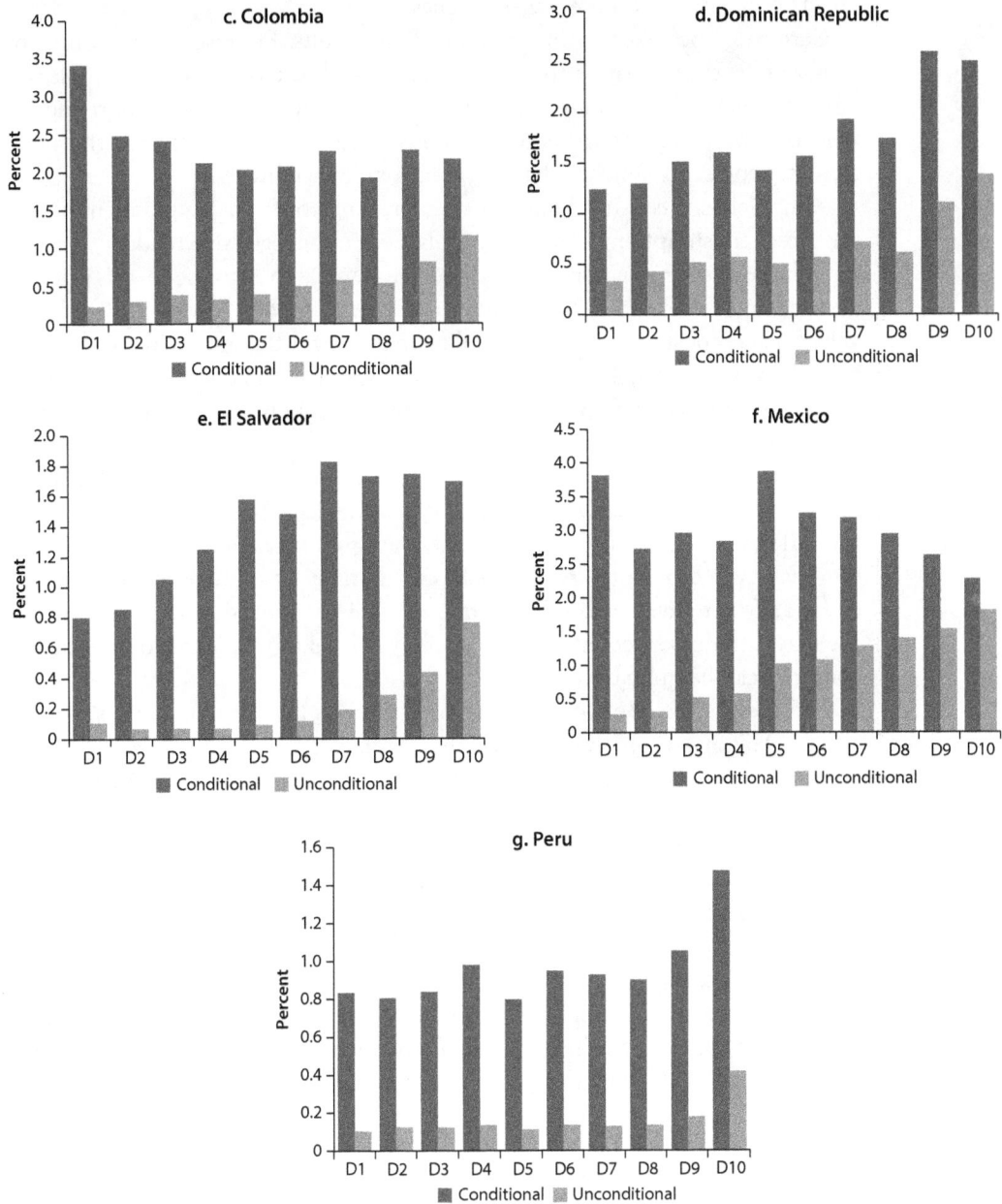

c. Colombia

d. Dominican Republic

e. El Salvador

f. Mexico

g. Peru

Source: Calculations using national household expenditure surveys and each country's input-output (IO) matrixes. Surveys are for the following years: Bolivia, 2012; Brazil, 2008; Colombia, 2009; Dominican Republic, 2008; El Salvador, 2001; Mexico, 2012; Peru, 2011.
Note: The figures show the average conditional and unconditional effects by income decile of a US$.25 per liter price shock in transport fuels (gasoline and diesel). Conditional effects are conditional on positive consumption of transport fuels. Unconditional effects include all households.

deciles (blue bars of figure 5.2) than for the richer deciles. However, for the Dominican Republic, El Salvador, and Peru the converse is true, with the larger effects concentrated on the richest deciles.

Care must be taken in interpreting these results. Because the measures of welfare effects are relative to the level of expenditure of a household, a larger effect for poorer households does not imply that they would spend more money than richer households (in absolute terms). Instead, it suggests that the price increase represents a higher share of their total expenditures.

Nonetheless, the results highlight an interesting policy challenge. Even though the direct consumption of transport fuels has a clear regressive tendency (richer households consume gasoline and diesel the most), an increase in the price of gasoline and diesel can have a substantial impact on poor households that own vehicles and consume these fuels. Policy makers will therefore have to balance the relatively low impact on poorer households as a whole with the significant impact on the relatively few lower-income households that consume gasoline and diesel.

The magnitude of the direct effect per decile across countries ranges from 0.07 percent in El Salvador to 1.81 percent in Mexico. Three factors may explain the differences between countries. First, the proportional change can be quite different, even though the price shock is the same in absolute terms (an increase of US$.25 per liter). Second, the rate of vehicle ownership and intensity of use may differ across countries. And, third, the level of total expenditures across countries can be quite different.

The average direct effect across income levels is highest in countries such as Brazil (0.62 percent), the Dominican Republic (0.74 percent), and Mexico (1.00 percent), even though they do not have the largest proportional price shocks. Instead, these countries have the highest vehicle ownership rates in our sample, measuring 27.5 percent, 26.0 percent, and 38.2 percent of the total population, respectively. By the same logic, countries such as El Salvador and Peru, which have the lowest vehicle ownership rates in the study sample (10.8 percent and 16.6 percent, respectively) have the lowest direct effects. This is in spite of having a price shock comparable to that of Mexico and 50 percent higher than those of Brazil and the Dominican Republic. The results therefore suggest that the rate of vehicle ownership is key to understanding the difference in magnitudes of the direct effect across countries.[3]

Indirect Effects

Households consume gasoline and diesel not only directly through their purchases at gas stations, but also indirectly. Because gasoline and diesel are used in many production processes, the rise in the cost of these inputs will likely translate into higher prices for many goods—assuming full pass-through to consumers. Moreover, transport costs would likely increase substantially because the transport sector is an intensive user of these fuels. Although the level of cost increases will vary depending on the energy-intensiveness of different goods and services, policy makers should expect consumers to face higher prices.

In what follows, we account for these indirect effects and assess the differences across the expenditure distribution.

Households across the expenditure distribution have very different consumption baskets. More important, these baskets differ in how much gasoline and diesel is embedded in them, which is a key determinant of how much they will be affected by increasing costs. To illustrate the significant differences in the consumption patterns of the poor and rich, figure 5.3 contrasts the average consumption basket of the second-poorest decile (decile 2) with that of the second-richest decile (decile 9). As expected, the consumption basket of the poorer households is dominated by food purchases. The basket of richer households has important contributions from the value of housing (because they reside in larger and more expensive dwellings) and expenditures on health and education, as well as more luxury expenses such as travel and restaurants.

Figure 5.3 Average Consumption Baskets of Income Decile 2 versus Income Decile 9, Selected LAC Countries
percent

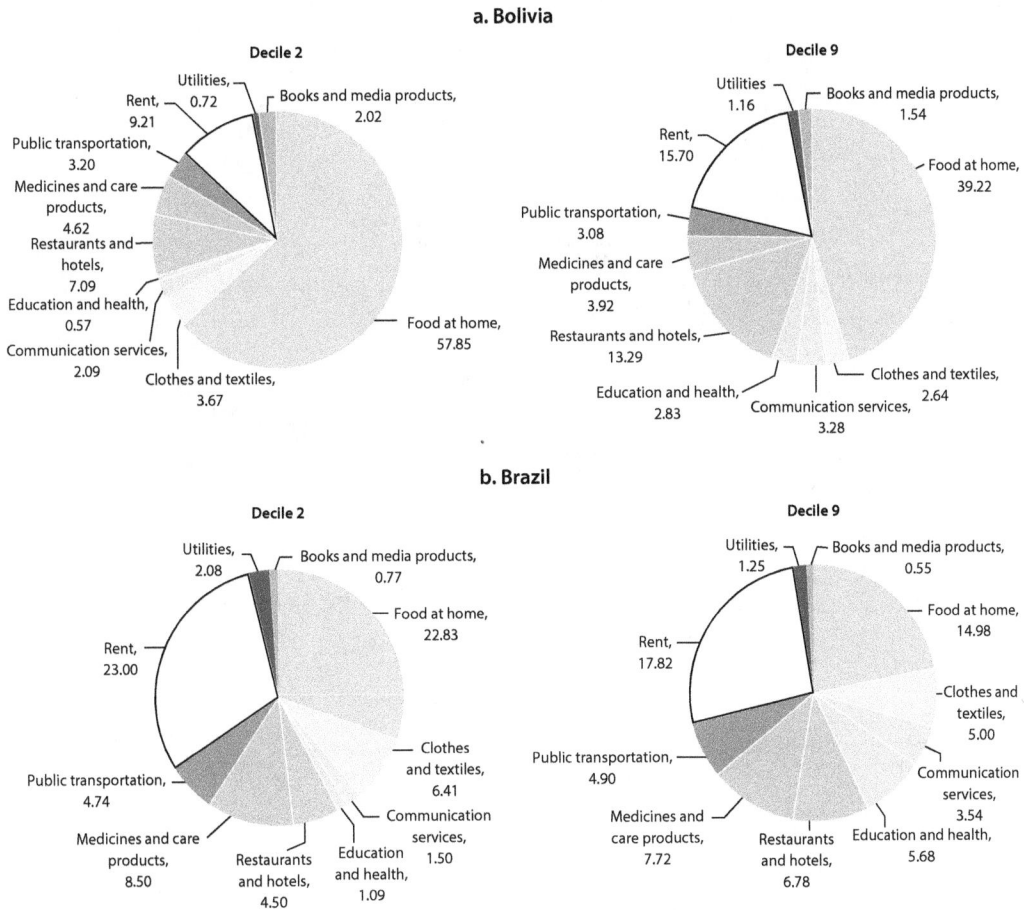

a. Bolivia

b. Brazil

figure continues next page

Figure 5.3 Average Consumption Baskets of Income Decile 2 versus Income Decile 9, Selected LAC Countries (continued)

c. Colombia

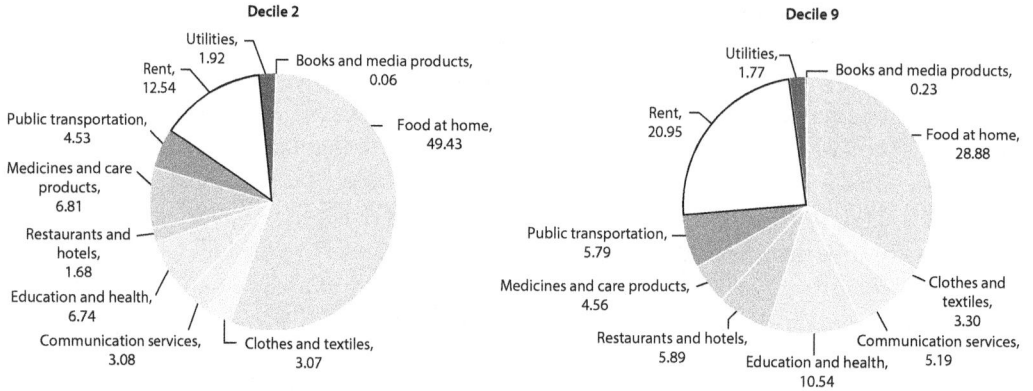

Decile 2

- Utilities, 1.92
- Rent, 12.54
- Public transportation, 4.53
- Medicines and care products, 6.81
- Restaurants and hotels, 1.68
- Education and health, 6.74
- Communication services, 3.08
- Books and media products, 0.06
- Food at home, 49.43
- Clothes and textiles, 3.07

Decile 9

- Utilities, 1.77
- Rent, 20.95
- Public transportation, 5.79
- Medicines and care products, 4.56
- Restaurants and hotels, 5.89
- Education and health, 10.54
- Books and media products, 0.23
- Food at home, 28.88
- Clothes and textiles, 3.30
- Communication services, 5.19

d. Dominican Republic

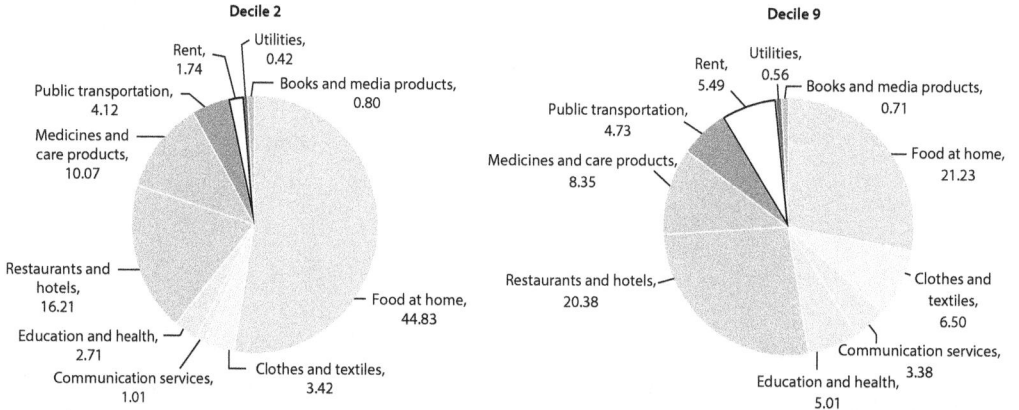

Decile 2

- Rent, 1.74
- Utilities, 0.42
- Public transportation, 4.12
- Medicines and care products, 10.07
- Restaurants and hotels, 16.21
- Education and health, 2.71
- Communication services, 1.01
- Books and media products, 0.80
- Food at home, 44.83
- Clothes and textiles, 3.42

Decile 9

- Rent, 5.49
- Utilities, 0.56
- Public transportation, 4.73
- Medicines and care products, 8.35
- Restaurants and hotels, 20.38
- Education and health, 5.01
- Books and media products, 0.71
- Food at home, 21.23
- Clothes and textiles, 6.50
- Communication services, 3.38

e. El Salvador

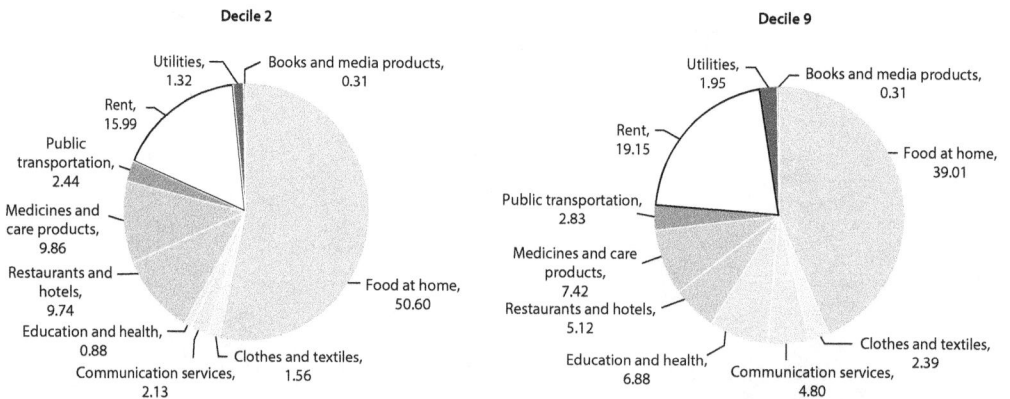

Decile 2

- Utilities, 1.32
- Rent, 15.99
- Public transportation, 2.44
- Medicines and care products, 9.86
- Restaurants and hotels, 9.74
- Education and health, 0.88
- Communication services, 2.13
- Books and media products, 0.31
- Food at home, 50.60
- Clothes and textiles, 1.56

Decile 9

- Utilities, 1.95
- Rent, 19.15
- Public transportation, 2.83
- Medicines and care products, 7.42
- Restaurants and hotels, 5.12
- Education and health, 6.88
- Communication services, 4.80
- Books and media products, 0.31
- Food at home, 39.01
- Clothes and textiles, 2.39

figure continues next page

Figure 5.3 Average Consumption Baskets of Income Decile 2 versus Income Decile 9, Selected LAC Countries *(continued)*

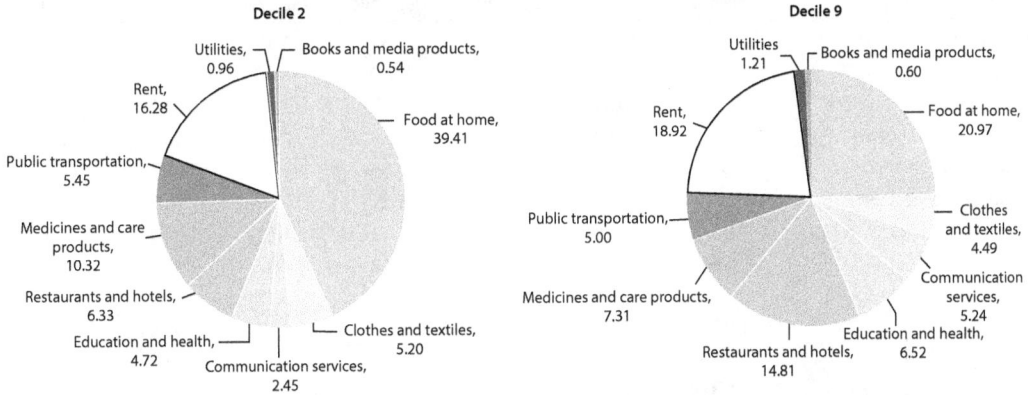

f. Mexico

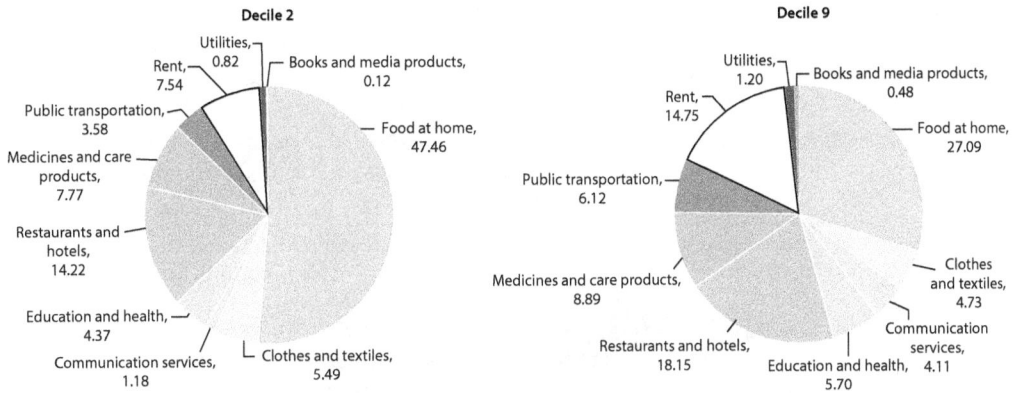

g. Peru

Source: Calculations using national household expenditure surveys for the following years: Bolivia, 2012; Brazil, 2008; Colombia, 2009; Dominican Republic, 2008; El Salvador, 2011; Mexico, 2012; Peru, 2011.

The effects of price increases in gasoline and diesel on the prices of consumer goods will vary, depending on how much gasoline and diesel are directly and indirectly embedded in their production processes. Table 5.2 lists the consumption categories most affected by an increase of US$.25 per liter in gasoline and diesel. Our calculation assumes that the higher costs resulting from the US$.25 increase are fully passed through to the consumer.

Discernable patterns emerge across countries. Of note is the consistency across the sample of the consumption categories that face the highest price increases. At the top of the list in every country are public transportation, medicines and care products (including personal care and cleaning products), as well as metal and crystal ware products (including pots, pans, and silverware). A similar consistency characterizes the least-affected categories. Dominant items here are rent, services (personal and professional), communication services (such as telephone, Internet, and cable TV), and entertainment and culture, as well as education and health.

Energy Pricing Policies for Inclusive Growth in Latin America and the Caribbean
http://dx.doi.org/10.1596/978-1-4648-1111-1

Table 5.2 Percentage Increase in Expenditure Categories from a US$.25 per Liter Price Shock in Transport Fuels, Selected LAC Countries

Expenditure category	Bolivia	Brazil	Colombia	Dominican Republic	El Salvador	Mexico	Peru
Food at home	3.35	1.49	0.84	0.83	1.55	0.70	1.78
Clothes and textiles	3.04	0.92	0.82	1.10	1.77	0.46	0.77
Communication services	1.70	0.39	0.50	0.12	1.23	0.52	0.41
Education and health	2.10	0.58	0.57	0.58	0.94	0.25	0.94
Restaurants and hotels	2.35	0.88	0.51	0.88	0.61	0.27	0.96
Metal and crystal ware	5.31	1.58	1.75	1.32	2.69	1.27	2.00
Medicines and care products	3.54	2.26	1.44	1.12	1.27	0.79	1.06
Services	1.61	0.62	0.33	0.12	0.01	0.31	0.57
Transport (public)	13.00	4.04	6.40	4.32	5.33	2.55	8.45
Rent	0.30	0.07	0.07	0.04	0.27	0.08	0.11
Utilities	5.19	0.99	0.41	7.81	7.47	3.41	3.56
Vehicle maintenance	5.41	0.74	0.83	0.83	1.30	0.14	1.48
Entertainment and culture	1.61	0.62	0.33	0.12	0.01	0.31	0.57
Books and media products	0.00	0.78	1.00	0.85	0.52	0.42	0.61
Tobacco	4.15	1.21	0.97	0.93	1.17	0.49	1.37
Electricity	5.19	0.99	0.41	7.81	7.47	3.41	3.56
Transport fuels	51.16	18.34	30.40	19.90	26.73	32.27	33.29
Cooking and heating fuels	0.00	0.00	0.00	0.00	0.00	0.00	0.00
Other fuels	2.56	1.45	1.25	0.85	0.56	0.67	0.96
Other	1.61	0.62	0.33	0.12	0.01	0.31	0.57

Source: Calculations using national household expenditure surveys and each country's input-output (IO) matrixes. Surveys are for the following years: Bolivia, 2012; Brazil, 2008; Colombia, 2009; Dominican Republic, 2008; El Salvador, 2011; Mexico, 2012; Peru, 2011.

The most important difference across countries arises for those countries where the electricity generation matrix is dominated by fossil fuels, such as El Salvador and the Dominican Republic. In these countries, electricity and utilities show up as the expenditure categories with the highest price increases.[4]

The results are what one might expect. Sectors that use gasoline and diesel intensively (such as public transport) or use petroleum derivatives as inputs (such as the chemical and pharmaceutical industries) or have energy-intensive production processes (such as metal production) are affected the most. Services that are mostly labor-intensive and do not use petroleum products as inputs are little affected.

Higher gasoline and diesel prices have indirect effects on all segments of the distribution by a similar magnitude. Although there are differences in the average size of the impact across countries, with the exception of Bolivia and Mexico, countries present a similar inverted U-shaped pattern across the expenditure distribution (see figure 5.4)—that is, the indirect effects are largest for the middle deciles and diminish toward the poorest and richest deciles. The pattern in Bolivia and Mexico is slightly different, with an indirect effect that monotonically diminishes with the level of expenditure.

Even though the size of the indirect effect is similar across the expenditure distribution, the composition of that effect varies greatly by income level.

For poorer households, the effect is dominated by higher food prices and, to a lesser extent, by the higher costs of public transportation. Meanwhile, for households in the middle of the distribution the impact of higher food expenses loses relevance. Instead, cost increases for public transport, restaurants, and hotels gain significance. In the Dominican Republic and El Salvador, cost increases for electricity emerge as important factors as well. At the other end of the distribution, the impact of higher food prices on rich households is about one-half to one-third the size of that on the poorest decile.

The magnitude of the indirect effect per decile across countries ranges from 0.53 percent in Mexico to 3.21 percent in Bolivia. The large effect measured for Bolivia is not surprising because Bolivia experienced the largest proportional price shock in the sample. What is puzzling is the significantly lower indirect effects in Colombia and Mexico. In spite of observing price shocks comparable to those in El Salvador and Peru, and almost twice as large as those in Brazil and

Figure 5.4 Indirect Effects of a US$.25 per Liter Price Shock in Transport Fuels, by Income Decile, Selected LAC Countries

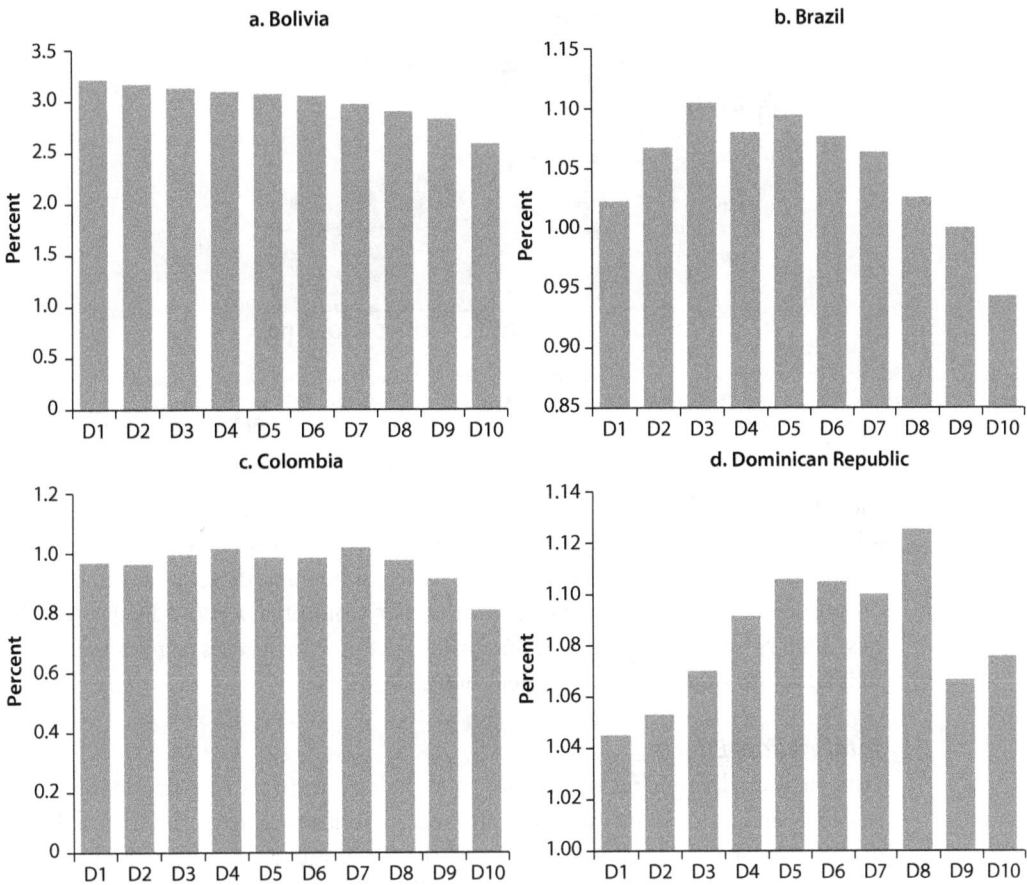

figure continues next page

Energy Pricing Policies for Inclusive Growth in Latin America and the Caribbean
http://dx.doi.org/10.1596/978-1-4648-1111-1

Figure 5.4 Indirect Effects of a US$.25 per Liter Price Shock in Transport Fuels, by Income Decile, Selected LAC Countries *(continued)*

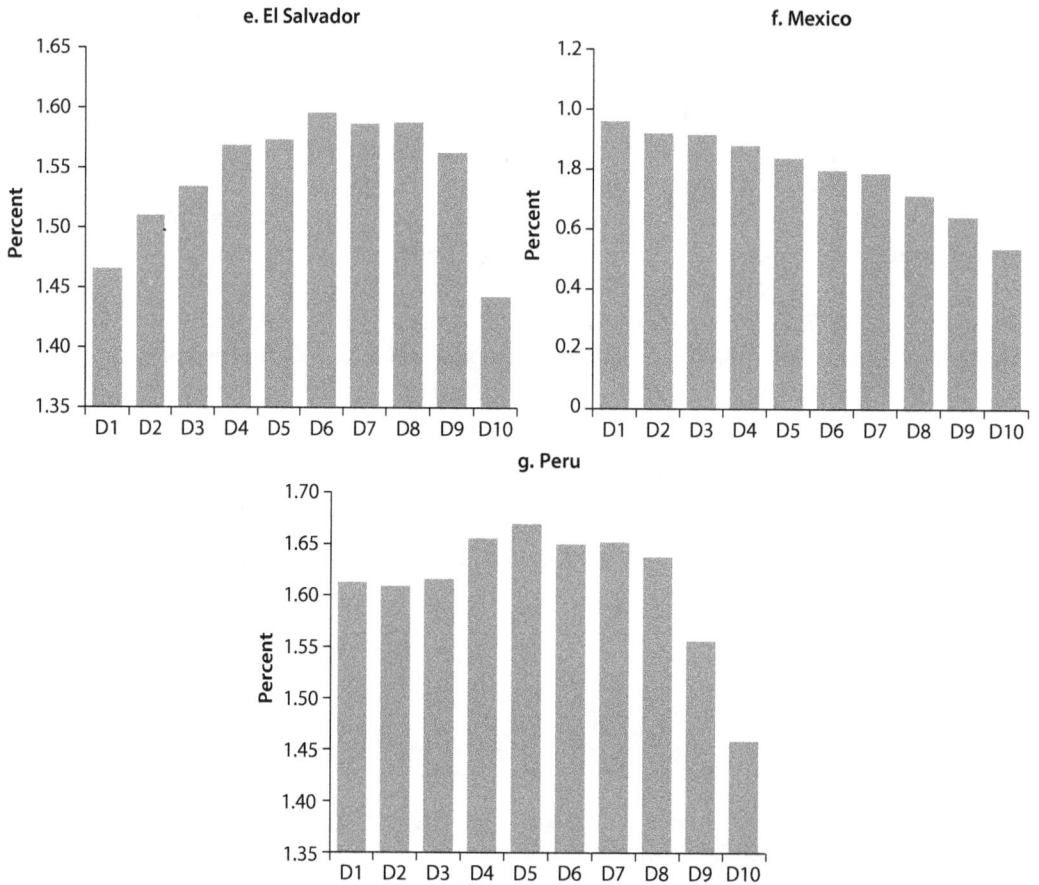

e. El Salvador

f. Mexico

g. Peru

Source: Calculations using national household expenditure surveys and each country's input-output (IO) matrixes. Surveys are for the following years: Bolivia, 2012; Brazil, 2008; Colombia, 2009; Dominican Republic, 2008; El Salvador, 2011; Mexico, 2012; Peru, 2011.
Note: The figures show the average indirect effects by income decile of a US$.25 per liter price shock in transport fuels (gasoline and diesel).

the Dominican Republic, householders in Colombia and Mexico have been affected much less by indirect effects. The reason for this in Mexico may relate to the country's transport sector, which uses gasoline and diesel in a much lower proportion relative to the other countries. For Colombia, a more likely hypothesis centers on the food manufacturing sector, which uses significantly less refined oil products than any other country.[5]

Total Welfare Effects

Our analysis demonstrates that the total welfare impacts of price increases in gasoline and diesel are widespread and significant across the expenditure distribution. Households at the top of the distribution are affected the most, whereas the composition of direct and indirect effects varies with income levels. Accounting for indirect effects raises the welfare loss estimates significantly and substantially changes the distributional profile of these losses. Similarly, conditioning on households that actually purchase these fuels increases the estimated welfare impact and changes the distributional implications.

With the exception of Bolivia, the total unconditional average welfare effect for householders hovers between 1 percent and 2.5 percent of their total expenditures. For Bolivia, the effect is between 3.45 percent and 3.7 percent, with the highest impact concentrated in the middle of the distribution. The large impacts observed in Bolivia—driven by large indirect effects—are mostly explained by the magnitude of the proportional price shock. Perhaps the most surprising results are those of Colombia and Mexico, where the total effects are relatively small because of the unexpectedly low indirect effects.

Consistent with previous literature, we find that indirect effects significantly boost welfare losses and markedly alter the distributional profile of these losses. In fact, with the exception of Mexico, the indirect effect is actually the largest contributor to welfare loss. Moreover, the indirect channel affects the poorest deciles the most, whereas its importance diminishes toward the richer deciles (see figure 5.5). Notably, in Brazil, Colombia, the Dominican Republic,

Figure 5.5 Total Effects of a US$.25 per Liter Price Shock in Transport Fuels, by Income Decile, Selected LAC Countries

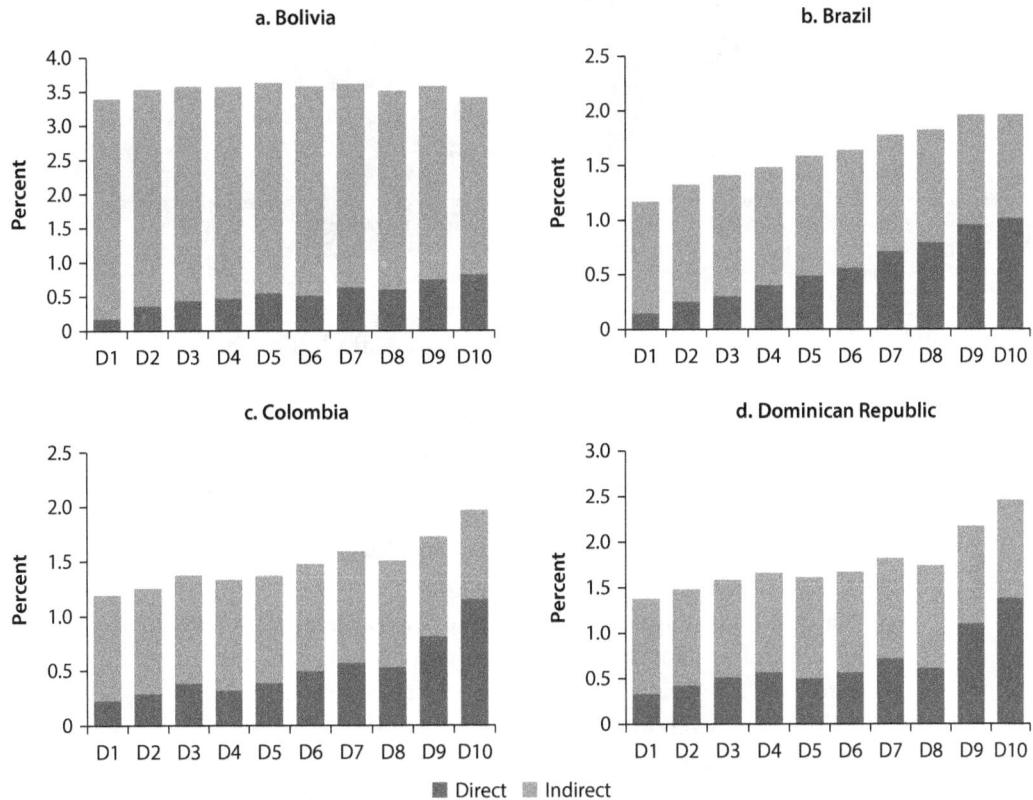

figure continues next page

Figure 5.5 Total Effects of a US$.25 per Liter Price Shock in Transport Fuels, by Income Decile, Selected LAC Countries *(continued)*

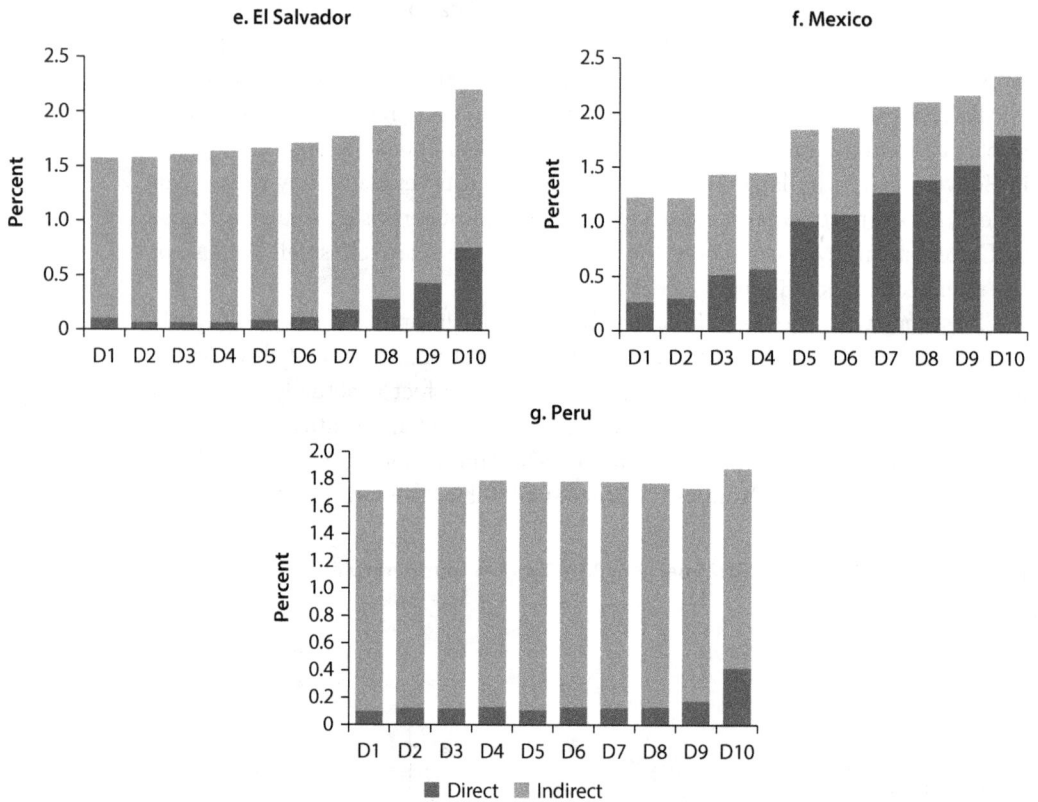

e. El Salvador

f. Mexico

g. Peru

■ Direct   ▒ Indirect

Source: Calculations using national household expenditure surveys and each country's input-output (IO) matrixes. Surveys are for the following years: Bolivia, 2012; Brazil, 2008; Colombia, 2009; Dominican Republic, 2008; El Salvador, 2011; Mexico, 2012; Peru, 2011.
Note: The figures show the average unconditional direct and indirect effects by income decile of a US$.25 per liter price shock in transport fuels (gasoline and diesel). Unconditional effects include all households.

and Mexico, the richest households are more affected by their direct consumption of gasoline and diesel than by the indirect effect.

After accounting for both the direct and indirect effects, we find that subsidies for gasoline and diesel are relatively regressive. The exceptions are Bolivia and Peru, where the combined effects are neutral (neither relatively regressive nor progressive). Conversely, subsidy removal would be relatively progressive (or neutral for Bolivia and Peru) because higher-income households will bear a higher proportional increase in expenditures.

Figure 5.6 illustrates how the subsidy benefits relative to household expenditures are distributed through the population—a Lorenz curve for the relative benefits of the subsidy. The x-axis charts the accumulated share of households sorted by their levels of expenditure per capita, and the y-axis plots the cumulative share of relative benefits. A relatively neutral subsidy—that is, one in which each household obtains the same benefits proportional to their level of expenditure—is represented by the 45° line. A curve above (below) the line represents a relatively progressive (regressive) subsidy, thereby indicating that lower-income households accrue relatively more (fewer) benefits than higher-income households.

For all countries in our sample, the direct effect is always relatively regressive and more regressive than the total effect (see figure 5.6). This finding implies that, relative to their level of expenditure, richer households are disproportionately capturing the direct benefits of lower gasoline and diesel prices. In general, however, the indirect effect is either neutral (meaning that households across the distribution are accruing a share of benefits that is commensurate with their level of expenditures) or relatively progressive (implying that lower-income households are benefitting more in relative terms than their richer counterparts).

Figure 5.6 Relative Lorenz Curves of a US$.25 per Liter Price Shock in Transport Fuels, Selected LAC Countries

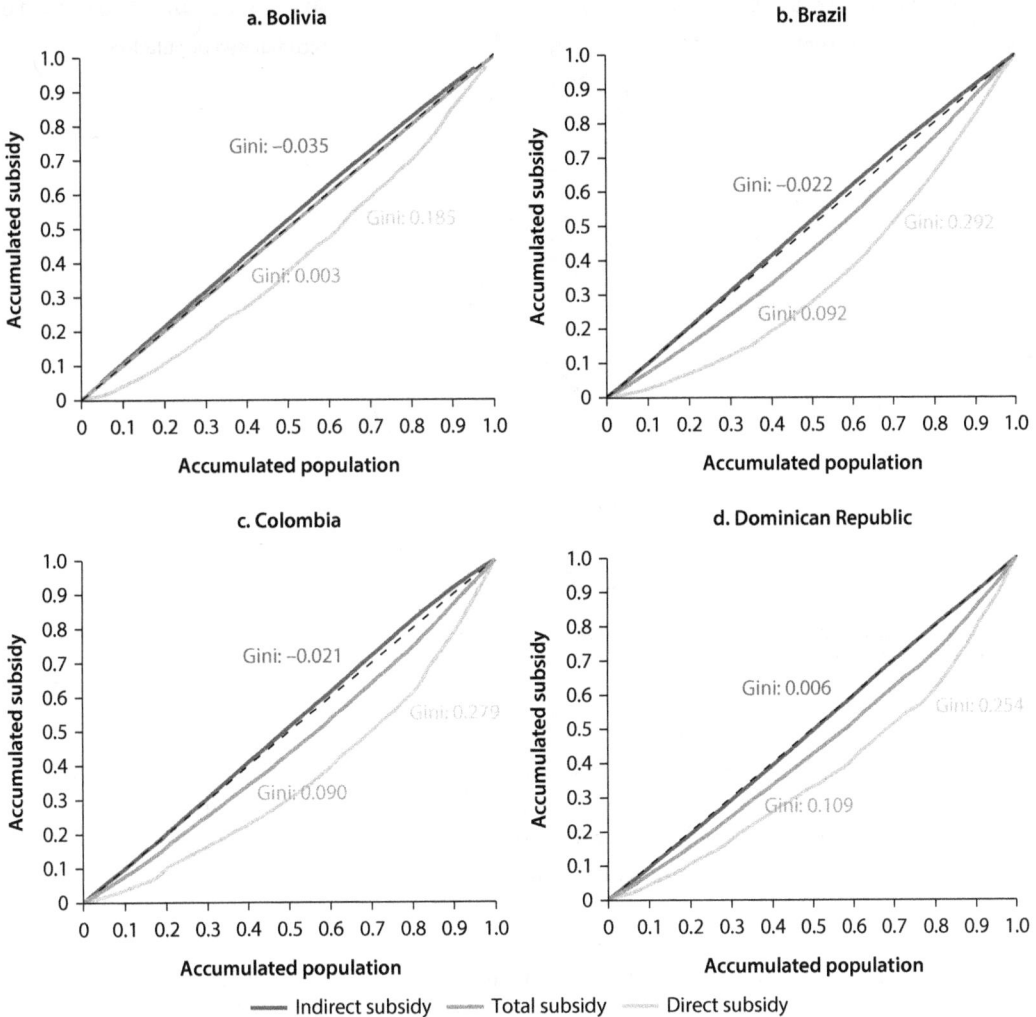

figure continues next page

Figure 5.6 Relative Lorenz Curves of a US$.25 per Liter Price Shock in Transport Fuels, Selected LAC Countries *(continued)*

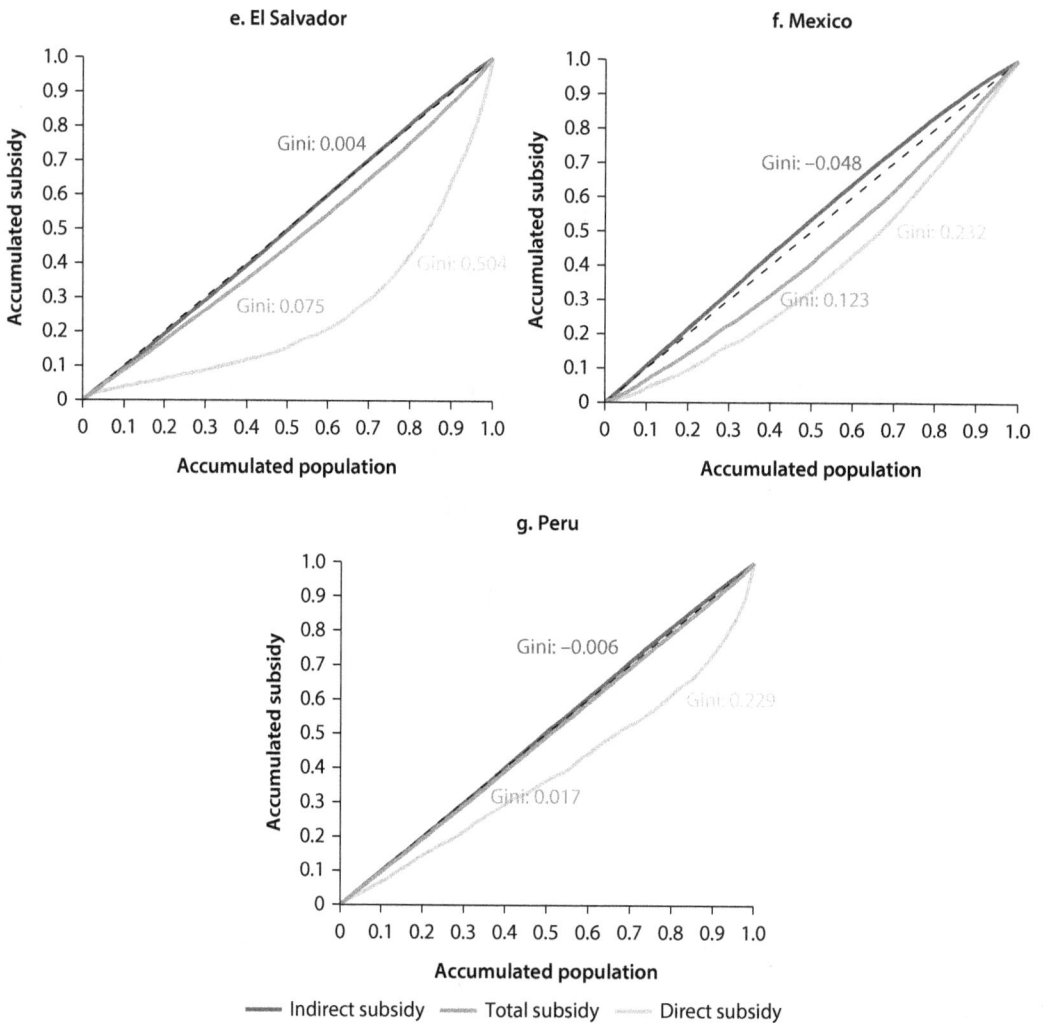

e. El Salvador

f. Mexico

g. Peru

——— Indirect subsidy   ——— Total subsidy   ········· Direct subsidy

Source: Calculations using national household expenditure surveys and each country's input-output (IO) matrixes. Surveys are for the following years: Bolivia, 2012; Brazil, 2008; Colombia, 2009; Dominican Republic, 2008; El Salvador, 2011; Mexico, 2012; Peru, 2011.
Note: The figures show the Lorenz curves for the relative impact of a US$.25 per liter price shock in transport fuels (gasoline and diesel) and the quasi-Gini coefficients.

Put simply, wealthier households have the highest direct gains, although poorer households are also benefitting indirectly. Nonetheless, in most countries the indirect benefits for the poorer deciles do not sufficiently counteract the direct benefits captured by richer households, rendering a relatively regressive total effect.

Furthermore, analysis of the share of total absolute benefits enjoyed by low-income households reveals that subsidies for gasoline and diesel are clearly favoring the richest households. In all countries and for all effects (direct, indirect, and total), the results show a distribution that is absolutely regressive, as seen in figure 5.7. Thus households at the top of the distribution are reaping most of the benefits.

Across countries, the direct effects are the most regressive. The indirect effects are slightly less regressive in the absolute sense, but even in this case households in the top decile are capturing most of the indirect benefits as well.

Comparing across countries (see figure 5.8), gasoline and diesel subsidies are the most regressive in countries with the highest rates of vehicle ownership (in both the absolute and relative dimensions). For example, in Brazil where vehicle ownership is relatively high (27.5 percent), the top 20 percent of households receive 58.7 percent of the benefits, and the bottom 40 percent receive only

Figure 5.7 Absolute Lorenz Curves of a US$.25 per Liter Price Shock in Transport Fuels, Selected LAC Countries

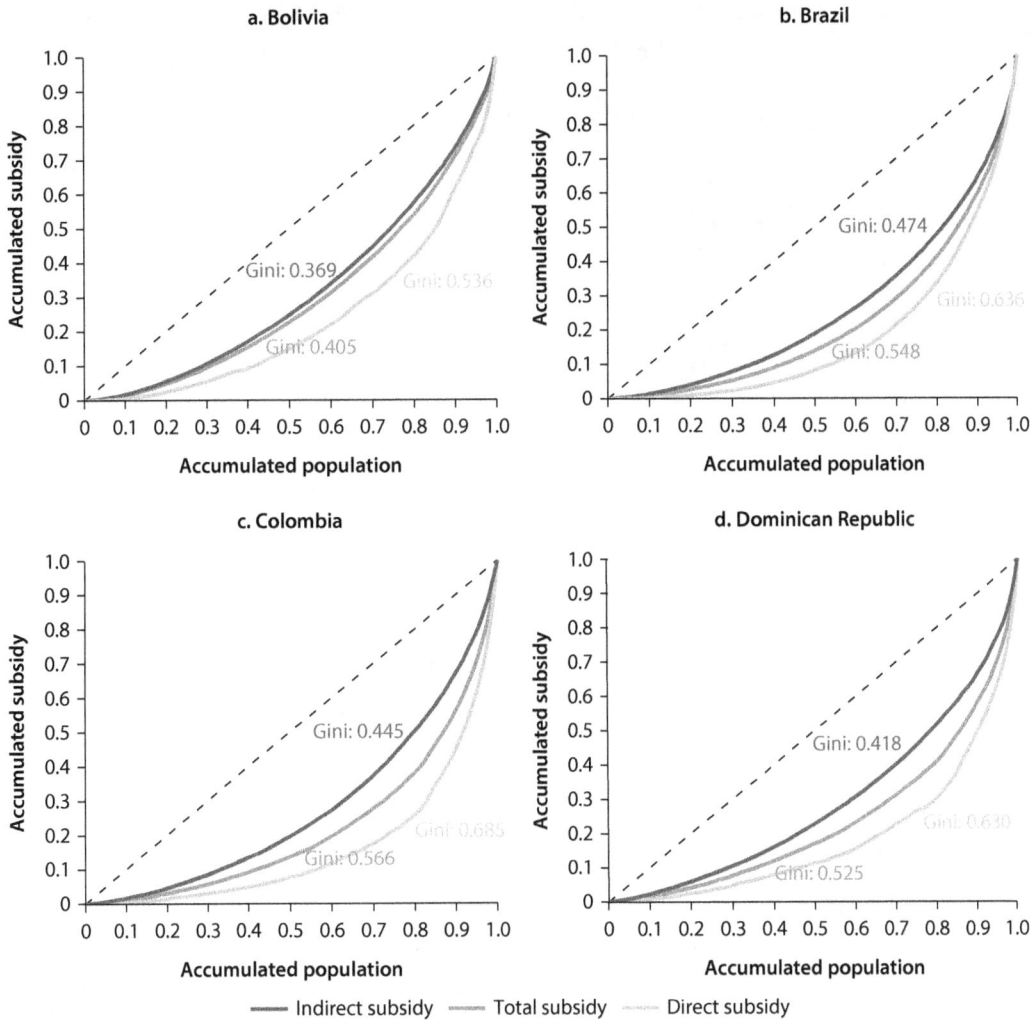

a. Bolivia

Gini: 0.369
Gini: 0.536
Gini: 0.405

b. Brazil

Gini: 0.474
Gini: 0.636
Gini: 0.548

c. Colombia

Gini: 0.445
Gini: 0.685
Gini: 0.566

d. Dominican Republic

Gini: 0.418
Gini: 0.630
Gini: 0.525

Indirect subsidy —— Total subsidy —— Direct subsidy

figure continues next page

Energy Pricing Policies for Inclusive Growth in Latin America and the Caribbean
http://dx.doi.org/10.1596/978-1-4648-1111-1

Figure 5.7 Absolute Lorenz Curves of a US$.25 per Liter Price Shock in Transport Fuels, Selected LAC Countries *(continued)*

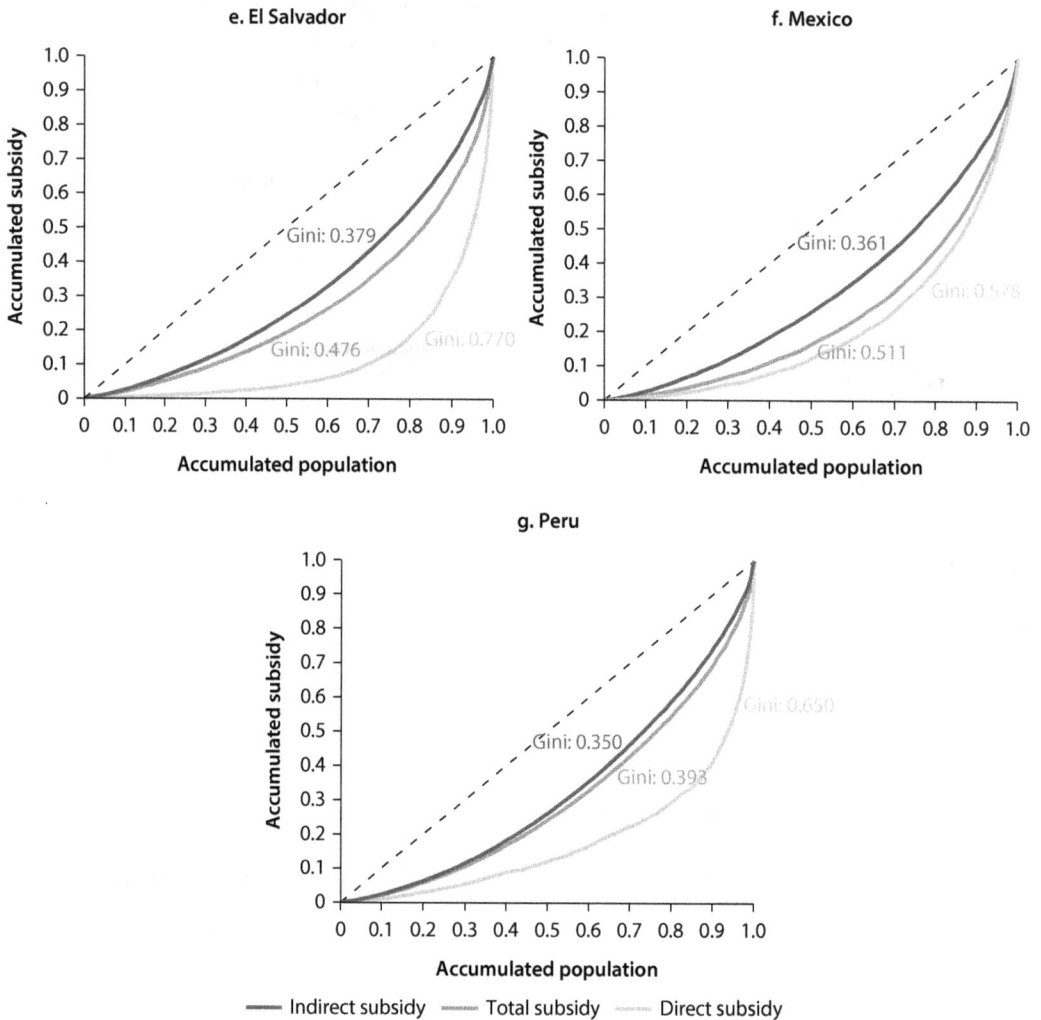

Source: Calculations using national household expenditure surveys and each country's input-output (IO) matrixes. Surveys are for the following years: Bolivia, 2012; Brazil, 2008; Colombia, 2009; Dominican Republic, 2008; El Salvador, 2011; Mexico, 2012; Peru, 2011.
Note: The figures show the Lorenz curves for the absolute impact of a US$.25 per liter price shock in transport fuels (gasoline and diesel) and the quasi-Gini coefficients.

9 percent—a ratio of 6.5. Yet in Peru, which has the lowest vehicle ownership in the study sample, the ratio is about 2.7 because the top 20 percent of households receive 45.4 percent of the benefits and the bottom 40 percent capture only 16.9 percent. This correlation is explained by the fact that vehicle ownership is generally more common across the richest segments of the population. Thus higher rates of vehicle ownership generally imply that more higher-income households have vehicles, and therefore these households will benefit directly from cheaper transport fuels. In countries with lower vehicle ownership such as

Figure 5.8 Relative and Absolute Lorenz Curves of a US$.25 per Liter Price Shock in Transport Fuels

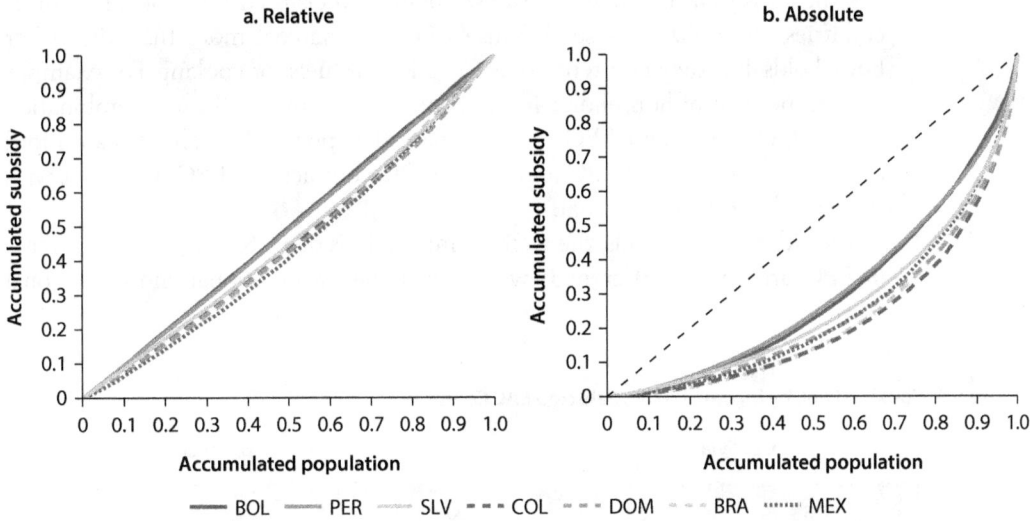

Source: Calculations using national household expenditure surveys and each country's input-output (IO) matrixes. Surveys are for the following years: Bolivia, 2012; Brazil, 2008; Colombia, 2009; Dominican Republic, 2008; El Salvador, 2011; Mexico, 2012; Peru, 2011.
Note: The figures show the Lorenz curves for the relative and absolute impacts of a US$.25 per liter price shock in transport fuels (gasoline and diesel). BOL = Bolivia; BRA = Brazil; COL = Colombia; DOM = Dominican Republic; MEX = Mexico; PER = Peru; SLV = El Salvador.

Bolivia and Peru, the most important benefits arise from indirect effects. As described earlier, these tend to be more evenly distributed across households.

Distributional Impacts of a Price Increase in Liquefied Petroleum Gas

This section turns to analyzing the impact on households across the expenditure distribution of increasing the price of LPG, a fuel used mainly for cooking, by US$.25 cents per liter.

Direct Effects

The use of LPG varies considerably across the expenditure distribution. Although this fuel is certainly used more widely than transport fuels, there is still a pro-rich bias in that poorer households use alternative—nonpetroleum-based—fuels for cooking.

Households in the poorest decile, with the exception of Brazil, mostly use nonpetroleum-based fuels (generally wood and charcoal) as their main cooking fuel (see figure 5.9). In moving toward the richest deciles, the penetration of petroleum-based cooking fuels rises significantly, with LPG being the most important. Although this trend is present in all countries, there is considerable variation across countries. At one extreme, in El Salvador 80 percent of households in the poorest decile use wood as their main cooking fuel. At the other extreme, less than 30 percent of those in the same demographic group in Brazil use only wood or charcoal.

Although most household surveys ask only about the main fuel used for cooking, the surveys in Brazil and Peru ask about all fuels used for cooking. In these countries, the poorer households use wood or charcoal more than the richer households, but they also tend to use a variety of fuels for cooking. For example, over 35 percent of households in the poorest decile of Brazil use a combination of wood, charcoal, and LPG. We speculate that poorer households can afford only limited amounts of LPG and thus need to complement LPG purchases with the use of wood and charcoal.

As with transport fuels, the welfare implications of a US$.25 per liter increase in LPG are vividly different if we contrast the unconditional and conditional

Figure 5.9 Cooking Fuel, by Income Decile, Selected LAC Countries

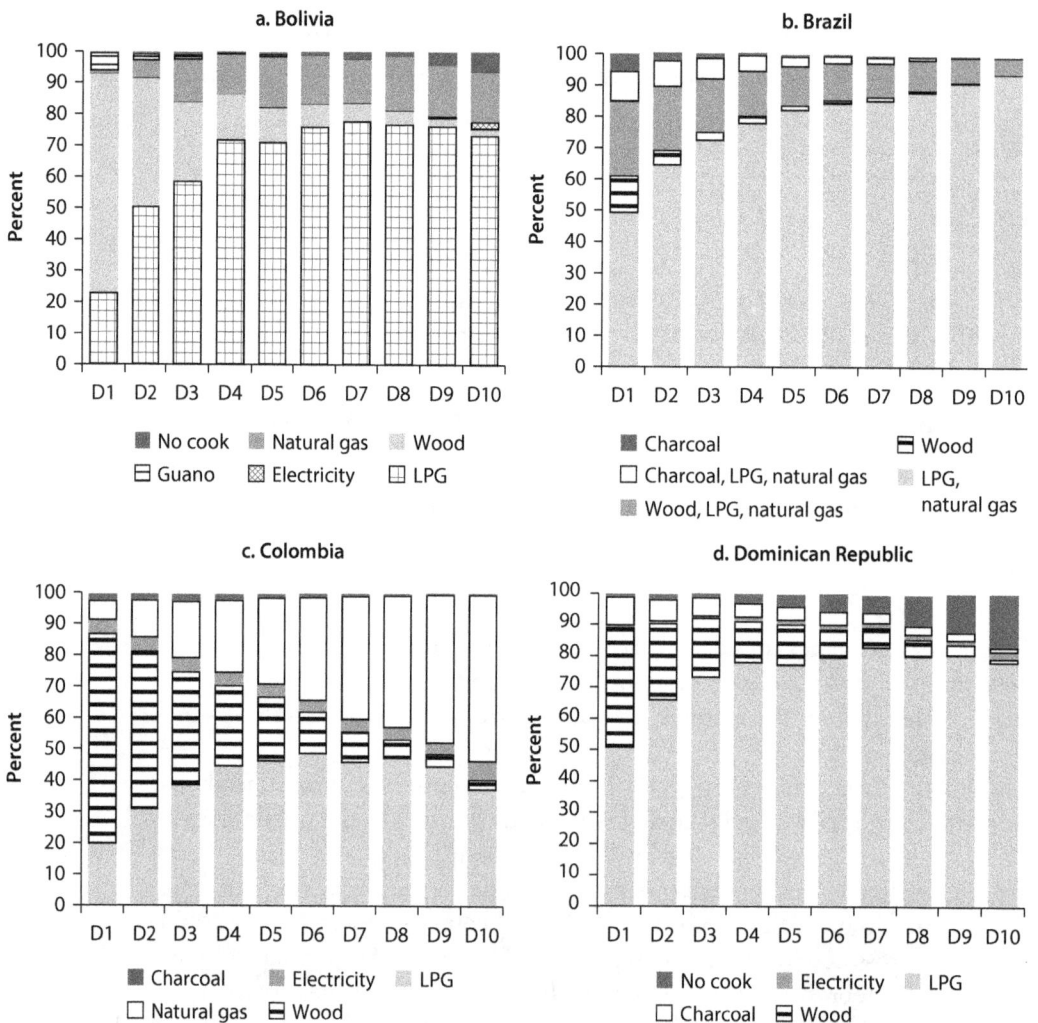

Figure 5.9 Cooking Fuel, by Income Decile, Selected LAC Countries *(continued)*

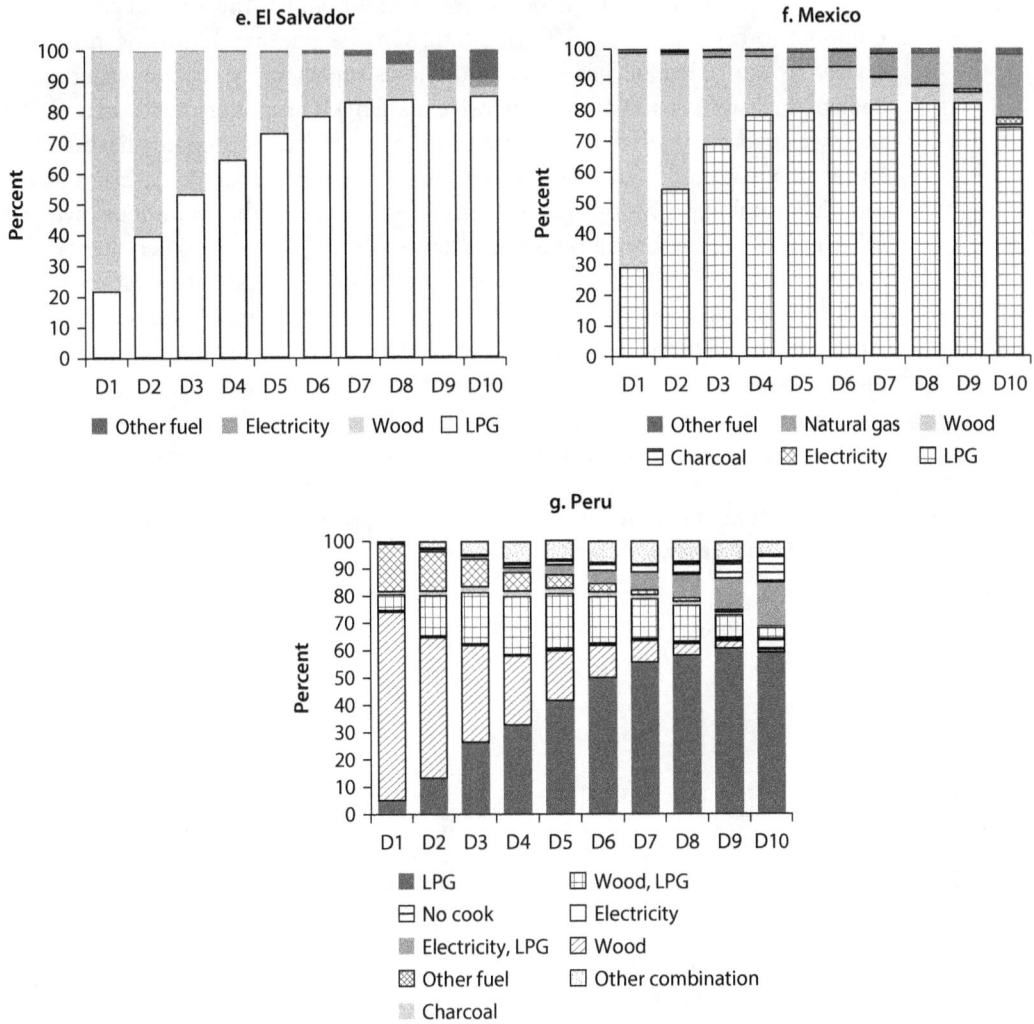

e. El Salvador

■ Other fuel ▨ Electricity ▨ Wood □ LPG

f. Mexico

■ Other fuel ▨ Natural gas ▨ Wood
⊟ Charcoal ⊠ Electricity ⊞ LPG

g. Peru

■ LPG   ⊞ Wood, LPG
⊟ No cook   □ Electricity
▨ Electricity, LPG ▨ Wood
⊠ Other fuel   □ Other combination
▨ Charcoal

Source: Calculations using national household expenditure surveys. Surveys are for the following years: Bolivia, 2012; Brazil, 2008; Colombia, 2009; Dominican Republic, 2008; El Salvador, 2011; Mexico, 2012; Peru, 2011.
Note: The figures show the average percentage of each cooking fuel by income decile. LPG = liquefied petroleum gas.

average direct welfare effects within a decile. Figure 5.10 presents both measures of direct welfare effects. Naturally, the difference in magnitude is more pronounced in the poorest deciles where more households rely on wood or charcoal as their main cooking fuel. As one moves across the expenditure distribution, the difference between the conditional and unconditional measures becomes less important because more households rely on LPG for cooking.

For the unconditional direct effect, two distinct patterns emerge across countries. Bolivia, Brazil, Colombia, and the Dominican Republic exhibit effects that are monotonically decreasing with the level of expenditure—with the exception of the first decile. Meanwhile, El Salvador, Mexico, and Peru exhibit an inverted U-shaped pattern across the expenditure distribution. For this second group, two opposing forces may explain the pattern. On the one hand, moving from the poorer to the richer deciles more households report using LPG as their main cooking fuel (referred to as the "extensive margin"). This pattern tends to increase the measured average effect. On the other hand, richer households have higher levels of expenditure. Therefore, as long as LPG expenditures increase at a lower rate than total expenditures (as evidenced by the size of the conditional average effect), LPG purchases will represent a lower share of the total expenditures (the "intensive margin"), thereby dampening the measured average effect. At the lower end of the distribution, the gains through the extensive margin tend to dominate. At the other end of the distribution, the declines through the intensive margin are more important, resulting in the inverted U-shaped pattern in figure 5.10.

Our results also indicate that for households that actually purchase LPG, the welfare effects clearly decrease with expenditure levels. In other words, because

Figure 5.10 Conditional and Unconditional Direct Effects of a US$.25 per Liter Price Shock in Liquefied Petroleum Gas (LPG), by Income Decile, Selected LAC Countries

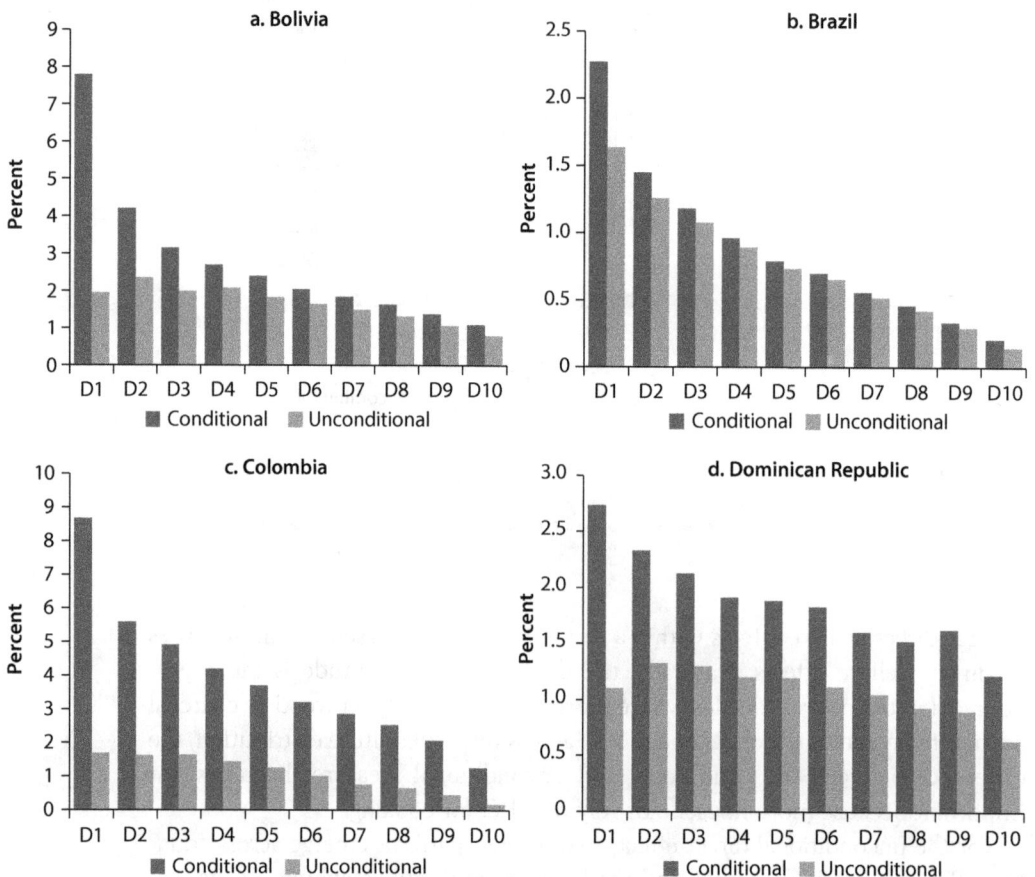

figure continues next page

Figure 5.10 Conditional and Unconditional Direct Effects of a US$.25 per Liter Price Shock in Liquefied Petroleum Gas (LPG), by Income Decile, Selected LAC Countries *(continued)*

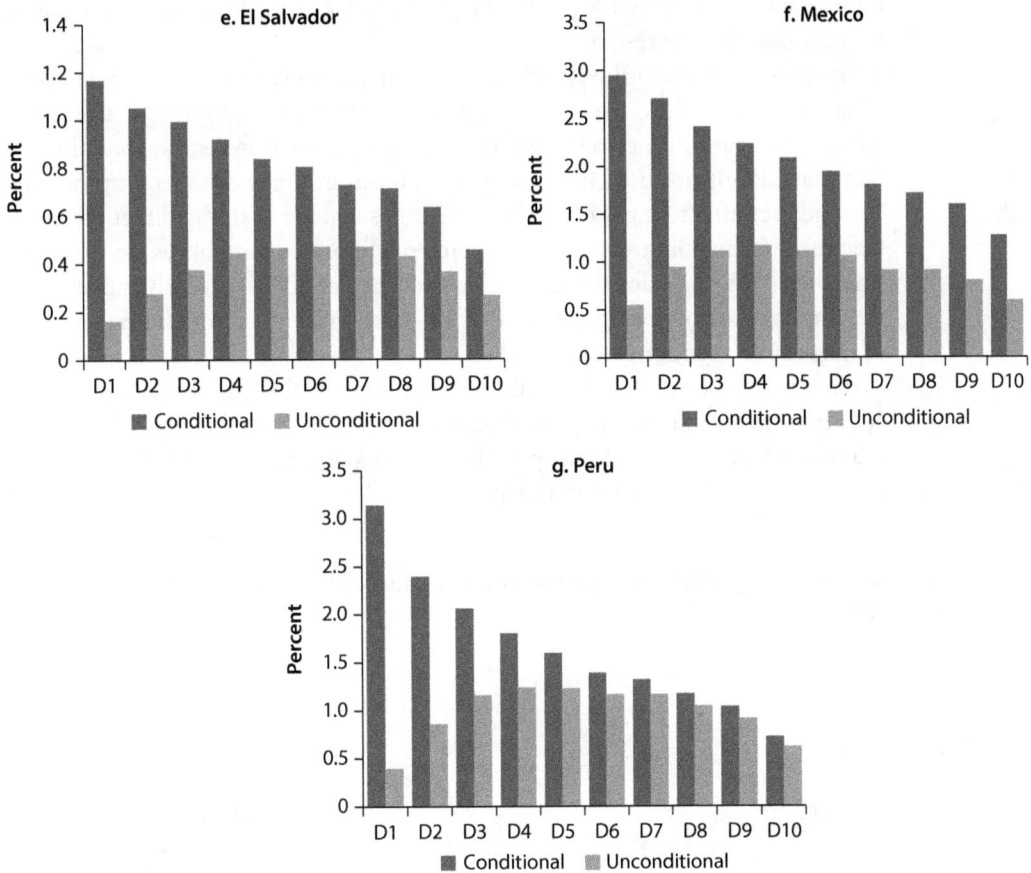

Source: Calculations using national household expenditure surveys and each country's input-output (IO) matrixes. Surveys are for the following years: Bolivia, 2012; Brazil, 2008; Colombia, 2009; Dominican Republic, 2008; El Salvador, 2011; Mexico, 2012; Peru, 2011.
Note: The figures show the average conditional and unconditional effects by income decile of a US$.25 per liter price shock in LPG. Conditional effects are conditional on positive consumption of transport fuels. Unconditional effects include all households.

a lower share of the expenditures of richer households is devoted to LPG consumption, these households are less affected by the price increase. Nonetheless, for poorer households that actually consume LPG, the effects can be quite significant—sometimes as much as three times the unconditional average effect.

Indirect Effects

For LPG, the indirect effects are relatively small, especially when compared with those associated with transport fuels. Unfortunately, some data limitations may influence the results. The IO matrixes published by national statistical agencies generally do not distinguish LPG production as a separate sector, including them instead in the "manufacture of oil refined products" sector. We complement the data from the IO matrixes with statistics from the Energy Statistics Database of the United Nations Statistics Division to estimate the expenditures on LPG by industry.

However, the Energy Statistics Database provides different levels of detail for each country. In many cases, then, we had to make some strong assumptions in order to impute the consumption level of LPG by industry. With those caveats in mind, we now describe the results.

In sharp contrast to the effects observed in transport fuels, the indirect effect of an increase in LPG pales in comparison with the magnitude of the direct effect. This reason is not only the data limitations described earlier, but also the fact that LPG is not used intensively as an input in many production processes. The indirect effect ranges from about 92 times smaller than the direct effect for Colombia (where industry and commerce rely more on natural gas) to 2.5 times smaller for Brazil. As for the impact on the price of all other goods and services, the implications of raising the price of LPG are evidently a much smaller order of magnitude than for gasoline and diesel.

There are no clear distributional patterns of the indirect effect across countries. In Colombia, the Dominican Republic, El Salvador, and Mexico, the indirect effect increases slightly with the level of expenditures of households (see figure 5.11). However, in Brazil and Peru the pattern is an inverted U shape,

Figure 5.11 Indirect Effects of a US$.25 per Liter Price Shock in Liquefied Petroleum Gas (LPG), by Income Decile, Selected LAC Countries

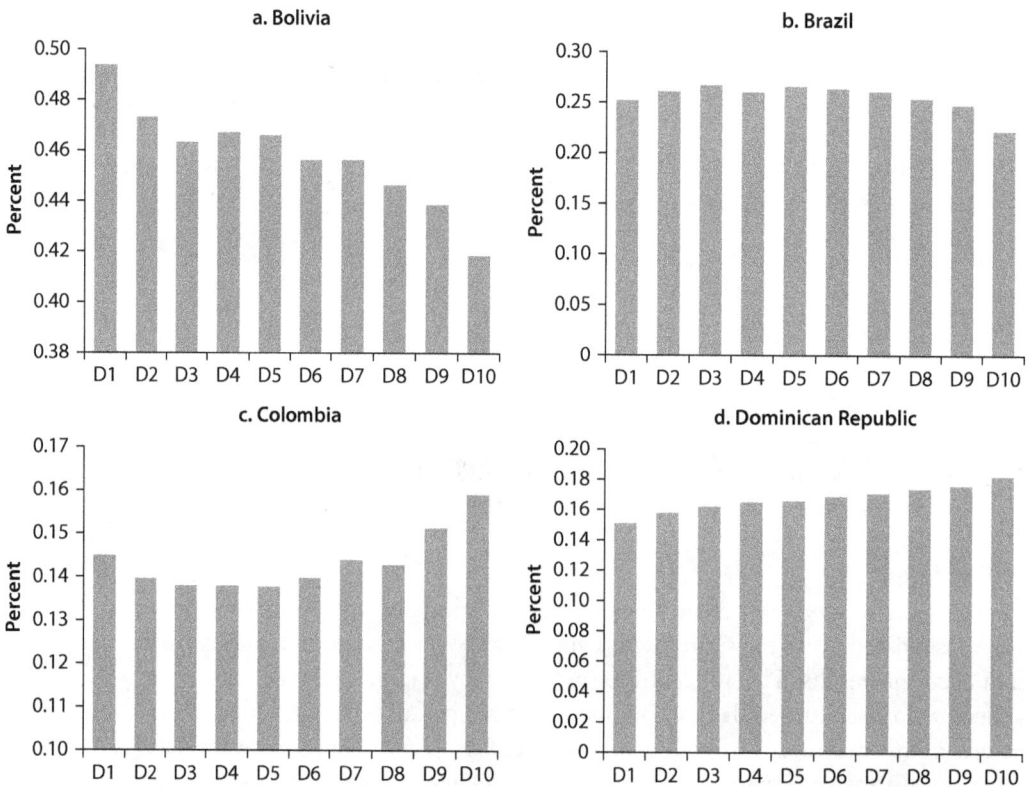

Figure 5.11 Indirect Effects of a US$.25 per Liter Price Shock in Liquefied Petroleum Gas (LPG), by Income Decile, Selected LAC Countries (continued)

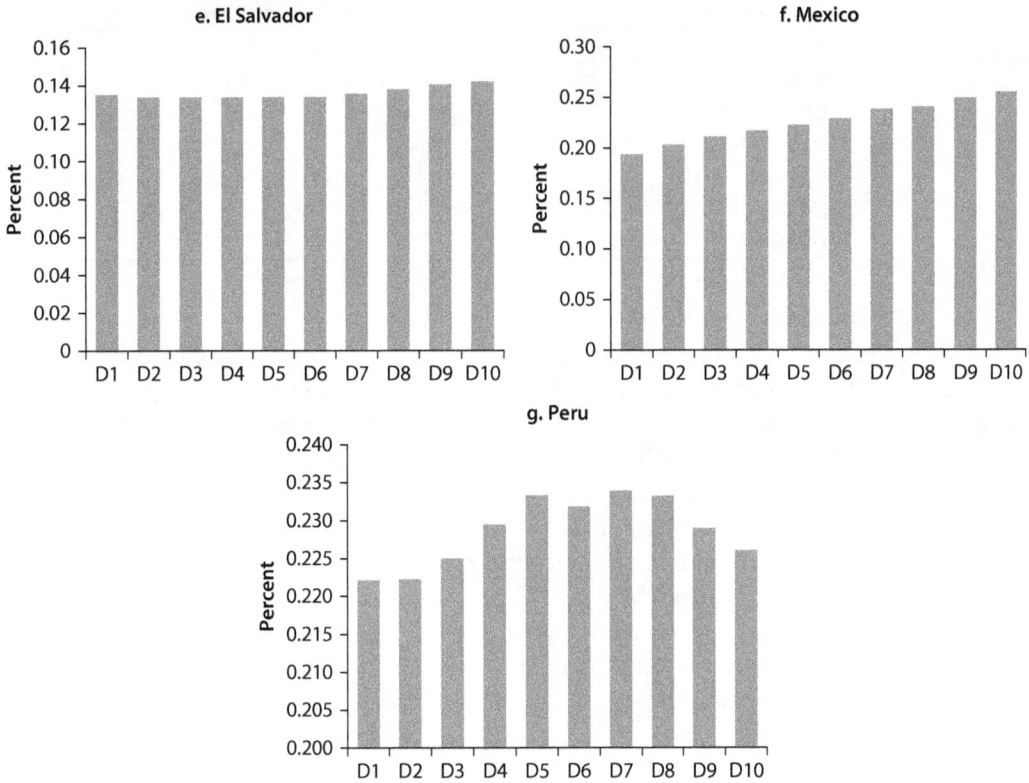

e. El Salvador

f. Mexico

g. Peru

Source: Calculations using national household expenditure surveys and each country's input-output (IO) matrixes. Surveys are for the following years: Bolivia, 2012; Brazil, 2008; Colombia, 2009; Dominican Republic, 2008; El Salvador, 2011; Mexico, 2012; Peru, 2011.
Note: The figures show the average indirect effect by income decile of a US$.25 per liter price shock in LPG.

whereas in Bolivia the effects decrease with wealth. In any case, the differences across deciles are never above 0.1 percent of their expenditure levels, suggesting that the effects are actually very small and similar across the expenditure distribution.

Total Welfare Effect

Ultimately, even for a product like LPG, which is used by poorer households much more intensively than gasoline or diesel, the majority of the benefits are captured by the richest households. The richest 20 percent of households receive 34 percent of the benefits on average, while the bottom 40 percent of households receive only 23 percent. In contrast to transport fuels, the welfare implications mostly derive from the direct effect on households because the indirect effects are relatively minor.

For a US$.25 per liter increase in the price of LPG, the total welfare effect ranges from 0.39 percent in El Salvador to 2.15 percent in Bolivia.

Table 5.3 Percentage Increase in Liquefied Petroleum Gas (LPG) Prices from a US$.25 per Liter Price Shock, Selected LAC Countries

Country (reference year)	Percentage increase in LPG price
Bolivia (2012)	183.4
Brazil (2008)	34.9
Colombia (2009)	69.2
Dominican Republic (2008)	54.4
El Salvador (2011)	37.0
Mexico (2012)	56.7
Peru (2011)	71.7

Source: World Bank data.

Because the indirect effects are relatively insignificant, the explanation for the differences in the welfare effect across countries must be sought elsewhere. The primary causes are the differences in the proportional size of the price increase and the different patterns of direct LPG consumption. As seen in table 5.3, the proportional increase in prices can be dramatically different, ranging from 34.9 percent in Brazil to 183.4 percent in Bolivia.

For LPG, the total welfare effect is essentially determined by the direct effects, which represent about 75 percent of the total welfare effect for Bolivia and Brazil and as much as 99 percent for Colombia. The indirect effects are relatively small, and they do not present clear distributional patterns. Moreover, the differences in the magnitude of the indirect effects across the expenditure distribution are very small. Once again, this finding implies that the distributional profile of the welfare impacts will essentially mirror the distributional profile of the direct effects.

Two interesting patterns emerge in considering the relative distributional profile of subsidies for LPG. In one set of countries—Bolivia, Brazil, Colombia, and the Dominican Republic—the distribution of benefits is relatively progressive. Poorer households are capturing a higher share of benefits relative to their expenditures. In another set of countries—El Salvador, Mexico, and Peru—most of the benefits are captured by households in the middle of the distribution. This clear demarcation in the findings suggests that poorer households capture a lower share of benefits (the Lorenz curve is below the 45° line). Yet, the fact that the Lorenz curve crosses the 45° line around the middle of the distribution (see figure 5.12) suggests a different outcome might emerge were the price of LPG increased. In such a scenario, lower-income households in the first set of countries would be the hardest hit in terms of relative welfare losses, whereas for the second set it would be the households in the middle of the distribution.

Even for LPG, which is more widely used across the expenditure distribution, the higher-income households capture most of the absolute benefits. Our results indicate that subsidies for LPG are absolutely regressive for all countries (see figure 5.12). The most unequal case is that of El Salvador, where the top 20 percent of households capture 38.5 percent of the benefits, and the bottom 40 percent of households capture only 17 percent of the benefits—a ratio of 2.26.

Figure 5.12 Relative and Absolute Lorenz Curves of a US$.25 per Liter Price Shock in Liquefied Petroleum Gas (LPG), Selected LAC Countries

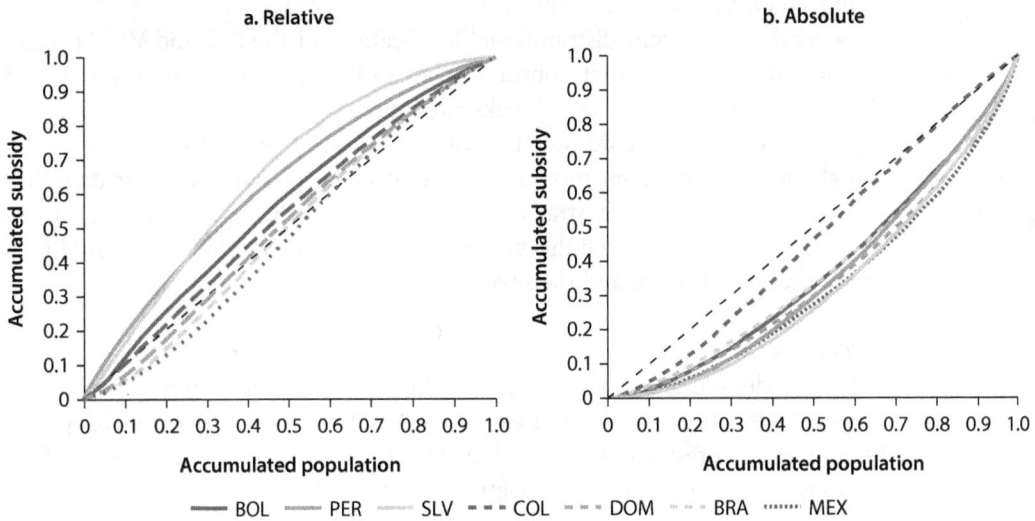

Source: Calculations using national household expenditure surveys and each country's input-output (IO) matrixes. Surveys are for the following years: Bolivia, 2012; Brazil, 2008; Colombia, 2009; Dominican Republic, 2008; El Salvador, 2011; Mexico, 2012; Peru, 2011.
Note: The figures show the Lorenz curves for the relative and absolute impacts of a US$.25 per liter price shock in LPG and the quasi-Gini coefficients. Relative (quasi-Gini) coefficients: Bolivia (BOL), –0.128; Brazil (BRA), –0.250; Colombia (COL), –0.299; Dominican Republic (DOM), –0.070; El Salvador (SLV), 0.045; Mexico (MEX), –0.021; Peru (PER), 0.007. Absolute (quasi-Gini) coefficients: Bolivia, 0.254; Brazil, 0.251; Colombia, 0.067; Dominican Republic, 0.299; El Salvador, 0.341; Mexico, 0.340; Peru, 0.290.

Excluding the special case of Colombia, the most even distribution of subsidy benefits is found in Brazil, where the richest 20 percent reap 34.5 percent of the benefits, and the poorest 40 percent receive 24.2 percent—a ratio of 1.42. We pinpoint Colombia as a special case because the penetration of LPG actually decreases for the higher-income households—about half of these households use natural gas.

Distributional Impacts of a Price Increase in Electricity

Analyzing the distributional impacts of increasing electricity prices presents some unique challenges. In contrast to the fuels just analyzed, electricity generally does not have a unique price. The LAC countries have instituted pricing policies that include volume differentiated tariff (VDT) schemes and increasing block tariff (IBT) pricing, as well as flat unit rates and targeted discounts.

This section analyzes the distributional impacts of the tariff schedules of El Salvador, Mexico, and the Dominican Republic. Each is different. El Salvador's can be categorized as a two-part tariff with a fixed charge that varies by consumption block and an IBT scheme for variable charges. Mexico's can be best described as a mixture of an IBT scheme with a VDT for high consumption—also known as the DAC or Domésticas de Alto Consumo (Residential High Consumption) tariff. Finally, the Dominican Republic has a two-part tariff comprising a fixed charge that changes with the total consumption level and a variable charge pricing scheme that is similar to Mexico's (that is, a combination of IBT and a VDT for high consumption).

Beyond the different distributional implications of the IBT and VDT pricing mechanisms,[6] the fact that consumers have different consumption levels and thus may end up in different blocks presents an analytical challenge. Because expenditure surveys record only the total expenditure on electricity, calculating actual kilowatt-hour consumption and the block tariff or unit price that applies to each household is not straightforward. Thus we have developed country-specific algorithms to match the expenditure level of a household to a tariff block or applicable unit price and then back out its kilowatt-hour consumption.

Direct Effects

In the last decade, access to electricity in El Salvador has improved substantially. In the 2011 national household expenditure survey, over 75 percent of households in the poorest decile reported having access to electricity, and the number of households with access to electricity increased rapidly with income level (see figure 5.13). In addition, a small fraction of households reported having access to electricity, but they reported no expenditures. Because this fraction is small and decreases with the expenditure level, this anomaly is most likely explained by a combination of misreporting and electricity theft.[7] In any case, a large majority of households have access and consume electricity. In this respect, consumption patterns differ from those of fuels.

Figure 5.13 Access to Electricity and Electricity Theft, by Income Decile: El Salvador, 2011

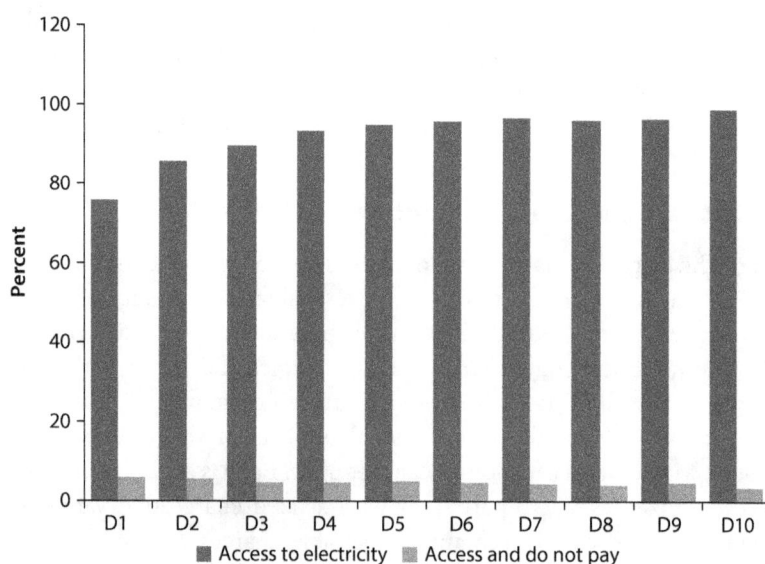

Source: Calculations using El Salvador's 2011 national household expenditure survey.
Note: The figure shows the average percentage of households with access to electricity and the average percentage of households that have zero expenditures and positive consumption of electricity by income decile.

For Mexico, we had to restrict our distributional analysis to the Federal District (Distrito Federal, DF) because of data limitations (although we do continue to refer to Mexico's results in our discussion for the sake of consistency with the rest of this report). The Federal District is the capital city of Mexico and has the legal status of a state. We used the 2010 expenditure survey, ENIGH, because it is representative at the state level for the DF.

In 2009, Luz y Fuerza del Centro, the utility company serving the DF, was absorbed by the Federal Electric Commission (Comisión Federal de Electricidad). This change is relevant because Luz y Fuerza del Centro was notoriously inefficient, with very high nontechnical losses stemming from electricity theft, failure to bill consumption properly, and metering errors,[8] which were rampant in its concession area.

Although access to electricity is largely universal in Mexico (see figure 5.14), many households with access to electricity do not pay for their consumption. This problem is found especially among the poorest deciles (reaching over 16 percent of households in decile 1), but it is still relevant to the richest deciles as well. Because it is not possible to infer the actual consumption for these households, we cannot assess the benefits of this implicit subsidy, and the distributional impact is therefore unclear. However, it is possible to say that a substantial number of the poorest households in Mexico are consuming electricity at no cost. The subsequent losses to the system are

Figure 5.14 Access to Electricity and Electricity Theft, by Income Decile: Mexico, 2012

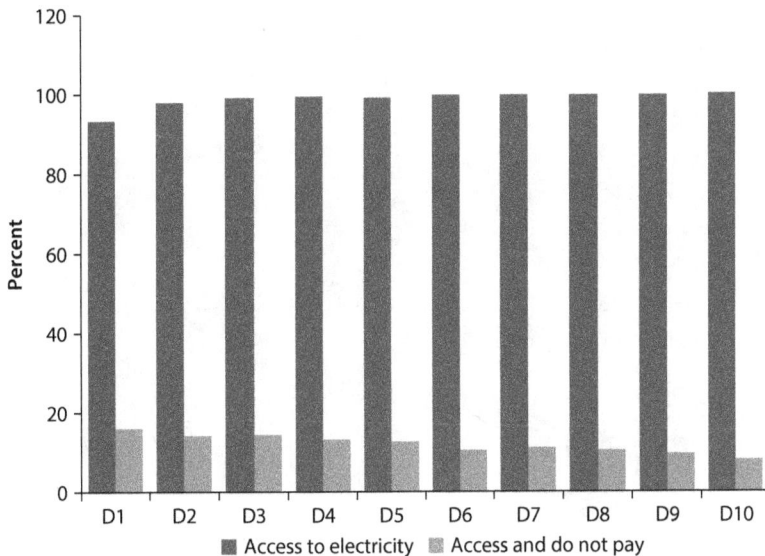

Source: Calculations using Mexico's 2012 national household expenditure survey.
Note: The figure shows the average percentage of households with access to electricity and the average percentage of households that have zero expenditures and positive consumption of electricity by income decile.

Energy Pricing Policies for Inclusive Growth in Latin America and the Caribbean
http://dx.doi.org/10.1596/978-1-4648-1111-1

absorbed by the utility company, passed on to paying consumers, or covered by the government. This analysis is based on households that report positive expenditures. The possible bias introduced should be considered when observing the results.

In the Dominican Republic, access to electricity has improved substantially over the last decade, with only about 14 percent of households in the bottom decile lacking access. As in El Salvador, the number of households without access to electricity declines with income (figure 5.15). However, the issue of nonpayment of electricity is severe and widespread across the income distribution. Although the proportion of households that do not pay for their electricity consumption declines with income, the numbers are staggering. About 55 percent of households in the bottom half of the distribution do not pay for their consumption, and even 33 percent of the richest households avoid payment. Again, because it is impossible to know how much electricity these households are consuming, we cannot assess the distributional impacts of this implicit subsidy. Thus as we did for the Federal District of Mexico, we based our analysis on the households that do pay for electricity, and once again caution the reader on the possible biases that may result.

El Salvador has implemented an IBT mechanism that is divided into four blocks. It heavily subsidizes householders who consume less than 100 kilowatt-hours.

Figure 5.15 Access to Electricity and Electricity Theft, by Income Decile: Dominican Republic, 2008

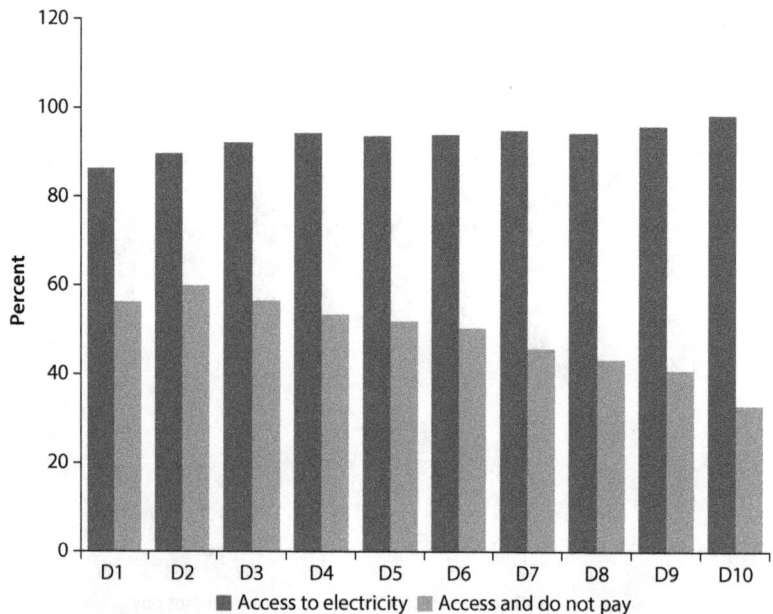

Source: Calculations using the Dominican Republic's 2008 national household expenditure survey.
Note: The figure shows the average percentage of households with access to electricity and the average percentage of households that have zero expenditures and positive consumption of electricity by income decile.

Table 5.4 Tariff Schedule, by Block: El Salvador, 2011
U.S. dollars

Tariff block	Fixed tariff	Variable tariff
Less than 50 kWh	0.803	0.086
More than 50 kWh and less than 100 kWh	0.809	0.089
More than 100 kWh and less than 200 kWh	0.880	0.198
More than 200 kWh	0.883	0.222

Source: World Bank data.
Note: kWh = kilowatt-hour.

Table 5.5 Tariff Schedule (Tariff 1), by Block: Mexico, 2012
U.S. dollars

Tariff block	Fixed tariff	Variable tariff
Less than 75 kWh	0	0.055
More than 75 kWh and less than 140 kWh	0	0.060
More than 140 kWh and less than 250 kWh	0	0.132
More than 250 kWh	5.630	0.276

Source: Regulatory Direction of Hydrocarbons and Mines, Ministry of Finance (El Salvador).
Note: kWh = kilowatt-hour.

Table 5.4 presents the tariff schedule with the defined blocks and the unit prices for each block in force during 2011. The tariff has two parts: a fixed charge that corresponds to marketing or commercialization charges[9] and a variable charge that is the sum of the distribution and energy charges. Like the unit price, the fixed charge is also specific to each consumption block, with higher-volume blocks paying higher fixed charges as well as higher unit prices. In addition, consumers in the first block have a minimum charge equivalent to 30 kilowatt-hours.

Mexico employs a pricing mechanism that combines an IBT approach with a separate two-part tariff for high-volume consumers (the DAC). Table 5.5 features the prices of the first three blocks, which are part of the IBT scheme. It also shows the separate rate for high-volume consumers. As the findings clearly show, the amount of the subsidy per unit decreases per block. In addition, the DAC tariff is priced above the average cost in order to cross-subsidize the lower-volume consumers.

The Dominican Republic has a somewhat hybrid tariff between those of El Salvador and Mexico. Like El Salvador, it has a two-part tariff composed of a fixed charge and a variable charge. The fixed charge depends on the total electricity consumed, with higher-volume consumers paying higher fixed charges. The variable charge is similar to that of Mexico; an IBT structure is divided into five blocks and a separate block for high-volume consumers (more than 700 kilowatt-hours per month) with a VDT pricing scheme (table 5.6).

Our analysis suggests that although a large proportion of the poorest households are in fact benefitting from subsidized tariffs, there is significant leakage

Table 5.6 Tariff Schedule, by Block: Dominican Republic, 2008
U.S. dollars

Tariff block	Fixed tariff	Variable tariff
Less than 75 kWh	0.659	0.089
More than 75 kWh and less than 125 kWh	2.121	0.089
More than 125 kWh and less than 175 kWh	2.897	0.089
More than 175 kWh and less than 300 kWh	3.116	0.112
More than 300 kWh and less than 700 kWh	3.116	0.201
More than 700 kWh	3.116	0.246

Source: Energy Information System, Ministry of Energy (Mexico).
Note: kWh = kilowatt-hour.

toward higher-income households. In El Salvador, a large majority of households in the poorest deciles have consumption levels that fall into the first two blocks, which are heavily subsidized. However, significant leakage is evident toward the higher-income households. Notably, almost 40 percent of households in decile 10 fall into these blocks (see figure 5.16). The third consumption block (100–199 kilowatt-hours per month) is also subsidized,[10] albeit at lower levels. This subsidy disproportionately benefits higher-income households. According to the regulatory agency of Mexico, the DAC tariff is the only tariff that is not subsidized. Among the households that pay for electricity in Mexico, the results show that less than 5 percent actually pay the DAC tariff (see figure 5.16). This implies that 95 percent of households have their electricity consumption subsidized. Similarly, the tariff structure in the Dominican Republic was originally conceived to allow for cross-subsidies from high-volume residential consumers to consumers in the 0–300 kilowatt-hours per month tranche. However, less than 4 percent of households actually have such high consumption levels.

As expected, consumption of electricity increases with the income level of households. Figure 5.17 plots the average kilowatt-hour consumption by decile conditional on positive expenditures. For El Salvador, because access to electricity is relatively high and most consumers pay for their electricity consumption, the line reflects accurately the average consumption per decile. For the Dominican Republic and Mexico, there is a significant gap between conditional and unconditional average consumption because of the high levels of nonpayment. Even so, we still find similar patterns in all three countries: an increase in consumption of electricity in relation to household wealth, with consumption for households in the richest decile spiking significantly.

In what follows, we simulate the welfare effects of increasing the price per kilowatt-hour by US$.05 cents for all tariff blocks. The price shock affects the variable charge but leaves the fixed charge (where it exists) at its original levels. The magnitude and distributional impacts of the direct effect are plotted in figure 5.18.

Different distributional patterns appear in each case. For El Salvador, the average unconditional direct effect hovers between 0.75 percent and 1.13 percent. It increases up to decile 8, after which it decreases. This pattern suggests that a reduction in the subsidy (a price increase) for electricity will affect households in

Figure 5.16 Percentage of Households in Each Tariff Block, by Income Decile: Dominican Republic, El Salvador, and Mexico

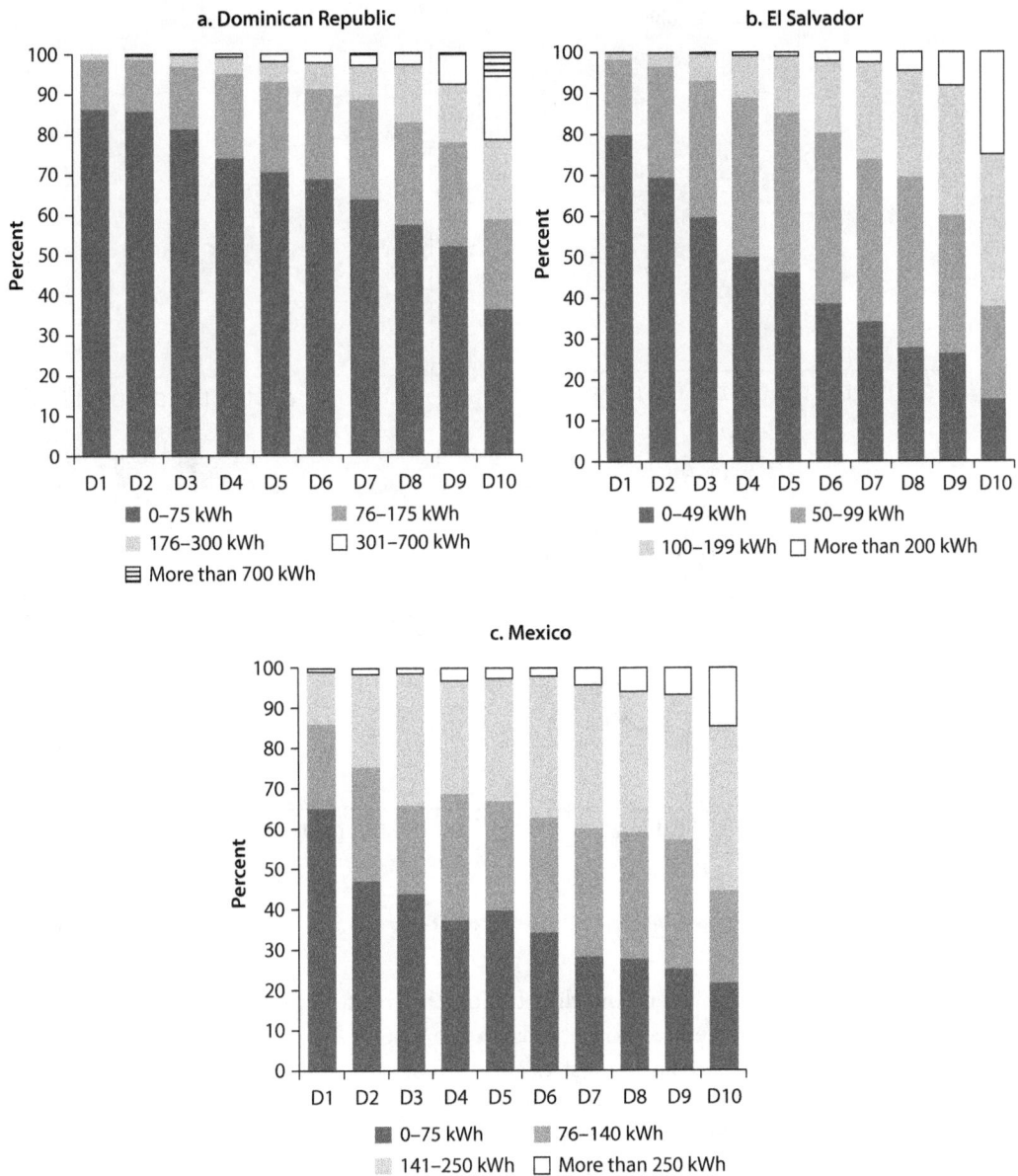

a. Dominican Republic

0–75 kWh
76–175 kWh
176–300 kWh
301–700 kWh
More than 700 kWh

b. El Salvador

0–49 kWh
50–99 kWh
100–199 kWh
More than 200 kWh

c. Mexico

0–75 kWh
76–140 kWh
141–250 kWh
More than 250 kWh

Source: Calculations using national household expenditure surveys for the following years: Dominican Republic, 2008; El Salvador, 2011; Mexico, 2010 and 2012.
Note: The figures show the average percentage of households in each electricity tariff block by income decile. kWh = kilowatt-hour.

Energy Pricing Policies for Inclusive Growth in Latin America and the Caribbean
http://dx.doi.org/10.1596/978-1-4648-1111-1

Figure 5.17 Average Electricity Consumption, by Income Decile for Positive Expenditures: Dominican Republic, El Salvador, and Mexico

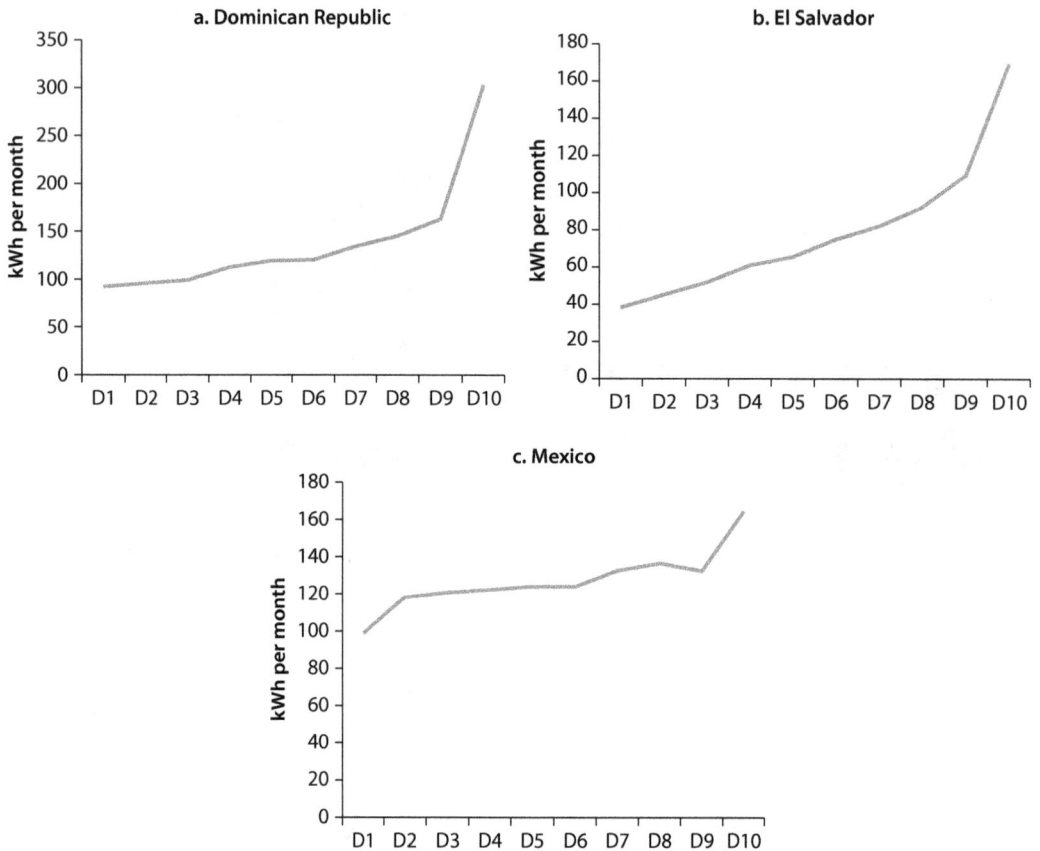

a. Dominican Republic

b. El Salvador

c. Mexico

Source: Calculations using national household expenditure surveys for the following years: Dominican Republic, 2008; El Salvador, 2011; Mexico, 2010 and 2012.
Note: The figures show the average consumption of electricity by income decile of households with positive expenditures on electricity. kWh = kilowatt-hour.

the middle and top of the distribution the most, with a smaller impact on the bottom 40 percent of households. In the case of Mexico, the effect ranges from 2.23 percent for the poorest decile to 0.60 percent for the richest decile. The distributional profile in Mexico is radically different than that of El Salvador, with the direct effect decreasing rather than increasing with household income levels. This implies that the lower-income households would be relatively more affected by a reduction in the electricity subsidy. Finally, in the Dominican Republic the measured direct effect increases with the income level of households—with the exception of the poorest decile. The Dominican Republic's higher-income households will therefore be most affected by a price increase in electricity.

The distributional profile of the conditional average direct effect is similar to the unconditional average direct effect. The only exception is the Dominican Republic. Meanwhile, the poorer deciles have the sharpest increases in the measured effect, as expected. In El Salvador, however, households in the middle of

Figure 5.18 Average Conditional and Unconditional Direct Effects of a US$.05 per Kilowatt-Hour Price Shock in Electricity, by Income Decile: Dominican Republic, El Salvador, and Mexico

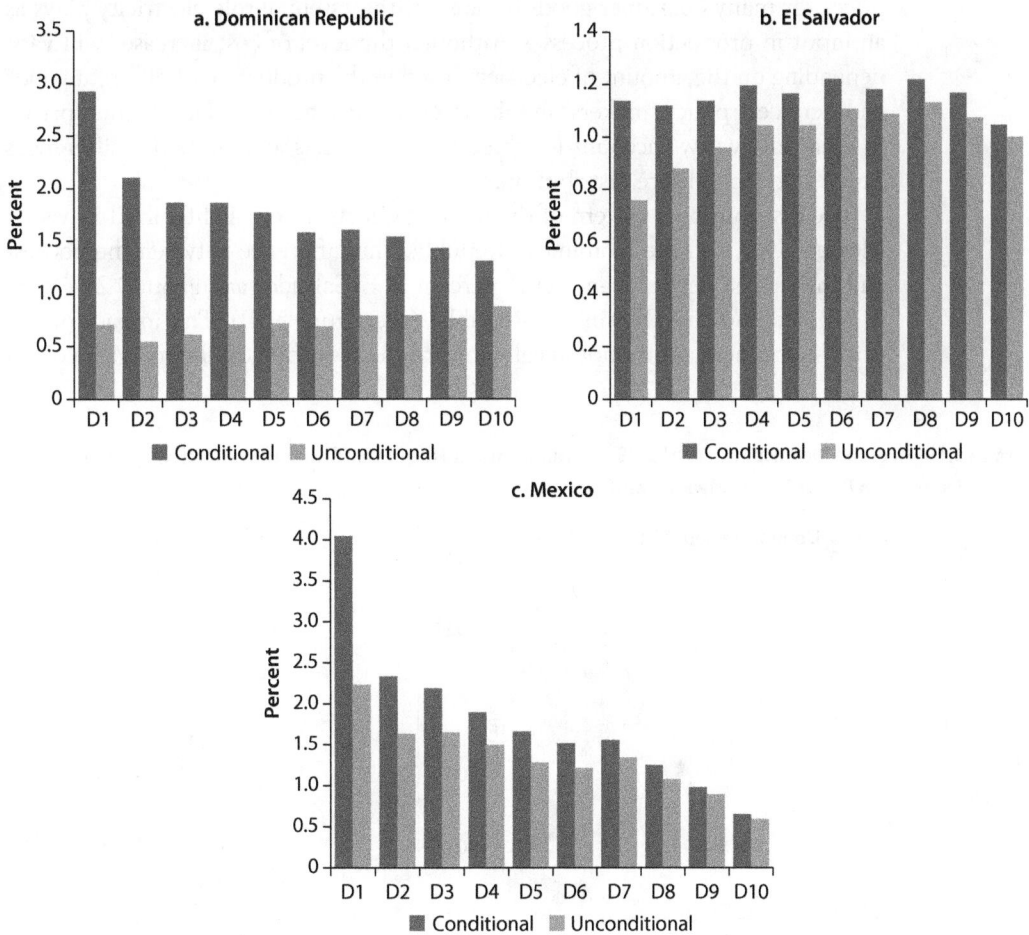

Source: Calculations using national household expenditure surveys and each country's input-output (IO) matrixes. Surveys are for the following years: Dominican Republic, 2008; El Salvador, 2011; Mexico, 2010 and 2012.
Note: The figures show the average conditional and unconditional effects by income decile of a US$.05 per kilowatt-hour price shock in electricity. Conditional effects are conditional on positive consumption of electricity. Unconditional effects include all households.

the distribution continue to suffer the largest direct welfare effects. On the other hand, when restricting the analysis to households that report positive expenditures on electricity, the richest households in the distribution emerge with the smallest welfare impacts. For Mexico, the measured direct effect almost doubles for the poorest decile, but the general distributional pattern is not significantly different from the unconditional average. Once again, this suggests that the hardest hit in relative terms will be the poorest households that pay for their electricity. In the Dominican Republic, when we focus on households that report a positive expenditure, the pattern changes substantially. With this measure, the poorest households become the most affected by a price increase, with the effect diminishing in line with household income levels.

Energy Pricing Policies for Inclusive Growth in Latin America and the Caribbean
http://dx.doi.org/10.1596/978-1-4648-1111-1

Indirect Effects

As previously noted, hikes in electricity prices are likely to translate into higher prices for many consumer goods because of the essential role electricity plays as an input in production processes. Although the level of cost increases will vary, depending on the amount of electricity used in the production of different goods and services, policy makers should expect consumers to face higher prices. In what follows, we account for these indirect effects and assess the differences across the expenditure distribution.

The distributional pattern of the indirect effects differs slightly in El Salvador, Mexico (DF), and the Dominican Republic: the difference between the poorest and the richest deciles is about 0.1 percent in El Salvador and about 0.2 percent in Mexico and the Dominican Republic (see figure 5.19). The impact of the price shock increases monotonically toward the wealthier deciles for Mexico and

Figure 5.19 Average Indirect Effects of a US$.05 per Kilowatt-Hour Price Shock in Electricity, by Income Decile: Dominican Republic, El Salvador, and Mexico

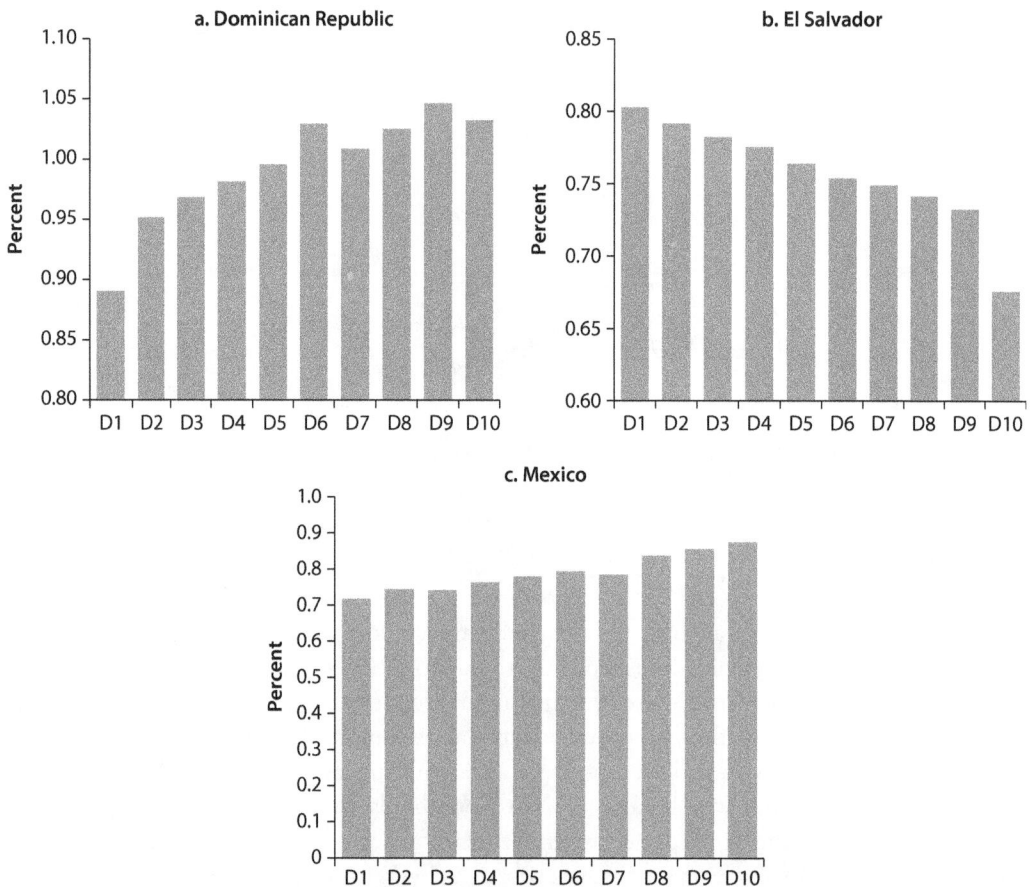

Source: Calculations using national household expenditure surveys and each country's input-output (IO) matrixes. Surveys are for the following years: Dominican Republic, 2008; El Salvador, 2011; Mexico, 2010 and 2012.

the Dominican Republic, while monotonically decreasing for El Salvador—that is, the richest households in Mexico and the Dominican Republic would be most affected by the indirect effects, whereas the opposite would be true in El Salvador.

In addition to the different distributional profile of the welfare losses, important differences can be found in the composition of the indirect effect. Yet, the indirect effect actually displays a similar pattern across the income distribution. In El Salvador, most of the indirect welfare losses arise from the increase in the price of food, chemical, and pharmaceutical products. Because food accounts for a large proportion of the overall budget of Salvadoran households, any price increase in foodstuffs has a large impact on households. As one moves across the distribution toward the higher-income households, items such as clothes and textiles, as well as education, health, and communication services, begin to form an important part of the indirect losses. For Mexico and the Dominican Republic, food is a major explanatory factor in welfare losses for lower-income households. Rising expenditures in restaurants and hotels, clothes and textiles, education, and health and communication services explain the bulk of the impact on higher-income households. Thus across the income distribution the three countries display similar patterns. For El Salvador, this effect tends to dominate because food expenditures represent a relatively larger proportion of total household expenditures.

Total Welfare Effect

The total welfare impacts are widespread and significant across the expenditure distribution, although the distributional patterns are different for each country. Accounting for indirect effects raises the welfare loss estimates significantly, but it does not alter the distributional profile of these losses. For El Salvador and Mexico, focusing on households that actually purchase electricity increases the estimated welfare impact but does not change the distributional implications. For the Dominican Republic, focusing on households that actually pay for electricity significantly changes the distributional profile of the welfare losses because of the widespread practice in the country of not paying for electricity consumption.

For El Salvador, the total welfare effect ranges from 1.5 percent to 1.9 percent of the total expenditures. Although the differences across the expenditure distribution are not large, it is households in the middle of the distribution that have the highest welfare losses. For Mexico, the largest impact is on the poorest households, with a measured effect of almost 3 percent. By contrast, the richest households suffer the least, with a less than 1.5 percent loss. Similar to El Salvador, the welfare loss in the Dominican Republic hovers around 1.6 percent and 2 percent. However, the welfare loss increases with the income level of households.

The difference in magnitude between countries is mainly due to two factors: the electricity consumption patterns across countries and the difference in the proportional magnitude of the price shock. For Mexico, the two factors go in the same direction. Electricity consumption is relatively high, and, because of the

large subsidy for electricity, a $US.05 per kilowatt-hour price increase is propor-
tionally large. In El Salvador, electricity consumption is much lower than in
Mexico or the Dominican Republic. Yet, the price increase is similar to that of
Mexico in the sense that it represents a large proportional increase. By contrast,
in the Dominican Republic electricity consumption is high, but the price
increase is lower proportionally.

Electricity subsidies are relatively neutral in the Dominican Republic and
El Salvador and relatively progressive in Mexico. In other words, the removal of
subsidies in the Dominican Republic and El Salvador will affect all households in
a relatively similar fashion (see figure 5.20). But in Mexico the removal of subsidies will affect the lower-income households relatively more than the higher-
income households. For El Salvador and the Dominican Republic, a relatively
neutral total welfare effect goes hand in hand with the relatively neutral direct
and indirect effects. In Mexico, the direct effect is relatively progressive, whereas
the indirect effect is slightly regressive. However, because the direct effect is relatively larger, the total welfare effect balances on the progressive side.

When we look at the distribution of the absolute benefits of electricity
subsidies, a shift occurs—they are regressive in all countries. El Salvador and the
Dominican Republic, despite the differences in their levels of consumption, have
similar distributional patterns. Both the indirect and direct effects are absolutely
regressive and with a similar distribution of benefits. Clearly, the total welfare
effect is similarly regressive in nature—that is, the removal of electricity subsidies

Figure 5.20 Relative and Absolute Lorenz Curves of a US$.05 per Kilowatt-Hour Price Shock in Electricity: Dominican Republic, El Salvador, and Mexico

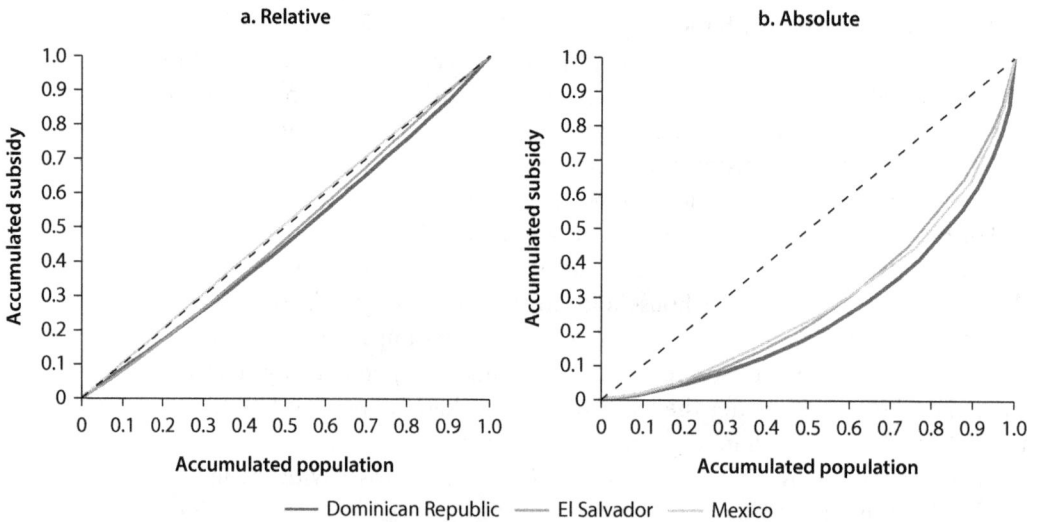

Source: Calculations using national household expenditure surveys and each country's input-output (IO) matrixes. Surveys are for the following years: Dominican Republic, 2008; El Salvador, 2011; Mexico, 2010 and 2012.
Note: The figures show Lorenz curves for the relative and absolute impacts of a US$.05 per kilowatt-hour price shock in electricity and the quasi-Gini coefficients. Relative (quasi-Gini) coefficients: Dominican Republic, 0.072; El Salvador, 0.048; Mexico, −0.011. Absolute (quasi-Gini) coefficients: Dominican Republic, 0.495; El Salvador, 0.415; Mexico, 0.418.

will have a greater impact on higher-income households. In Mexico, the subsidies are also absolutely regressive, but they display a different pattern. There, although the direct benefits are regressive, they are less so than the indirect effects. In any case, the impact of removing electricity subsidies would fall most heavily on higher-income households.

Notes

1. We should also note, as a mitigating factor, that many studies have found that the demand for oil is highly price-inelastic, mostly because there are few if any close substitutes for petroleum products. Nonetheless, even if demand for petroleum products were inelastic, the approach does not account for substitution effects on all other goods and services whose prices are changing.

2. These results are not discussed further in the report but can be made available upon request.

3. A comparison of Brazil and the Dominican Republic, which have similar vehicle ownership rates and similar proportional price shocks, reveals that the effect is more important in the Dominican Republic, where households spend a larger share of their expenditures on gasoline and diesel. Two reasons may account for this larger share: either households in the Dominican Republic use their vehicles more intensively than Brazilian households and thus consume more gasoline or diesel, or, because the Dominican Republic is poorer than Brazil and thus has lower levels of total expenditure, the same amount of gasoline would represent a higher share of expenditure there than in Brazil.

4. Unfortunately, IO tables do not separate electricity from other utility services such as water, sewage, and garbage collection. Although it is unlikely that water, sewage, and garbage collection will be affected in the same way as electricity production, we have opted for increasing their prices as well in order to present the maximum possible welfare effects on households.

5. We cannot provide a full explanation for these facts. They are directly related to the technical coefficients of the IO tables published by each country's statistical agency. In Mexico, the technical coefficient for oil-refined products in transport is 0.07, whereas it ranges from 0.18 to 0.24 in all other countries. For Colombia, the technical coefficient for oil-refined products in the manufacture of food products is 0.003, whereas it ranges from 0.006 to 0.027 in the other countries.

6. As described in chapter 2, VDT pricing schemes see the price per unit paid by the consumer increase for all units consumed in line with their total level of consumption. With the VDT approach, the per unit price is the same for all consumers in a certain consumption block. For electricity, higher consumption blocks generally have higher unit prices. In the case of IBT pricing schemes, one defined unit price per block exists for all consumers regardless of their total level of consumption. Consumers pay a higher unit price only when they exceed the maximum consumption level set by their assigned block.

7. Illegal connections in which the users do not pay for the electricity consumed.

8. Metering anomalies include, but are not limited to, destroyed, burned, or jammed meters. Another problem is direct service with no billing—that is, service was disconnected by the Federal Electric Commission because of nonpayment, but the service remains connected or is reconnected by the consumer. Failure to bill consumption

properly includes a metered service that is not billed; a service with estimated reading in which the meter reads incorrectly; and the application of an incorrect tariff.

9. Fixed in accordance with SIGET Agreement 587-E-2012.

10. Prices have been fixed according to the January 2011 tariff. Thus they do not reflect the higher costs of electricity.

References

Coady, David, and David Newhouse. 2006. "Ghana: Evaluating the Fiscal and Social Costs of Increases in Domestic Fuel Prices." In *Poverty and Social Analysis of Reforms: Lessons and Examples from Implementation*, edited by A. Coudouel, A. Dani, and S. Paternostro. Washington, DC: World Bank.

Coady, D. P., M. El Said, R. Gillingham, K. Kpodar, P. A. Medas, and D. L. Newhouse. 2006. "The Magnitude and Distribution of Fuel Subsidies: Evidence from Bolivia, Ghana, Jordan, Mali, and Sri Lanka." IMF Working Paper WP/06/247, International Monetary Fund, Washington, DC.

Deaton, Angus. 1989. "Household Survey Data and Pricing Policies in Developing Countries." *World Bank Economic Review* 3 (2): 183–210.

Kpodar, Kangni. 2006. "Distributional Effects of Oil Price Changes on Household Expenditures: Evidence from Mali." IMF WP/06/91, International Monetary Fund, Washington, DC.

Porto, Guido G. 2006. "Using Survey Data to Assess the Distributional Effects of Trade Policy." *Journal of International Economics* 70 (1): 140–60.

CHAPTER 6

Considerations for Policy Makers

Introduction

Because international oil prices are currently low, it is the ideal time to reform energy pricing policies. The removal of subsidies under such circumstances would likely mean smaller impacts on the welfare of the poor and less serious effects on national competitiveness.

Oil prices are, however, beginning to inch back upward. If they continue on the same trajectory, policy makers may be encouraged to provide new energy subsidies or extend the existing subsidies—measures that are invariably popular with consumers because of the short-term protection they offer against rising prices. However, energy subsidies create large distortions, and in the long term they do little to help the poor and vulnerable. Furthermore, they can be accompanied by significant fiscal costs.

Although it is important for governments in the Latin America and the Caribbean (LAC) region to consider subsidy reforms, it is not recommended that such reforms be carried out in isolation. Rather, they should be considered together with alternative uses for saved resources and complementary reforms that can increase national economic efficiency and competitiveness.

Fairness for All: A Guiding Principle for Energy Pricing Reforms

Countries need to guide their energy policies toward transparent, fair, and automatic price adjustments that track international prices closely. At the same time, it is important that national governments across the LAC region strengthen their social safety nets. And they should aim infrastructure investments at industries that are vulnerable to the effects of higher oil and electricity prices.

If pricing policies in the energy market are to be guided by the principles of fairness, transparency, and the promotion of competition, then it is vital that all actors involved in the process are fully aware of the rules and regulations at all times. Furthermore, these rules and regulations should be fair to all actors, not just a privileged minority or a priority segment of the economy. The principle of universal fairness also requires that such rules and regulations be effectively enforced by the pertinent authorities. Thus it is imperative that regulatory agencies be well equipped, well funded, and well staffed.

Transparency is a prerogative that is all too often overlooked in many LAC countries. Price distortions arising from subsidies and other government interventions are frequently absorbed by state-owned enterprises (SOEs) without explicit disclosure. This situation prevents the true cost of such measures from appearing in government budgets, and thus it effectively hides them from public taxpayers and private investors alike.

Fair and vigorous competition is a crucial building block of a fair-for-all pricing system. Failure to promote competition through the price-setting process encourages inefficiencies and makes oligopolistic collusion more likely, leading to higher prices for final consumers. By contrast, a truly competitive energy market results in greater efficiency and benefits consumers in the form of lower retail prices. The most appropriate means of achieving these benefits is a market-based (or deregulated) approach to pricing—the route adopted by the majority of developed economies. And yet, a market-based approach does not negate the need for reasonable technical, environmental, health, and safety regulations.

The LAC countries, however, may not meet all of the conditions required for a completely deregulated approach to pricing. Despite gradual privatization, the influence of SOEs over the refining and distribution stages remains high in most oil-producing countries (notably, Bolivia, Brazil, Colombia, and Mexico). Meanwhile, oil-importing countries are generally small economies in the Central America and the Caribbean subregion, where strong competition is often not the norm. As a consequence, it is difficult to describe any of the LAC countries in our study as implementing a pure, deregulated market-based approach to pricing. Moves to deregulate pricing should therefore be accompanied by strong antitrust monitoring and enforcement, in addition to investments and policies that promote competition.

Government interference in the price-setting process may be necessary in some countries. For example, when natural monopolies emerge (such as in the transport of oil or natural gas through pipelines), the government should intervene and regulate prices to avoid the higher prices and economic inefficiencies that would result from monopolistic behavior. Similarly, a purely market-based approach to pricing may have to be paired with strong antitrust monitoring and enforcement in small net importing economies, where the substantial up-front capital costs of import terminals and refineries may be a major barrier to entry. As an interim measure, policy makers may justifiably consider transparent regulatory frameworks that mimic the outcomes of well-functioning competitive markets. The most common such framework is that of pricing formulas based on import parity prices (IPPs) or, when the product is locally produced and exported, on export parity prices (EPPs).

A final consideration in creating a fair pricing system centers on the fiscal resources saved by subsidy reform. Ideally, these should be used to benefit the public by reverting them back to the economy through lower taxes or other public spending and investments. It is also reasonable for governments to consider easing the cost burdens on hard-hit sectors by introducing tailored interventions. Improvements in road infrastructure, for example, can help reduce the transport costs affected by high oil prices. Regulatory changes can play a role as well. Consider cargo trucks. The present rules prevent trucks in Central America from picking up goods in another country. Removing this regulation could bring high transport costs down. Likewise, policy makers may choose to consider targeted subsidies for vulnerable sections of the population. Such subsidies ensure that assistance reaches those who need it most, while reducing the fiscal cost of energy subsidies overall.

Fresh Thinking: Adopting an Engaging, Targeted, Apolitical Approach

Reform of energy pricing policy cannot be achieved through pure regulation alone. For regulatory reforms to be successful, education and informational campaigns have to alert citizens to the large inequities that arise from price controls and subsidies.

In addition, governments should expand and strengthen social safety nets. In this way, they can effectively deliver compensatory mechanisms that target the poor and vulnerable. Another way to protect the groups most at risk from higher energy costs is direct cash transfers. The advantage of such transfers over subsidies is the extent to which transfers can specifically target vulnerable householders. Direct cash transfers also avoid the unintended consequences that commonly follow government interference in the pricing process—consequences that can negatively affect the same poor and vulnerable individuals whom policy makers are endeavoring to protect.

In a similar vein, governments may also choose to make direct investments that will support industries highly likely to be negatively affected by the price impacts of oil shocks. Helping the transport sector by financing upgrades in road infrastructure and improvements in road safety are an illustrative case in point.

Last but not least, it is imperative that governments approach the question of energy pricing reform in a rational and objective manner. Energy prices are a highly emotive topic, especially when spikes in international oil prices lead to sudden increases in the final prices for domestic consumers. It is tempting for politicians to diverge from established policy commitments and intervene arbitrarily in the pricing process. Such politicization of the energy market may result in short-term political gains, but it creates instability in the market and potentially incurs major fiscal costs for the government. One solution is for governments to establish credible agencies that can control the pricing process for energy products independently of short-term political pressure. This "tie your hands" approach effectively depoliticizes the pricing process by removing the capacity of political actors to interfere.

Energy Pricing Policies for Inclusive Growth in Latin America and the Caribbean
http://dx.doi.org/10.1596/978-1-4648-1111-1

Hydrocarbons

Often policy makers have objectives beyond providing access to fuels at lower costs. One such objective relates to the notoriously volatile nature of international oil prices, which can create tensions and uncertainty in some economies. When domestic prices are completely delinked from international prices, these kinds of policies require important fiscal outlays, especially if they are sustained for long periods. Policy makers should also consider other knock-on effects such as the impact of sustained price distortions on the investment decisions of private firms. If prices charged by private producers are kept artificially low, they may well be dissuaded from investing in exploration. Thus a strategy that seeks to lower prices in the short run may jeopardize the level of oil production in the future. Another consideration is the impact that underpricing fuels has on consumption. Artificially low fuel prices can lead to higher consumption, which has important long-term consequences on the environment and public health (via higher rates of air pollution).

Parity Prices: Frequency and Duration

Although it is conceptually simple and appealing, at least five policy issues need to be considered when designing a pricing formula framework.

First, authorities need to define the frequency of price adjustments. In general, the lower the frequency, the more stable prices will be in the short run. The longer policy makers wait to make adjustments, however, the higher the probability they will have to make a large adjustment. In our sample, the Dominican Republic, El Salvador, Haiti (before 2011), and Honduras have weekly adjustment of prices, whereas Colombia, Mexico, and Peru have monthly adjustments.

Second, governments must determine the averaging period for the reference price. Policy makers should consider that longer averaging periods yield less volatile domestic prices. On the flip side, longer averaging periods may lead to a lower correlation with international prices, and they also come with the risk of larger departures from cost recovery, potentially leading to large financing gaps. The LAC countries have taken different approaches. For example, the Dominican Republic sets the prices on Monday of every week, referencing the price of Tuesday of the previous week. Honduras uses a moving average of 22 days. Table 6.1 presents the details of reference price determination in IPP pricing formulas for seven of the nine sample countries.

Third, policy makers should also consider the market structure throughout the entire production chain (see Kojima 2016). In countries in which there is one state-run refinery but many private fuel retailers, it makes sense to regulate prices at the wholesale level rather than at the retail level. Relevant examples in our sample are Bolivia, Brazil, and Colombia. On the other hand, if importers or refineries are vertically integrated with downstream retailers, prices should be regulated at the retail level such as for gasoline and diesel in Mexico.

Fourth, policy makers need to consider whether to impose price caps or fix price levels. Price caps set the maximum prices that can be charged by retailers,

Table 6.1 Reference Price Determination in IPP Pricing Formulas, Selected LAC Countries

Country	Gasoline	Diesel	LPG	Kerosene	Natural gas
Colombia	Prices are set monthly, based on parity price tendencies of the last 60 days.	Prices are set monthly, based on parity price tendencies of the last 60 days.	—	—	—
Dominican Republic	Prices are set each Monday (referencing price on Tuesday of the previous week).	Prices are set each Monday (referencing price on Tuesday of the previous week).	Prices are set each Monday (referencing price on Tuesday of the previous week).	—	Prices are set weekly.
El Salvador	Prices are set weekly as the average of the last 30 days.	Prices are set weekly as the average of the last 30 days.	Prices are set weekly as the average of the last 30 days.	—	—
Haiti	Prices frozen since 2011. Weekly adjustments before 2011.	Prices frozen since 2011. Weekly adjustments before 2011.	—	Prices frozen since 2011. Weekly adjustments before 2011.	—
Honduras	Prices are set weekly as the moving average of the last 22 days.	Prices are set weekly as the moving average of the last 22 days.	Prices are set weekly as the moving average of the last 22 days.	Prices are set weekly as the moving average of the last 22 days.	—
Mexico	Prices are set monthly.	Prices are set monthly.	Prices are set monthly.	Prices are set monthly.	—
Peru	Prices are set monthly.	Prices are set monthly.	Prices are set monthly.	—	—

Source: World Bank.
Note: IPP = import parity price; LPG = liquefied petroleum gas; — = not available.

but they have the advantage of allowing for price competition. Indeed, departures from the price cap can even indicate the level of competition in a particular market. On the downside, price caps increase the costs of monitoring, whereas fixing price levels facilitates monitoring the compliance of all agents. What fixing price levels fails to do, however, is to allow for competition in prices.

Finally, policy makers need to decide how to respond in the event of a sharp change in international energy prices. Governments frequently intervene in the regulatory framework based on IPP pricing formulas when the international oil price is volatile. The impacts of these interventions should be considered. Introducing price reductions through a specific additional line in the pricing formula may result in a significant fiscal cost, for example. On the positive side, this approach at least has the benefit of being transparent: the per unit subsidy is explicit, as are the fiscal outlays needed to finance the subsidy. This transparency helps with budgetary planning. And yet, this approach does nothing to change the rules or incentives for agents in the production and marketing chain. The opposite is true for less explicit measures such as the introduction of reductions in commercial margins or temporary suspensions of the formula. These measures, however, substantially affect the general business environment. They also obscure the magnitude of the implicit subsidy they create.

Energy Pricing Policies for Inclusive Growth in Latin America and the Caribbean
http://dx.doi.org/10.1596/978-1-4648-1111-1

Price Stabilization

We recognize that price stabilization is a popular strategy for making fuel prices in the domestic market more predictable, but we advise against its use in the hydrocarbons market because of the frequency with which governments are left with large fiscal costs and unpredictable fiscal outlays.

In theory, price stabilization funds are self-financing schemes that can moderate the transmission of international oil price volatility. In practice, however, oil prices need to follow a mean-reversion pattern for this to occur—that is, oil prices must revert to their average value in the short term, allowing for periods of revenue loss to be regularly compensated by periods of savings. These conditions were not in place in the late 2000s when international oil prices rose steadily. In the LAC countries that introduced price stabilization such as Colombia and Peru, this measure created a long under-recovery period with a growing deficit. Such a scenario means that governments end up unintentionally subsidizing fuels through exceptional loans and other stopgap measures.

There is one exception, however. Price stabilization funds could be a useful transitory mechanism toward automatic price adjustments in countries whose governments have instituted price freezing. In such cases, the funds should include automatic adjustments in their target levels (such as those used in Colombia and Peru after the fixed-price targets became too misaligned with international prices), which would allow prices to move gradually. Over time, governments should have wider bands, allowing more fluctuation in prices, until pricing policies can transition to automatic adjustments.

Unintended Consequences: Switching, Adulteration, and Smuggling

Policy makers need to be aware that intervening in the price-setting process and the tax structure in the hydrocarbons sector can affect market outcomes in ways that are not always intended. Well-intentioned policies can backfire or incur ballooning costs because consumers react and change their behavior in response to relative price changes.

One example is overconsumption. Governments may design their pricing strategies to promote certain products and incentivize switching from certain fuels to others. For example, the government of the Dominican Republic introduced a universal price subsidy for liquefied petroleum gas (LPG) to encourage consumers to switch cooking fuel from charcoal to LPG. As detailed earlier, the policy had the effect of encouraging overconsumption because consumers switched to using LPG for transport as well as for domestic purposes. The move not only resulted in a high fiscal cost to the government, but also led to a significant revenue loss as consumers moved away from taxed products such as diesel and gasoline as well.

Large differences in prices among substitutable fuels can also lead to problems of adulteration. This illegal activity manifests itself most commonly in mixing regular gasoline and premium gasoline and adding kerosene to diesel.

Countries that provide large subsidies for kerosene intended for cooking and lighting in households may soon confront the issue of automotive diesel being blended with kerosene. Such malpractice is difficult to monitor because a considerable amount of kerosene (about 30 percent of the total mix) can be added without immediate detection. Similarly, detecting the difference between premium and regular gasoline is difficult without authorities continually monitoring service station inventories.

Policy makers should also be aware that price-setting mechanisms can unintentionally lead to cross-border smuggling of hydrocarbon products. This occurs when a large price differential exists between two neighboring countries. The government that subsidizes fuel products bears the cost of all the subsidized fuel that ends up in a neighboring country. The receiving government also bears a cost, however, in the form of lost tax revenues. Evidence exists of smuggling from Ecuador and Bolivia to Peru. Meanwhile, when the subsidy for LPG was universal in El Salvador, large quantities of LPG cylinders were smuggled into neighboring Guatemala and Honduras. The most glaring example in the LAC region, however, is the smuggling across the República Bolivariana de Venezuela–Colombia frontier. The price of gasoline and diesel was more than 1,000 percent higher in Colombia than in the República Bolivariana de Venezuela, leading to a major smuggling problem. In 2010 over 14,000 barrels per day of fuel were smuggled into Colombia, equivalent to 10–15 percent of Colombia's total fuel consumption. This smuggling represents a loss of some US$225.2 million in tax revenues to the Colombian government (Garcia et al. 2010).

One of the ways to respond to the smuggling epidemic is to provide frontier zones with fuel tax exemptions. Such a measure effectively reduces the price differential with the neighboring state. Policy makers in Colombia adopted this approach to smuggled fuels from República Bolivariana de Venezuela. This type of policy is not unfamiliar in the LAC region, mirroring as it does certain geographically based tax exemptions that are used as a strategy for promoting development. Such exemptions can be problematic, however, because they create price differentials internally between regions, which in turn present a profitable arbitrage opportunity. In most cases, the problem of smuggling simply moves within the country's borders. Domestic smuggling is harder to monitor, leading to an increase in enforcement costs.

Electricity Market

As previously noted, the electricity market is best conceptualized as three separate yet interdependent segments: generation, transmission, and distribution and commercialization. Although there is a component of strategic planning by the government, the generation market can be thought of as a competitive market and regulated as such. The other two segments are natural monopolies, and so they should be closely regulated.

Electricity Sector: Investing in Infrastructure

Some LAC countries have privatized both their transmission networks and their distribution and commercialization sectors. Regardless of the underlying market structure, it is essential that revenues in these two segments allow not only for covering operation and maintenance costs but also for continual investment in technological modernization. Without such upkeep and investment, transmission lines are likely to become congested as demand for electricity grows.

In addition, underinvestment in these segments inevitably leads to the deterioration of transmission lines and distribution equipment, as well as ineffective metering, monitoring, and enforcement systems. Together, these problems can result in excessive technical and nontechnical losses, as detailed in chapter 3, and they can severely destabilize the finances of utilities if not kept in check. Faced with such circumstances, utility companies could struggle to even cover their operating expenses, let alone make investments in new distribution equipment and transmission lines. This lack of investment would in turn generate higher technical losses, resulting in a downward spiral as utilities prove unable to take other mitigating steps to reduce losses, such as installing more meters, rehabilitating grids, and prosecuting electricity theft.

LAC countries with high technical and nontechnical losses must therefore institute comprehensive reform packages (see box 6.1). These reforms must

---

Box 6.1 Action Plans to Reduce Power Losses in the Electricity Sector

Governments should draw up a comprehensive action plan to address power losses in a way that involves all key players, according to a handbook prepared by PA Consulting Group for the U.S. Agency for International Development (PA Consulting Group 2004). The key features of any such plan should include the following:

- *Market segmentation.* When total losses are high, large consumers usually account for a large fraction of the losses. Measures should be taken to reduce the incentive of large consumers to collude with the utility employees who read their meters—a common problem in Latin America and the Caribbean countries. This could be achieved through systematic field inspections of large consumers' premises by trusted employees.
- *Name and shame.* Publicizing the identity of large consumers that steal electricity can be an effective deterrent to others because these entities are generally sensitive to anything that will damage their public reputations.
- *Reliable information.* Providing reliable and transparent information within the company, as well as to the government and other external stakeholders, allows more accurate detection of problem areas and enables quicker corrective actions. Obtaining quality data may require an upgrade in management information systems.
- *Public commitment.* One way governments can prove their political commitment to combating electricity theft is to ensure that the prosecution of politically powerful individuals is not being impeded, should such an incident arise.

---

provide utilities with sufficient revenue for investments in new equipment, technology, and the monitoring of payments to reduce losses. If they do not, the vicious cycle of chronic deficits will only continue. Consumers could be asked to make up for the inefficiencies of utility companies in the form of higher electricity prices, but this is neither fair nor efficient.

Not all loss reduction measures are costly for utilities. Better management can achieve good results. Steps that utilities might consider include creating a large customer department as an organizational unit within the commercial department; undertaking a field assessment to determine which consumers are supplied by different voltage lines and who are the largest low-voltage customers; regularizing supply where needed; and establishing processes to systematically monitor the electricity supply to large customers and to take immediate corrective action when irregularities occur (for more, see Antmann 2016).

Policy makers should also consider adopting Advanced Metering Infrastructure (AMI). Such a telecommunications-based system enables electricity consumption to be metered remotely, which brings down metering costs dramatically. AMI has been used in countries such as Brazil, the Dominican Republic, and Honduras, where it has also been shown to be highly effective in detecting and discouraging theft. With AMI systems in place, consumers become aware that their utility can monitor their consumption at its convenience, which gives them the incentive to change their behavior. In addition, AMI can enhance electricity companies' internal anticorruption efforts by eliminating the need for field operatives (who are liable to collusion with large consumers). Finally, AMI enables utilities to introduce effective programs for prepaid consumption, which is generally a very good commercial option for low-income consumers.

In terms of electricity generation, the government has a role in strategically planning investment in new power plants. Countries that rely extensively on one technology for generation are exposed to substantial risks. For example, in economies in which fossil fuel–based generation plants dominate, price risks linked to the oil market can be high. Countries reliant on hydroelectric generation face problems when the hydrological conditions are not favorable. Although certain financial instruments may help hedge some of the short-term risks (see box 6.2), countries need to diversify their electricity generation matrix to minimize their exposure to any price shock.

As highlighted earlier, the LAC region has been an innovator in the use of auctions in new generation capacity, which has helped lower overall costs. We encourage further use of this approach both "in the market" and "for the market." These auctions can either be technology-specific or project-specific. They may even be used for full competitive procurement where all technologies compete head to head. Reasons to support the competitive auction approach include an abundance of conveniently located renewable resources, a significant decline in technology prices, and the built-in guarantee of the benefits of competition being passed on to the end user.

Energy Pricing Policies for Inclusive Growth in Latin America and the Caribbean
http://dx.doi.org/10.1596/978-1-4648-1111-1

Box 6.2 Uruguay Weather and Oil Price Insurance Transaction

UTE, Uruguay's state-owned public electric company, generates more than 80 percent of its energy needs from hydropower plants. When rainfall or accumulated water reserves are low, the state utility is forced to purchase fuels—typically oil derivatives and natural gas—to use as inputs for electricity production. When the price of oil is high, generation costs skyrocket. High oil prices not only damage UTE's bottom line, but also increase the prices for consumers and affect the national budget. In 2012 the cost of supplying electricity to meet national demand reached a record US$1.4 billion as drought conditions obliged the country to shift to fossil fuels. The original projected cost for the year was US$953 million.

In an effort to hedge against the financial exposure to low rainfall and high oil prices, UTE signed a US$450 million weather and oil price insurance contract with the World Bank in December 2013. At the time, this contract was the largest in the weather risk management market. It was also the first time that a public utility company had used this type of risk management tool.

The contract insured UTE for 18 months against drought and high oil prices. To measure the extent of a drought and potential insurance payout to UTE, daily rainfall data are collected at 39 weather stations spread throughout the river basins. If precipitation falls below the level set up as a trigger in the contract, UTE would receive a payout of up to US$450 million based on the severity of the drought and oil price levels. If oil prices are high, the payout would be larger to offset the high cost of fuel purchases.

This contract formed part of a broader framework of measures designed to reduce UTE's vulnerabilities to drought. Among the other steps were stabilization funds and contingent financing with private banks. This type of arrangement is replicable in other countries.

Pricing Policies

In terms of pricing policies for final consumers, the overarching principle is that, whatever the tariff structure chosen by a government, the revenue generated should cover all operating, maintenance, and generation costs, as well as investments in new, modern equipment.

Underpricing electricity is a widespread phenomenon in the LAC region's energy markets. The effect of this practice is significant yet more modest relative to fuels. Furthermore, unlike the fuels market, most electricity subsidies are directed toward the residential sector rather than industry users. Even so, the effects of underpricing should not be underestimated. In some countries, subsidies have served to exacerbate existing inefficiencies. In doing so, they have caused a further deterioration of service quality and led to an increase in consumer costs. This is another reason it is vital for policy makers to develop plans to restore electricity utilities to financial health as part of any subsidy reform process.

The most common tariff framework for end users is the increasing block tariff (IBT). IBT-based pricing mechanisms have important benefits for richer households, which tend to consume more electricity because all consumers, irrespective of income, are subsidized in the first block.

Even when the tariff is structured in a cross-subsidy framework, poor households may still end up consuming more than the subsidized consumption level because they have larger families or live in multifamily dwellings. Their consumption will also be inflated if, as is probable, they use older, less energy-efficient appliances.

Paying the full costs of electricity that their household consumes may not be possible for some low-income consumers. If helping the poor is a central concern for policy makers, one of the best tools is a flat price per kilowatt-hour, with a different pricing structure for households that qualify for the federal antipoverty programs. Brazil has implemented just such a policy with considerable success. For LAC countries that already have conditional cash transfer programs and expansive social safety nets in place, it would be relatively easy to follow Brazil's lead. An alternative option for policy makers could be to transfer the cash equivalent of a predetermined level of kilowatt-hour consumption to qualifying poor families. This would encourage such families to efficiently conserve electricity.

Additional Considerations

In what follows, we pose some additional considerations for policy makers wishing to design efficient, resilient electricity markets. On the one hand, it is important to introduce incentives and protections that enhance competition in all segments of the electricity production chain. On the other hand, any electricity system will be subject to risks—some technological in nature and some driven by conditions beyond the control of policy makers. The following sections highlight some tools for managing these risks.

Avoiding Collusion

Because the price that distributors pay generation companies for electricity is passed on to consumers, regulators have to consider how to avoid the threat of collusion. One common way of achieving this is to place ownership limits on private firms that have interests in more than one segment of the electricity production chain. In Colombia, for example, distributors are prohibited from having more than 25 percent ownership of a generation company (and vice versa). Another option is for governments to limit the amount of energy that can be purchased through long-term contracts. In the Dominican Republic, the limit is set at 80 percent. Under this kind of arrangement, the remainder of the energy demand has to be met by energy purchased in the spot market. The spot market consists of government auctions held by the regulatory agency at regular intervals. In Colombia, the auctions occur every hour, whereas in Peru the interval is every 15 minutes.

Anticipating a Rise in Small-Scale Renewable Power Generation

An important emerging consideration for policy makers is how an uptake in distributed, small-scale power generation may affect the retail electricity market.

As technology develops and the price for small-scale self-generation continues to drop, low-volume consumers could well join large industrial power users in the retail market. This would bring considerably more competition to the market. It would also threaten the basis of the cross-subsidy model as high-income residential and commercial customers who currently cross-subsidize poorer or lower-volume consumers begin to install their own photovoltaic panels or alternative renewable power microgenerators. Such an eventuality would inevitably see grid consumption fall, causing utilities to carefully consider whether their fixed costs are being properly recovered. If not, they may be required to review their pricing policies based on cross-subsidies. The LAC region is not alone in this dilemma: regulatory regimes in the developed world are in a similar state of flux as well.

Coping with Drought

For countries that rely heavily on hydroelectric generation, drought or extended dry spells can present major risks to the power network. Drought is not uncommon, and it may well become more frequent if global temperatures continue to rise. In Peru, for example, because of the hydrological cycle it is highly probable that the country will experience a dry season at least once every seven years. When water levels are low, countries usually turn to thermal power plants to meet demand. These plants tend to be fossil fuel–based, and the costs are dependent on the international prices of oil. This scenario played out in Brazil when it faced adverse hydrological conditions in 2012 and 2013. The result was a nine-fold increase in the marginal cost of the electricity system.[1] Uruguay also has a generation matrix based on hydroelectric generation, with thermal power generation as a backup. To hedge against the risk of facing high oil prices during a drought (which would increase generation costs dramatically), the country has developed a financial instrument in conjunction with the World Bank Group (see box 6.2).

Conclusions

A majority of countries in the LAC region are currently facing tight fiscal conditions and low growth prospects. The end of the commodity supercycle has brought in turn an end to the economic bonanza that helped lift millions of families out of poverty, and it has left governments with challenging fiscal positions. Moreover, governments in the LAC region tend to be small as measured by their revenue as a percentage of their gross domestic product. As a result, their limited government resources have to be used wisely and need to better target the poor and vulnerable. One important area for improvement is energy pricing policies, where eliminating expensive subsidies and price controls can create fiscal space (now and in the future) and allow governments to better direct their resources toward protecting lower-income households.

Although some governments have made progress in reforming energy subsidies, the temptation and political pressures to reintroduce them will resurface if oil prices shoot up again. Thus it is important for governments to implement energy pricing processes that are protected from these pressures. For one thing, they could consider delegating pricing authority to independent agencies that are shielded from political pressure. They also could institute automatic price adjustment policies that cannot be easily modified, thereby further distancing themselves from pricing decisions.

Although energy subsidies are an inefficient policy tool to protect the welfare of the poor, energy price increases can have a big impact on these households. It is important, then, that governments expand the coverage and depth of their social safety nets so they can provide relief for poor households if energy prices were to increase. Direct cash transfers or vouchers have been found to be a much more effective policy tool for protecting poor and vulnerable households from the negative welfare impacts of rising energy costs.

Note

1. As measured by the Balance Liquidation Value (Preco de Liquidacao de Diferencas, PLD).

References

Antmann, P. 2016. "Reducing Technical and Non-technical Losses in the Power Sector." Background paper for Energy Strategy Energy Unit, World Bank, Washington, DC.

Gracia, Orlando, et al. 2010. "La Ley de Fronteras y su efecto en el comercio de combustibles líquidos." *Oxford Review of Economic Policy* (January).

Kojima, Masami. 2016. "Fossil Fuel Subsidy and Pricing Policies: Recent Developing Country Experience." Policy Research Working Paper 7531, World Bank, Washington, DC.

PA Consulting Group. 2004. "Improving Power Distribution Company Operations to Accelerate Power Sector Reform." November. http://pdf.usaid.gov/pdf_docs/PNADJ549.pdf.

APPENDIX A

Fuels Fiscal Cost Tables

Table A.1 Bolivia—Gasoline

		Unit[a]	2009	2010	2011	2012	2013
A.	Supply cost[b]	$/liter	0.637	0.734	0.925	0.967	0.935
B1.	Taxes, excluding transfers (includes VAT and excise taxes)[c]	$/liter	0.193	0.193	0.193	0.193	0.193
B2.	Taxes (counterfactual): standard VAT	$/liter	0.095	0.110	0.138	0.145	0.140
C.	**Total reference benchmark price**						
C1	Total reference price without taxes	$/liter	0.637	0.734	0.925	0.967	0.935
C2	Total reference price at counterfactual tax rates	$/liter	0.733	0.844	1.063	1.112	1.074
D.	**End-user price**						
D1	End-user price, excluding all taxes	$/liter	0.271	0.271	0.271	0.271	0.271
D2	End-user price, including all taxes	$/liter	0.465	0.465	0.465	0.465	0.465
E.	**Price gap**						
E1	Price gap, excluding all taxes (C1 – D1)	$/liter	0.366	0.463	0.654	0.696	0.663
E2	Price gap, with fuel at counterfactual tax rates (C2 – D2)	$/liter	0.268	0.379	0.598	0.647	0.609
F.	**Consumption**						
F1	Volume	Million liters	911.63	1,008.57	1,102.87	1,174.48	1,289.43
G.	**Fiscal cost**						
G1	Fiscal cost, excluding all taxes (E1* F1)	$, millions	333.72	466.64	720.87	817.62	855.23
G2	Fiscal cost, with fuel at counterfactual tax rates (E2 * F1)	$, millions	244.15	382.10	659.90	760.14	785.79
G3	Fiscal expenditures (F1 * (E2 – E1))	$, millions	(89.57)	(84.54)	(60.98)	(57.48)	(69.44)

a. All prices and costs are expressed in current U.S. dollars.
b. CIF (cost, insurance, and freight) price + local distribution and retail margins + other costs.
c. VAT (value added tax) is $.04 per liter, and excise tax (IEDH) is $.15 per liter.

Table A.2 Bolivia—Diesel

		Unit[a]	2009	2010	2011	2012	2013
A.	Supply cost[b]	$/liter	0.589	0.713	0.942	0.970	0.940
B1.	Taxes, excluding transfers (includes VAT and excise taxes)[c]	$/liter	0.195	0.195	0.195	0.195	0.195
B2.	Taxes (counterfactual): standard VAT	$/liter	0.088	0.107	0.141	0.145	0.140
C.	Total reference benchmark price						
C1	Total reference price without taxes	$/liter	0.589	0.713	0.942	0.970	0.940
C2	Total reference price at counterfactual tax rates	$/liter	0.677	0.820	1.083	1.115	1.080
D.	End-user price						
D1	End-user price, excluding all taxes	$/liter	0.267	0.267	0.267	0.267	0.267
D2	End-user price, including all taxes	$/liter	0.462	0.462	0.462	0.462	0.462
E.	Price gap						
E1	Price gap, excluding all taxes (C1 − D1)	$/liter	0.321	0.446	0.675	0.702	0.673
E2	Price gap, with fuel at counterfactual tax rates (C2 − D2)	$/liter	0.214	0.357	0.621	0.652	0.618
F.	Consumption						
F1	Volume	Million liters	1,261.58	1,329.47	1,446.70	1,555.21	1,694.49
G.	Fiscal cost						
G1	Fiscal cost, excluding all taxes (E1 * F1)	$, millions	405.48	592.54	976.55	1,092.45	1,139.87
G2	Fiscal cost, with fuel at counterfactual tax rates (E2 * F1)	$, millions	270.54	475.02	898.23	1,014.63	1,047.55
G3	Fiscal expenditures (F1 * (E2 − E1))	$, millions	(134.95)	(117.52)	(78.32)	(77.82)	(92.32)

a. All prices and costs are expressed in current U.S. dollars.
b. CIF (cost, insurance, and freight) price + local distribution and retail margins + other costs.
c. VAT (value added tax) is $.04 per liter, and excise tax (IEDH) is $.15 per liter.

Table A.3 Bolivia—Liquefied Petroleum Gas (LPG)

		Unit[a]	2009	2010	2011	2012	2013
A.	Supply cost[b]	$/liter	0.464	0.552	0.634	0.525	0.536
B1.	Taxes, excluding transfers (includes VAT and excise taxes)[c]	$/liter	0.020	0.020	0.020	0.020	0.020
B2.	Taxes (counterfactual): standard VAT	$/liter	0.069	0.083	0.095	0.078	0.080
C.	Total reference benchmark price						
C1	Total reference price without taxes	$/liter	0.464	0.552	0.634	0.525	0.536
C2	Total reference price at counterfactual tax rates	$/liter	0.533	0.635	0.728	0.603	0.617
D.	End-user price						
D1	End-user price, excluding all taxes	$/liter	0.136	0.136	0.136	0.136	0.136
D2	End-user price, including all taxes	$/liter	0.157	0.157	0.157	0.157	0.157
E.	Price gap						
E1	Price gap, excluding all taxes (C1 − D1)	$/liter	0.328	0.416	0.497	0.388	0.400
E2	Price gap, with fuel at counterfactual tax rates (C2 − D2)	$/liter	0.376	0.478	0.572	0.446	0.460
F.	Consumption						
F1	Volume	Million liters	549.66	551.12	581.37	559.42	398.32

table continues next page

Table A.3 Bolivia—Liquefied Petroleum Gas (LPG) *(continued)*

	Unit[a]	2009	2010	2011	2012	2013
G.	**Fiscal cost**						
G1	Fiscal cost, excluding all taxes (E1 * F1)	$, millions	180.02	229.24	289.21	217.28	159.39
G2	Fiscal cost, with fuel at counterfactual tax rates (E2 * F1)	$, millions	206.91	263.47	332.41	249.73	183.20
G3	Fiscal expenditures (F1 * (E2 − E1))	$, millions	26.89	34.24	43.20	32.45	23.81

a. All prices and costs are expressed in current U.S. dollars.
b. CIF (cost, insurance, and freight) price + local distribution and retail margins + other costs.
c. VAT (value added tax) is $.02 per liter, and excise tax is zero.

Table A.4 Bolivia—Natural Gas (Power)

		Unit[a]	2009	2010	2011	2012	2013
A.	**Supply cost**[b]	$/MPC	5.350	6.500	8.390	10.170	9.860
B1.	**Taxes, excluding transfers (includes VAT and excise taxes)**[c]	$/MPC	0.544	0.661	0.853	1.034	1.003
B2.	**Taxes (counterfactual): standard VAT**	$/MPC	0.799	0.971	1.253	1.519	1.473
C.	**Total reference benchmark price**						
C1	Total reference price without taxes	$/MPC	5.350	6.500	8.390	10.170	9.860
C2	Total reference price at counterfactual tax rates	$/MPC	6.149	7.471	9.643	11.689	11.333
D.	**End-user price**						
D1	End-user price, excluding all taxes	$/MPC	1.180	1.180	1.180	1.180	1.180
D2	End-user price, including all taxes	$/MPC	1.300	1.300	1.300	1.300	1.300
E.	**Price gap**						
E1	Price gap, excluding all taxes (C1 − D1)	$/MPC	4.170	5.320	7.210	8.990	8.680
E2	Price gap, with fuel at counterfactual tax rates (C2 − D2)	$/MPC	4.849	6.171	8.343	10.389	10.033
F.	**Consumption**						
F1	Volume	Million MPCs	41.38	49.24	54.40	53.88	53.11
G.	**Fiscal cost**						
G1	Fiscal cost, excluding all taxes (E1 * F1)	$, millions	172.54	261.95	392.19	484.38	460.96
G2	Fiscal cost, with fuel at counterfactual tax rates (E2 * F1)	$, millions	200.65	303.86	453.84	559.78	532.82
G3	Fiscal expenditures (F1* (E2 − E1))	$, millions	28.11	41.91	61.65	75.40	71.86

a. All prices and costs are expressed in current U.S. dollars. MPC = thousand cubic feet.
b. Weighted export price to Brazil and Argentina.
c. VAT (value added tax) is 10.17 percent of the supply cost, and excise tax is zero.

Table A.5 Bolivia—Natural Gas (Households)

		Unit[a]	2009	2010	2011	2012	2013
A.	**Supply cost**[b]	$/MPC	5.350	6.500	8.390	10.170	9.860
B1.	**Taxes, excluding transfers (includes VAT and excise taxes)**[c]	$/MPC	0.412	0.500	0.645	0.782	0.758
B2.	**Taxes (counterfactual): standard VAT**	$/MPC	0.799	0.971	1.253	1.519	1.473
C.	**Total reference benchmark price**						
C1	Total reference price without taxes	$/MPC	5.350	6.500	8.390	10.170	9.860
C2	Total reference price at counterfactual tax rates	$/MPC	6.149	7.471	9.643	11.689	11.333

table continues next page

Table A.5 Bolivia—Natural Gas (Households) *(continued)*

		Unit[a]	2009	2010	2011	2012	2013
D.	**End-user price**						
D1	End-user price, excluding all taxes	$/MPC	0.910	0.910	0.910	0.910	0.910
D2	End-user price, including all taxes	$/MPC	0.980	0.980	0.980	0.980	0.980
E.	**Price gap**						
E1	Price gap, excluding all taxes (C1 − D1)	$/MPC	4.440	5.590	7.480	9.260	8.950
E2	Price gap, with fuel at counterfactual tax rates (C2 − D2)	$/MPC	5.169	6.491	8.663	10.709	10.353
F.	**Consumption**						
F1	Volume	Million MPCs	37.90	50.91	46.79	51.30	56.72
G.	**Fiscal cost**						
G1	Fiscal cost, excluding all taxes (E1 * F1)	$, millions	168.26	284.61	349.99	475.05	507.60
G2	Fiscal cost, with fuel at counterfactual tax rates (E2 * F1)	$, millions	195.90	330.49	405.36	549.41	587.18
G3	Fiscal expenditures (F1* (E2 − E1))	$, millions	27.64	45.88	55.37	74.36	79.58

a. All prices and costs are expressed in current U.S. dollars. MPC = thousand cubic feet.
b. Weighted export price to Brazil and Argentina.
c. VAT (value added tax) is 7.69 percent of the supply cost, and excise tax is zero.

Table A.6 Bolivia—Natural Gas (Direct)

		Unit[a]	2009	2010	2011	2012	2013
A.	**Supply cost[b]**	$/MPC	5.350	6.500	8.390	10.170	9.860
B1.	**Taxes, excluding transfers (includes VAT and excise taxes)[c]**	$/MPC	0.543	0.659	0.851	1.032	1.000
B2.	**Taxes (counterfactual): standard VAT**	$/MPC	0.799	0.971	1.253	1.519	1.473
C.	**Total reference benchmark price**						
C1	Total reference price without taxes	$/MPC	5.350	6.500	8.390	10.170	9.860
C2	Total reference price at counterfactual tax rates	$/MPC	6.149	7.471	9.643	11.689	11.333
D.	**End-user price**						
D1	End-user price, excluding all taxes	$/MPC	1.380	1.380	1.380	1.380	1.380
D2	End-user price, including all taxes	$/MPC	1.520	1.520	1.520	1.520	1.520
E.	**Price gap**						
E1	Price gap, excluding all taxes (C1 − D1)	$/MPC	3.970	5.120	7.010	8.790	8.480
E2	Price gap, with fuel at counterfactual tax rates (C2 − D2)	$/MPC	4.629	5.951	8.123	10.169	9.813
F.	**Consumption**						
F1	Volume	Million MPCs	6.96	7.48	7.48	7.22	8.12
G.	**Fiscal cost**						
G1	Fiscal cost, excluding all taxes (E1 * F1)	$, millions	27.63	38.28	52.41	63.45	68.86
G2	Fiscal cost, with fuel at counterfactual tax rates (E2 * F1)	$, millions	32.22	44.49	60.73	73.41	79.69
G3	Fiscal expenditures (F1 * (E2 − E1))	$, millions	4.59	6.21	8.32	9.96	10.83

a. All prices and costs are expressed in current U.S. dollars. MPC = thousand cubic feet.
b. Weighted export price to Brazil and Argentina.
c. VAT (value added tax) is 10.14 percent of the supply cost, and excise tax is zero.

Table A.7 Brazil—Gasoline

	Unit[a]	2008	2009	2010	2011	2012	2013
A.	**Supply cost[b]**	$/liter	1.020	0.815	0.973	1.209	1.164	1.125
B1.	**Taxes, excluding transfers (includes VAT and excise taxes)[c]**	$/liter	0.692	0.502	0.621	0.749	0.634	0.579
B2.	**Taxes (counterfactual): standard VAT**	$/liter	0.173	0.139	0.165	0.205	0.198	0.191
C.	**Total reference benchmark price**							
C1	Total reference price without taxes	$/liter	1.020	0.815	0.973	1.209	1.164	1.125
C2	Total reference price at counterfactual tax rates	$/liter	1.193	0.954	1.139	1.414	1.362	1.316
D.	**End-user price**							
D1	End-user price, excluding all taxes	$/liter	0.671	0.754	0.838	0.884	0.767	0.744
D2	End-user price, including all taxes	$/liter	1.363	1.256	1.459	1.632	1.401	1.323
E.	**Price gap**							
E1	Price gap, excluding all taxes (C1 − D1)	$/liter	0.348	0.061	0.136	0.325	0.396	0.381
E2	Price gap, with fuel at counterfactual tax rates (C2 − D2)	$/liter	−0.170	−0.302	−0.320	−0.218	−0.040	−0.007
F.	**Consumption**							
F1	Volume	Million liters	25,174.78	25,409.09	29,843.67	35,491.26	39,697.72	41,365.26
G.	**Fiscal cost**							
G1	Fiscal cost, excluding all taxes (E1 * F1)	$, millions	8,772.75	1,559.47	4,051.38	11,539.89	15,734.69	15,761.32
G2	Fiscal cost, with fuel at counterfactual tax rates (E2 * F1)	$, millions	(4,287.24)	(7,684.27)	(9,547.38)	(7,737.13)	(1,569.12)	(277.29)
G3	Fiscal expenditures (F1 * (E2 − E1))	$, millions	(13,059.99)	(9,243.74)	(13,598.76)	(19,277.02)	(17,303.81)	(16,038.61)

a. All prices and costs are expressed in current U.S. dollars.
b. CIF (cost, insurance, and freight) price + local distribution and retail margins.
c. VAT (value added tax, ICMS) is a weighted average from different ad valorem state taxes (PIS/COFINS), and excise tax (CIDE) is a unit tax.

Table A.8 Brazil—Diesel

	Unit[a]	2008	2009	2010	2011	2012	2013
A.	**Supply cost[b]**	$/liter	0.925	0.613	0.772	1.001	0.988	1.005
B1.	**Taxes, excluding transfers (includes VAT and excise taxes)[c]**	$/liter	0.423	0.309	0.398	0.491	0.438	0.409
B2.	**Taxes (counterfactual): standard VAT**	$/liter	0.157	0.104	0.131	0.170	0.168	0.171

table continues next page

Table A.8 Brazil—Diesel *(continued)*

		Unit[a]	2008	2009	2010	2011	2012	2013
C.	**Total reference benchmark price**							
C1	Total reference price without taxes	$/liter	0.925	0.613	0.772	1.001	0.988	1.005
C2	Total reference price at counterfactual tax rates	$/liter	1.082	0.718	0.903	1.172	1.156	1.175
D.	**End-user price**							
D1	End-user price, excluding all taxes	$/liter	0.687	0.721	0.740	0.720	0.630	0.709
D2	End-user price, including all taxes	$/liter	1.110	1.030	1.138	1.211	1.068	1.119
E.	**Price gap**							
E1	Price gap, excluding all taxes (C1 – D1)	$/liter	0.238	−0.108	0.032	0.282	0.357	0.295
E2	Price gap, with fuel at counterfactual tax rates (C2 – D2)	$/liter	−0.028	−0.313	−0.235	−0.039	0.087	0.057
F.	**Consumption**							
F1	Volume	Million liters	43,541.00	42,453.00	46,771.00	48,885.00	51,838.00	54,764.00
G.	**Fiscal cost**							
G1	Fiscal cost, excluding all taxes (E1 * F1)	$, millions	10,347.73	(4,588.43)	1,474.32	13,762.50	18,522.47	16,163.36
G2	Fiscal cost, with fuel at counterfactual tax rates (E2 * F1)	$, millions	(1,211.18)	(13,280.70)	(10,992.44)	(1,928.21)	4,522.13	3,112.12
G3	Fiscal expenditures (F1 * (E2 – E1))	$, millions	(11,558.91)	(8,692.27)	(12,466.77)	(15,690.71)	(14,000.33)	(13,051.24)

a. All prices and costs are expressed in current U.S. dollars.
b. CIF (cost, insurance, and freight) price + local distribution and retail margins.
c. VAT (value added tax, ICMS) is a weighted average from different ad valorem state taxes (PIS/COFINS), and excise tax (CIDE) is a unit tax.

Table A.9 Brazil—Kerosene

		Unit[a]	2008	2009	2010	2011	2012	2013
A.	**Supply cost**[b]	$/liter	0.850	0.478	0.617	0.860	0.874	0.806
B1.	**Taxes, excluding transfers (includes VAT and excise taxes)**[c]	$/liter	0.333	0.220	0.271	0.345	0.334	0.270
B2.	**Taxes (counterfactual): standard VAT**	$/liter	0.144	0.081	0.105	0.146	0.149	0.137
C.	**Total reference benchmark price**							
C1	Total reference price without taxes	$/liter	0.850	0.478	0.617	0.860	0.874	0.806
C2	Total reference price at counterfactual tax rates	$/liter	0.994	0.559	0.722	1.006	1.023	0.943

table continues next page

Table A.9 Brazil—Kerosene *(continued)*

		Unit[a]	2008	2009	2010	2011	2012	2013
D.	**End-user price**							
D1	End-user price, excluding all taxes	$/liter	0.807	0.446	0.583	0.796	0.815	0.838
D2	End-user price, including all taxes	$/liter	1.140	0.667	0.854	1.140	1.149	1.109
E.	**Price gap**							
E1	Price gap, excluding all taxes (C1 − D1)	$/liter	0.043	0.032	0.034	0.064	0.059	−0.032
E2	Price gap, with fuel at counterfactual tax rates (C2 − D2)	$/liter	−0.146	−0.107	−0.133	−0.135	−0.126	−0.165
F.	**Consumption**							
F1	Volume	Million liters	5,251.78	5,444.71	6,265.45	6,969.63	7,303.57	7,234.25
G.	**Fiscal cost**							
G1	Fiscal cost excluding all taxes (E1 * F1)	$, millions	224.60	172.66	212.00	445.63	433.03	(231.92)
G2	Fiscal cost, with fuels at counterfactual tax rates (E2 * F1)	$, millions	(765.79)	(585.07)	(830.74)	(938.98)	(922.46)	(1,194.40)
G3	Fiscal expenditures (F1 * (E2 − E1))	$, millions	(990.39)	(757.73)	(1,042.74)	(1,384.61)	(1,355.48)	(962.48)

a. All prices and costs are expressed in current U.S. dollars.
b. CIF (cost, insurance, and freight) price + local distribution and retail margins.
c. VAT (value added tax, ICMS) is a weighted average from different ad valorem state taxes, and excise tax (PIS/COFINS) is a unit tax.

Table A.10 Brazil—Liquefied Petroleum Gas (LPG)

		Unit[a]	2008	2009	2010	2011	2012	2013
A.	**Supply cost[b]**	$/liter	0.726	0.601	0.792	0.901	0.718	0.708
B1.	**Taxes, excluding transfers (includes VAT and excise taxes)[c]**	$/liter	0.170	0.146	0.184	0.205	0.170	0.164
B2.	**Taxes (counterfactual): standard VAT**	$/liter	0.123	0.102	0.135	0.153	0.122	0.120
C.	**Total reference benchmark price**							
C1	Total reference price without taxes	$/liter	0.726	0.601	0.792	0.901	0.718	0.708
C2	Total reference price at counterfactual tax rates	$/liter	0.849	0.703	0.927	1.054	0.840	0.828
D.	**End-user price**							
D1	End-user price, excluding all taxes	$/liter	0.580	0.602	0.723	0.752	0.665	0.628
D2	End-user price, including all taxes	$/liter	0.750	0.747	0.907	0.958	0.836	0.793
E.	**Price gap**							
E1	Price gap, excluding all taxes (C1 − D1)	$/liter	0.146	−0.001	0.069	0.148	0.053	0.080

table continues next page

Table A.10 Brazil—Liquefied Petroleum Gas (LPG) *(continued)*

		Unit[a]	2008	2009	2010	2011	2012	2013
E2	Price gap, with fuel at counterfactual tax rates (C2 − D2)	$/liter	0.099	−0.044	0.020	0.096	0.004	0.036
F.	**Consumption**							
F1	Volume	Million liters	12,415.00	12,187.00	12,604.00	13,903.00	13,131.00	13,608.00
G.	**Fiscal cost**							
G1	Fiscal cost, excluding all taxes (E1 * F1)	$, millions	1,809.88	(8.30)	867.18	2,059.74	692.00	1,082.61
G2	Fiscal cost, with fuel at counterfactual tax rates (E2 * F1)	$, millions	1,227.54	(538.26)	248.37	1,336.20	58.59	487.62
G3	Fiscal expenditures (F1 * (E2 − E1))	$, millions	(582.34)	(529.96)	(618.80)	(723.53)	(633.41)	(594.99)

a. All prices and costs are expressed in current U.S. dollars.
b. CIF (cost, insurance, and freight) price + local distribution and retail margins.
c. VAT (value added tax, ICMS) is a weighted average from different ad valorem state taxes, and excise tax (PIS/COFINS) is a unit tax.

Table A.11 Colombia—Gasoline

		Unit[a]	2008	2009	2010	2011	2012
A.	**Supply cost**[b]	$/liter	0.736	0.580	0.721	0.916	0.975
B1.	**Taxes, excluding transfers (includes VAT and excise taxes)**[c]	$/liter	0.266	0.229	0.269	0.303	0.315
B2.	**Taxes (counterfactual): standard VAT**	$/liter	0.118	0.093	0.115	0.147	0.156
C.	**Total reference benchmark price**						
C1	Total reference price without taxes	$/liter	0.736	0.580	0.721	0.916	0.975
C2	Total reference price at counterfactual tax rates	$/liter	0.854	0.673	0.837	1.062	1.131
D.	**End-user price**						
D1	End-user price, excluding all taxes	$/liter	0.653	0.598	0.729	0.866	0.961
D2	End-user price, including all taxes	$/liter	0.963	0.883	1.059	1.208	1.299
E.	**Price gap**						
E1	Price gap, excluding all taxes (C1 − D1)	$/liter	0.083	−0.018	−0.008	0.050	0.014
E2	Price gap, with fuel at counterfactual tax rates (C2 − D2)	$/liter	−0.109	−0.210	−0.222	−0.145	−0.168
F.	**Consumption**						
F1	Volume	Million liters	3,957.67	3,812.59	4,038.91	4,137.56	4,282.64
G.	**Fiscal cost**						
G1	Fiscal cost, excluding all taxes (E1 * F1)	$, millions	327.49	(70.39)	(31.91)	206.20	60.38
G2	Fiscal cost, with fuel at counterfactual tax rates (E2 * F1)	$, millions	(431.68)	(802.49)	(897.36)	(600.43)	(718.88)
G3	Fiscal expenditures (F1 * (E2 − E1))	$, millions	(759.17)	(732.10)	(865.45)	(806.63)	(779.26)

a. All prices and costs are expressed in current U.S. dollars.
b. CIF (cost, insurance, freight) price + local distribution and retail margins + other costs.
c. VAT (value added tax) includes global tax, and it is 16 percent of the supply cost. Excise tax (regional tax) is a unit tax.

Table A.12 Colombia—Diesel

	Unit[a]	2008	2009	2010	2011	2012
A. Supply cost[b]	$/liter	0.871	0.570	0.757	0.994	0.982
B1. Taxes, excluding transfers (includes VAT and excise taxes)[c]	$/liter	0.173	0.122	0.154	0.191	0.190
B2. Taxes (counterfactual): standard VAT	$/liter	0.139	0.091	0.121	0.159	0.157
C. Total reference benchmark price						
C1 Total reference price without taxes	$/liter	0.871	0.570	0.757	0.994	0.982
C2 Total reference price at counterfactual tax rates	$/liter	1.010	0.662	0.878	1.153	1.139
D. End-user price						
D1 End-user price, excluding all taxes	$/liter	0.624	0.571	0.720	0.875	0.949
D2 End-user price, including all taxes	$/liter	0.803	0.732	0.896	1.054	1.128
E. Price gap						
E1 Price gap, excluding all taxes (C1 − D1)	$/liter	0.247	0.000	0.037	0.119	0.033
E2 Price gap, with fuel at counterfactual tax rates (C2 − D2)	$/liter	0.208	−0.071	−0.018	0.099	0.011
F. Consumption						
F1 Volume	Million liters	5,379.41	5,640.55	6,093.19	6,690.90	6,824.37
G. Fiscal cost						
G1 Fiscal cost, excluding all taxes (E1 * F1)	$, millions	1,329.29	(2.48)	223.62	796.23	227.07
G2 Fiscal cost, with fuel at counterfactual tax rates (E2 * F1)	$, millions	1,116.64	(399.53)	(109.02)	660.71	74.22
G3 Fiscal expenditures (F1 * (E2 − E1))	$, millions	(212.65)	(397.04)	(332.65)	(135.52)	(152.85)

a. All prices and costs are expressed in current U.S. dollars.
b. CIF (cost, insurance, and freight) price + local distribution and retail margins + other costs.
c. VAT (value added tax) includes global tax, and it is 16 percent of the supply cost. Excise tax (regional tax) is a unit tax.

Table A.13 Dominican Republic—Gasoline

	Unit[a]	2008	2009	2010	2011	2012	2013
A. Supply cost[b]	$/liter	0.838	0.588	0.710	0.930	0.938	0.976
B1. Taxes, excluding transfers (includes VAT and excise taxes)[c]	$/liter	0.426	0.378	0.392	0.466	0.498	0.495
B2. Taxes (counterfactual): standard VAT	$/liter	0.134	0.094	0.114	0.149	0.169	0.176
C. Total reference benchmark price							
C1 Total reference price without taxes	$/liter	0.838	0.588	0.710	0.930	0.938	0.976
C2 Total reference price at counterfactual tax rates	$/liter	0.972	0.683	0.823	1.079	1.107	1.151
D. End-user price							
D1 End-user price, excluding all taxes	$/liter	0.844	0.589	0.717	0.932	0.939	0.976
D2 End-user price, including all taxes	$/liter	1.271	0.967	1.110	1.397	1.437	1.471
E. Price gap							
E1 Price gap, excluding all taxes (C1 − D1)	$/liter	−0.007	0.000	−0.008	−0.002	−0.001	0.000
E2 Price gap, with fuel at counterfactual tax rates (C2 − D2)	$/liter	−0.299	−0.284	−0.286	−0.319	−0.330	−0.320

table continues next page

Energy Pricing Policies for Inclusive Growth in Latin America and the Caribbean
http://dx.doi.org/10.1596/978-1-4648-1111-1

Table A.13 Dominican Republic—Gasoline *(continued)*

	Unit[a]	2008	2009	2010	2011	2012	2013
F.	**Consumption**						
F1	Volume	Million liters	457.66	551.91	785.85	724.53	621.19	819.54
G.	**Fiscal cost**						
G1	Fiscal cost, excluding all taxes (E1 * F1)	$, millions	(3.08)	(0.09)	(6.19)	(1.16)	(0.55)	(0.26)
G2	Fiscal cost, with fuel at counterfactual tax rates (E2 * F1)	$, millions	(136.90)	(156.91)	(225.04)	(230.82)	(204.81)	(262.34)
G3	Fiscal expenditures (F1 * (E2 − E1))	$, millions	(133.82)	(156.82)	(218.85)	(229.65)	(204.26)	(262.08)

a. All prices and costs are expressed in current U.S. dollars.
b. CIF (cost, insurance, and freight) price + local distribution and retail margins.
c. VAT (value added tax) is zero, and excise tax is a mix of unit tax and ad valorem.

Table A.14 Dominican Republic—Diesel

	Unit[a]	2008	2009	2010	2011	2012	2013
A.	**Supply cost[b]**	$/liter	0.896	0.618	0.732	0.993	1.036	1.033
B1.	**Taxes, excluding transfers (includes VAT and excise taxes)[c]**	$/liter	0.293	0.238	0.256	0.314	0.335	0.335
B2.	**Taxes (counterfactual): standard VAT**	$/liter	0.143	0.099	0.117	0.159	0.186	0.186
C.	**Total reference benchmark price**						
C1	Total reference price without taxes	$/liter	0.896	0.618	0.732	0.993	1.036	1.033
C2	Total reference price at counterfactual tax rates	$/liter	1.039	0.717	0.849	1.152	1.222	1.219
D.	**End-user price**						
D1	End-user price, excluding all taxes	$/liter	0.896	0.618	0.734	0.996	1.036	1.033
D2	End-user price, including all taxes	$/liter	1.189	0.856	0.991	1.310	1.371	1.368
E.	**Price gap**						
E1	Price gap, excluding all taxes (C1 − D1)	$/liter	0.000	0.000	−0.003	−0.003	0.000	0.000
E2	Price gap, with fuel at counterfactual tax rates (C2 − D2)	$/liter	−0.150	−0.139	−0.142	−0.158	−0.149	−0.150
F.	**Consumption**						
F1	Volume	Million liters	228.07	237.23	291.33	269.60	432.75	286.78
G.	**Fiscal cost**						
G1	Fiscal cost, excluding all taxes (E1 * F1)	$, millions	0.00	0.00	(0.77)	(0.71)	0.00	0.00
G2	Fiscal cost, with fuel at counterfactual tax rates (E2 * F1)	$, millions	(34.20)	(32.94)	(41.31)	(42.62)	(64.52)	(42.90)
G3	Fiscal expenditures (F1 * (E2 − E1))	$, millions	(34.20)	(32.94)	(40.54)	(41.91)	(64.52)	(42.90)

a. All prices and costs are expressed in current U.S. dollars.
b. CIF (cost, insurance, and freight) price + local distribution and retail margins.
c. VAT (value added tax) is zero, and excise tax is a mix of unit tax and ad valorem.

Table A.15 Dominican Republic—Liquefied Petroleum Gas (LPG)

		Unit[a]	2008	2009	2010	2011	2012	2013
A.	Supply cost[b]	$/liter	0.416	0.404	0.501	0.628	0.580	0.572
B1.	Taxes, excluding transfers (includes VAT and excise taxes)[c]	$/liter	0.048	0.047	0.062	0.080	0.071	0.061
B2.	Taxes (counterfactual): standard VAT	$/liter	0.067	0.065	0.080	0.101	0.104	0.103
C.	Total reference benchmark price							
C1	Total reference price without taxes	$/liter	0.416	0.404	0.501	0.628	0.580	0.572
C2	Total reference price at counterfactual tax rates	$/liter	0.482	0.469	0.581	0.729	0.684	0.675
D.	End-user price							
D1	End-user price, excluding all taxes	$/liter	0.417	0.405	0.503	0.628	0.579	0.571
D2	End-user price, including all taxes	$/liter	0.465	0.452	0.565	0.708	0.650	0.631
E.	Price gap							
E1	Price gap, excluding all taxes (C1 − D1)	$/liter	−0.002	−0.001	−0.002	0.001	0.001	0.001
E2	Price gap, with fuel at counterfactual tax rates (C2 − D2)	$/liter	0.017	0.017	0.016	0.021	0.034	0.044
F.	Consumption							
F1	Volume	Million liters	1,542.55	1,564.51	1,556.94	1,284.01	1,352.91	1,459.28
G.	Fiscal cost							
G1	Fiscal cost, excluding all taxes (E1 * F1)	$, millions	(2.41)	(1.26)	(3.42)	1.05	0.77	1.79
G2	Fiscal cost, with fuel at counterfactual tax rates (E2 * F1)	$, millions	26.76	26.81	24.24	27.00	46.56	63.52
G3	Fiscal expenditures (F1 * (E2 − E1))	$, millions	29.17	28.06	27.66	25.95	45.79	61.73

a. All prices and costs are expressed in current U.S. dollars.
b. CIF (cost, influence, and freight) price + local distribution and retail margins.
c. VAT (value added tax) is zero, and excise tax is a mix of unit tax and ad valorem.

Table A.16 El Salvador—Gasoline

		Unit[a]	2011	2012	2013
A.	Supply cost[b]	$/liter	0.827	0.845	0.813
B1.	Taxes, excluding transfers (includes VAT and excise taxes)[c]	$/liter	0.210	0.241	0.237
B2.	Taxes (counterfactual): standard VAT	$/liter	0.108	0.110	0.106
C.	Total reference benchmark price				
C1	Total reference price without taxes	$/liter	0.827	0.845	0.813
C2	Total reference price at counterfactual tax rates	$/liter	0.935	0.955	0.919
D.	End-user price				
D1	End-user price, excluding all taxes	$/liter	0.854	0.866	0.839
D2	End-user price, including all taxes	$/liter	1.064	1.107	1.076
E.	Price gap				
E1	Price gap excluding all taxes (C1 − D1)	$/liter	−0.026	−0.021	−0.026
E2	Price gap, with fuel at counterfactual tax rates (C2 − D2)	$/liter	−0.129	−0.152	−0.157
F.	Consumption				
F1	Volume	Million liters	401.70	397.24	410.83

table continues next page

Table A.16 El Salvador—Gasoline (continued)

		Unit[a]	2011	2012	2013
G.	**Fiscal cost**				
G1	Fiscal cost, excluding all taxes (E1 * F1)	$ million	(10.62)	(8.46)	(10.73)
G2	Fiscal cost, with fuel at counterfactual tax rates (E2 * F1)	$ million	(51.82)	(60.41)	(64.70)
G3	Fiscal expenditures (F1 * (E2 − E1))	$ million	(41.20)	(51.95)	(53.98)

a. All prices and costs are expressed in current U.S. dollars.
b. CIF (cost, insurance, and freight) price + local distribution and retail margins + other costs.
c. VAT (value added tax) is zero, and excise tax is a unit tax.

Table A.17 El Salvador—Diesel

		Unit[a]	2011	2012	2013
A.	**Supply cost[b]**	$/liter	0.857	0.881	0.856
B1.	**Taxes, excluding transfers (includes VAT and excise taxes)[c]**	$/liter	0.194	0.197	0.195
B2.	**Taxes (counterfactual): standard VAT**	$/liter	0.111	0.114	0.111
C.	**Total reference benchmark price**				
C1	Total reference price without taxes	$/liter	0.857	0.881	0.856
C2	Total reference price at counterfactual tax rates	$/liter	0.968	0.995	0.967
D.	**End-user price**				
D1	End-user price, excluding all taxes	$/liter	0.874	0.897	0.877
D2	End-user price, including all taxes	$/liter	1.068	1.094	1.072
E.	**Price gap**				
E1	Price gap, excluding all taxes (C1 − D1)	$/liter	−0.018	−0.016	−0.021
E2	Price gap, with fuel at counterfactual tax rates (C2 − D2)	$/liter	−0.100	−0.099	−0.105
F.	**Consumption**				
F1	Volume	Million liters	715.31	699.60	706.86
G.	**Fiscal cost**				
G1	Fiscal cost, excluding all taxes (E1 * F1)	$, millions	(12.54)	(11.40)	(15.05)
G2	Fiscal cost, with fuel at counterfactual tax rates (E2 * F1)	$, millions	(71.66)	(69.20)	(73.91)
G3	Fiscal expenditures (F1 * (E2 − E1))	$, millions	(59.11)	(57.80)	(58.86)

a. All prices and costs are expressed in current U.S. dollars.
b. CIF (cost, insurance, and freight) price + local distribution and retail margins + other costs.
c. VAT (value added tax) is zero, and excise tax is a unit tax.

Table A.18 El Salvador—Liquefied Petroleum Gas (LPG)

		Unit[a]	2011	2012	2013
A.	**Supply cost[b]**	$/liter	0.600	0.480	0.486
B1.	**Taxes, excluding transfers (includes VAT and excise taxes)[c]**	$/liter	0.080	0.068	0.064
B2.	**Taxes (counterfactual): standard VAT**	$/liter	0.078	0.062	0.063
C.	**Total reference benchmark price**				
C1	Total reference price without taxes	$/liter	0.600	0.480	0.486
C2	Total reference price at counterfactual tax rates	$/liter	0.678	0.542	0.550
D.	**End-user price**				
D1	End-user price, excluding all taxes	$/liter	0.210	0.104	0.078
D2	End-user price, including all taxes	$/liter	0.291	0.172	0.142
E.	**Price gap**				
E1	Price gap, excluding all taxes (C1 − D1)	$/liter	0.390	0.376	0.408
E2	Price gap, with fuel at counterfactual tax rates (C2 − D2)	$/liter	0.387	0.371	0.408

table continues next page

Table A.18 El Salvador—Liquefied Petroleum Gas (LPG) *(continued)*

		Unit[a]	2011	2012	2013
F.	**Consumption**				
F1	Volume	Million liters	397.74	403.16	463.27
G.	**Fiscal cost**				
G1	Fiscal cost, excluding all taxes (E1 * F1)	$, millions	154.96	151.54	189.15
G2	Fiscal cost, with fuel at counterfactual tax rates (E2 * F1)	$, millions	154.05	149.40	188.87
G3	Fiscal expenditures (F1 * (E2 − E1))	$, millions	(0.91)	(2.14)	(0.28)

a. All prices and costs are expressed in current U.S. dollars.
b. CIF (cost, insurance, and freight) price + local distribution and retail margins + other costs.
c. VAT (value added tax) is zero, and excise tax is a unit tax.

Table A.19 Haiti—Gasoline

		Unit[a]	2008	2009	2010	2011	2012	2013
A.	**Supply cost[b]**	$/liter	1.054	0.806	0.930	1.152	1.181	1.160
B1.	**Taxes, excluding transfers (includes VAT and excise taxes)[c]**	$/liter	0.552	0.394	0.470	0.602	0.621	0.600
B2.	**Taxes (counterfactual): standard VAT**	$/liter	0.000	0.000	0.000	0.000	0.000	0.000
C.	**Total reference benchmark price**							
C1	Total reference price without taxes	$/liter	1.054	0.806	0.930	1.152	1.181	1.160
C2	Total reference price at counterfactual tax rates	$/liter	1.054	0.806	0.930	1.152	1.181	1.160
D.	**End-user price**							
D1	End-user price, excluding all taxes	$/liter	0.869	0.637	0.721	0.703	0.639	0.616
D2	End-user price, including all taxes	$/liter	1.421	1.030	1.191	1.305	1.260	1.215
E.	**Price gap**							
E1	Price gap, excluding all taxes (C1 − D1)	$/liter	0.185	0.169	0.209	0.449	0.542	0.544
E2	Price gap, with fuel at counterfactual tax rates (C2 − D2)	$/liter	−0.367	−0.225	−0.262	−0.153	−0.079	−0.055
F.	**Consumption**							
F1	Volume	Million liters	96.57	110.69	119.75	124.90	141.71	117.14
G.	**Fiscal cost**							
G1	Fiscal cost, excluding all taxes (E1 * F1)	$, millions	17.86	18.71	24.99	56.09	76.75	63.74
G2	Fiscal cost, with fuel at counterfactual tax rates (E2 * F1)	$, millions	(35.46)	(24.86)	(31.32)	(19.14)	(11.23)	(6.50)
G3	Fiscal expenditures (F1* (E2 − E1))	$, millions	(53.32)	(43.57)	(56.31)	(75.23)	(87.98)	(70.24)

a. All prices and costs are expressed in current U.S. dollars.
b. CIF (cost, insurance, and freight) price + local distribution and retail margins.
c. VAT (valued added tax) is zero, and excise tax is a unit tax.

Table A.20 Haiti—Diesel

		Unit[a]	2008	2009	2010	2011	2012	2013
A.	**Supply cost[b]**	$/liter	1.210	0.837	0.975	1.202	1.239	1.220
B1.	**Taxes, excluding transfers (includes VAT and excise taxes)[c]**	$/liter	0.092	0.074	0.082	0.092	0.092	0.090
B2.	**Taxes (counterfactual): standard VAT**	$/liter	0.000	0.000	0.000	0.000	0.000	0.000

table continues next page

Table A.20 Haiti—Diesel (continued)

	Unit[a]	2008	2009	2010	2011	2012	2013
C.	**Total reference benchmark price**						
C1	Total reference price without taxes	$/liter	1.210	0.837	0.975	1.202	1.239	1.220
C2	Total reference price at counterfactual tax rates	$/liter	1.210	0.837	0.975	1.202	1.239	1.220
D.	**End-user price**						
D1	End-user price, excluding all taxes	$/liter	0.914	0.613	0.761	0.925	0.927	0.896
D2	End-user price, including all taxes	$/liter	1.006	0.687	0.843	1.017	1.020	0.985
E.	**Price gap**						
E1	Price gap, excluding all taxes (C1 − D1)	$/liter	0.296	0.225	0.214	0.277	0.312	0.325
E2	Price gap, with fuel at counterfactual tax rates (C2 − D2)	$/liter	0.203	0.151	0.132	0.185	0.219	0.235
F.	**Consumption**						
F1	Volume	Million liters	435.08	461.16	530.09	530.17	535.70	586.41
G.	**Fiscal cost**						
G1	Fiscal cost, excluding all taxes (E1 * F1)	$, millions	128.73	103.55	113.43	147.06	166.99	190.54
G2	Fiscal cost, with fuel at counterfactual tax rates (E2 * F1)	$, millions	88.50	69.44	70.02	98.04	117.46	137.87
G3	Fiscal expenditures (F1 * (E2 − E1))	$, millions	(40.23)	(34.11)	(43.41)	(49.02)	(49.53)	(52.67)

a. All prices and costs are expressed in current U.S. dollars.
b. CIF (cost, insurance, and freight) price + local distribution and retail margins.
c. VAT (value added tax) is zero, and excise tax is a unit tax.

Table A.21 Haiti—Kerosene

	Unit[a]	2008	2009	2010	2011	2012	2013
A.	**Supply cost[b]**	$/liter	0.948	0.596	0.731	0.953	0.968	0.929
B1.	**Taxes, excluding transfers (includes VAT and excise taxes)[c]**	$/liter	0.066	0.047	0.054	0.066	0.066	0.064
B2.	**Taxes (counterfactual): standard VAT**	$/liter	0.000	0.000	0.000	0.000	0.000	0.000
C.	**Total reference benchmark price**						
C1	Total reference price without taxes	$/liter	0.948	0.596	0.731	0.953	0.968	0.929
C2	Total reference price at counterfactual tax rates	$/liter	0.948	0.596	0.731	0.953	0.968	0.929
D.	**End-user price**						
D1	End-user price, excluding all taxes	$/liter	0.919	0.592	0.701	0.931	0.948	0.914
D2	End-user price, including all taxes	$/liter	0.985	0.639	0.756	0.996	1.014	0.979
E.	**Price gap**						
E1	Price gap, excluding all taxes (C1 − D1)	$/liter	0.030	0.004	0.030	0.022	0.020	0.015
E2	Price gap, with fuel at counterfactual tax rates (C2 − D2)	$/liter	−0.037	−0.043	−0.024	−0.044	−0.046	−0.050
F.	**Consumption**						
F1	Volume	Million liters	81.73	92.23	99.73	99.07	82.88	78.67

table continues next page

Table A.21 Haiti—Kerosene *(continued)*

	Unit[a]	2008	2009	2010	2011	2012	2013
G.	**Fiscal cost**						
G1	Fiscal cost, excluding all taxes (E1 * F1)	$, millions	2.42	0.36	3.00	2.18	1.65	1.15
G2	Fiscal cost, with fuel at counterfactual tax rates (E2 * F1)	$, millions	(3.02)	(3.96)	(2.43)	(4.33)	(3.83)	(3.90)
G3	Fiscal expenditures (F1 * (E2 − E1))	$, millions	(5.43)	(4.32)	(5.42)	(6.51)	(5.48)	(5.06)

a. All prices and costs are expressed in current U.S. dollars.
b. CIF (cost, insurance, and freight) price + local distribution and retail margins.
c. VAT (value added tax) is zero, and excise tax is a unit tax.

Table A.22 Honduras—Gasoline, Premium

		Unit[a]	2008	2009	2010	2011	2012	2013
A.	**Supply cost[b]**	$/liter	0.789	0.538	0.683	0.901	0.963	0.950
B1.	**Taxes, excluding transfers (includes VAT and excise taxes)[c]**	$/liter	0.305	0.305	0.305	0.305	0.305	0.305
B2.	**Taxes (counterfactual): standard VAT**	$/liter	0.095	0.065	0.082	0.108	0.116	0.143
C.	**Total reference benchmark price**							
C1	Total reference price without taxes	$/liter	0.789	0.538	0.683	0.901	0.963	0.950
C2	Total reference price at counterfactual tax rates	$/liter	0.884	0.603	0.765	1.009	1.078	1.093
D.	**End-user price**							
D1	End-user price, excluding all taxes	$/liter	0.687	0.502	0.642	0.859	0.938	0.969
D2	End-user price, including all taxes	$/liter	0.992	0.807	0.947	1.164	1.243	1.274
E.	**Price gap**							
E1	Price gap, excluding all taxes (C1 − D1)	$/liter	0.102	0.036	0.041	0.042	0.025	−0.019
E2	Price gap, with fuel at counterfactual tax rates (C2 − D2)	$/liter	−0.109	−0.204	−0.183	−0.155	−0.165	−0.182
F.	**Consumption**							
F1	Volume	Million liters	348.64	381.16	421.26	388.91	410.95	406.20
G.	**Fiscal cost**							
G1	Fiscal cost, excluding all taxes (E1 * F1)	$, millions	35.47	13.84	17.22	16.48	10.17	(7.59)
G2	Fiscal cost, with fuel at counterfactual tax rates (E2 * F1)	$, millions	(37.98)	(77.93)	(76.91)	(60.23)	(67.84)	(73.73)
G3	Fiscal expenditures (F1 * (E2 − E1))	$, millions	(73.46)	(91.77)	(94.13)	(76.71)	(78.01)	(66.14)

a. All prices and costs are expressed in current U.S. dollars.
b. CIF (cost, insurance, and freight) price + local distribution and retail margins.
c. VAT (value added tax) is zero, and excise tax is a unit tax.

Table A.23 Honduras—Gasoline, Regular

		Unit[a]	2008	2009	2010	2011	2012	2013
A.	**Supply cost[b]**	$/liter	0.780	0.529	0.660	0.865	0.920	0.902
B1.	**Taxes, excluding transfers (includes VAT and excise taxes)[c]**	$/liter	0.262	0.262	0.262	0.262	0.262	0.262
B2.	**Taxes (counterfactual): standard VAT**	$/liter	0.094	0.064	0.079	0.104	0.110	0.135
C.	**Total reference benchmark price**							
C1	Total reference price without taxes	$/liter	0.780	0.529	0.660	0.865	0.920	0.902

table continues next page

Table A.23 Honduras—Gasoline, Regular *(continued)*

		Unit[a]	2008	2009	2010	2011	2012	2013
C2	Total reference price at counterfactual tax rates	$/liter	0.874	0.593	0.739	0.969	1.030	1.038
D.	**End-user price**							
D1	End-user price, excluding all taxes	$/liter	0.761	0.581	0.728	0.954	1.011	0.990
D2	End-user price, including all taxes	$/liter	1.023	0.843	0.990	1.216	1.273	1.251
E.	**Price gap**							
E1	Price gap, excluding all taxes (C1 − D1)	$/liter	0.019	−0.052	−0.068	−0.089	−0.091	−0.087
E2	Price gap, with fuel at counterfactual tax rates (C2 − D2)	$/liter	−0.149	−0.250	−0.251	−0.247	−0.242	−0.214
F.	**Consumption**							
F1	Volume	Million liters	256.79	268.86	304.03	298.40	310.42	301.18
G.	**Fiscal cost**							
G1	Fiscal cost, excluding all taxes (E1 * F1)	$, millions	4.93	(13.92)	(20.68)	(26.56)	(28.21)	(26.27)
G2	Fiscal cost, with fuel at counterfactual tax rates (E2 * F1)	$, millions	(38.29)	(67.27)	(76.25)	(73.75)	(75.26)	(64.40)
G3	Fiscal expenditures (F1 * (E2 − E1))	$, millions	(43.22)	(53.35)	(55.57)	(47.19)	(47.05)	(38.13)

a. All prices and costs are expressed in current U.S. dollars.
b. CIF (cost, insurance, and freight) price + local distribution and retail margins.
c. VAT (value added tax) is zero, and excise tax is a unit tax.

Table A.24 Honduras—Diesel

		Unit[a]	2008	2009	2010	2011	2012	2013
A.	**Supply cost[b]**	$/liter	0.867	0.568	0.709	0.928	0.945	0.916
B1.	**Taxes, excluding transfers (includes VAT and excise taxes)[c]**	$/liter	0.161	0.161	0.161	0.161	0.161	0.161
B2.	**Taxes (counterfactual): standard VAT**	$/liter	0.104	0.068	0.085	0.111	0.113	0.137
C.	**Total reference benchmark price**							
C1	Total reference price without taxes	$/liter	0.867	0.568	0.709	0.928	0.945	0.916
C2	Total reference price at counterfactual tax rates	$/liter	0.971	0.636	0.794	1.039	1.059	1.053
D.	**End-user price**							
D1	End-user price, excluding all taxes	$/liter	0.806	0.589	0.739	0.976	0.998	0.958
D2	End-user price, including all taxes	$/liter	0.967	0.750	0.900	1.137	1.159	1.119
E.	**Price gap**							
E1	Price gap, excluding all taxes (C1 − D1)	$/liter	0.061	−0.021	−0.030	−0.048	−0.053	−0.042
E2	Price gap, with fuel at counterfactual tax rates (C2 − D2)	$/liter	0.004	−0.115	−0.107	−0.098	−0.101	−0.066
F.	**Consumption**							
F1	Volume	Million liters	873.77	821.07	876.28	853.36	901.09	956.00

table continues next page

Table A.24 Honduras—Diesel *(continued)*

	Unit[a]	2008	2009	2010	2011	2012	2013
G.	**Fiscal cost**							
G1	Fiscal cost, excluding all taxes (E1 * F1)	$, millions	53.36	(17.57)	(26.57)	(40.78)	(47.68)	(40.23)
G2	Fiscal cost, with fuel at counterfactual tax rates (E2 * F1)	$, millions	3.32	(94.09)	(93.42)	(83.40)	(90.83)	(63.14)
G3	Fiscal expenditures (F1 * (E2 − E1))	$, millions	(50.04)	(76.52)	(66.84)	(42.62)	(43.14)	(22.91)

a. All prices and costs are expressed in current U.S. dollars.
b. CIF (cost, insurance, and freight) price + local distribution and retail margins.
c. VAT (value added tax) is zero, and excise tax is a unit tax.

Table A.25 Honduras—Diesel (Other)

	Unit[a]	2008	2009	2010	2011	2012	2013
A.	**Supply cost[b]**	$/liter	0.800	0.501	0.637	0.853	0.856	0.820
B1.	**Taxes, excluding transfers (includes VAT and excise taxes)[c]**	$/liter	0.161	0.161	0.161	0.161	0.161	0.161
B2.	**Taxes (counterfactual): standard VAT**	$/liter	0.096	0.060	0.076	0.102	0.103	0.123
C.	**Total reference benchmark price**							
C1	Total reference price without taxes	$/liter	0.800	0.501	0.637	0.853	0.856	0.820
C2	Total reference price at counterfactual tax rates	$/liter	0.896	0.561	0.713	0.955	0.959	0.943
D.	**End-user price**							
D1	End-user price, excluding all taxes	$/liter	0.806	0.589	0.739	0.976	0.998	0.958
D2	End-user price, including all taxes	$/liter	0.967	0.750	0.900	1.137	1.159	1.119
E.	**Price gap**							
E1	Price gap, excluding all taxes (C1 − D1)	$/liter	−0.006	−0.088	−0.102	−0.123	−0.142	−0.138
E2	Price gap, with fuel at counterfactual tax rates (C2 − D2)	$/liter	−0.071	−0.189	−0.187	−0.182	−0.201	−0.176
F.	**Consumption**							
F1	Volume	Million liters	311.58	282.82	295.10	283.80	299.35	308.77
G.	**Fiscal cost**							
G1	Fiscal cost, excluding all taxes (E1 * F1)	$, millions	(1.81)	(24.96)	(30.08)	(34.88)	(42.51)	(42.48)
G2	Fiscal cost, with fuel at counterfactual tax rates (E2 * F1)	$, millions	(22.15)	(53.59)	(55.13)	(51.61)	(60.04)	(54.30)
G3	Fiscal expenditures (F1 * (E2 − E1))	$, millions	(20.34)	(28.63)	(25.05)	(16.73)	(17.53)	(11.82)

a. All prices and costs are expressed in current U.S. dollars.
b. CIF (cost, insurance, and freight) price + local distribution and retail margins.
c. VAT (value added tax) is zero, and excise tax is a unit tax.

Table A.26 Honduras—Liquefied Petroleum Gas (LPG)

	Unit[a]	2008	2009	2010	2011	2012	2013
A.	**Supply cost[b]**	$/liter	0.536	0.370	0.468	0.537	0.451	0.448
B1.	**Taxes, excluding transfers (includes VAT and excise taxes)[c]**	$/liter	0.032	0.032	0.032	0.032	0.032	0.032
B2.	**Taxes (counterfactual): standard VAT**	$/liter	0.064	0.044	0.056	0.064	0.054	0.067
C.	**Total reference benchmark price**							
C1	Total reference price without taxes	$/liter	0.536	0.370	0.468	0.537	0.451	0.448
C2	Total reference price at counterfactual tax rates	$/liter	0.601	0.415	0.524	0.602	0.505	0.515

table continues next page

Energy Pricing Policies for Inclusive Growth in Latin America and the Caribbean
http://dx.doi.org/10.1596/978-1-4648-1111-1

Table A.26 Honduras—Liquefied Petroleum Gas (LPG) *(continued)*

		Unit[a]	2008	2009	2010	2011	2012	2013
D.	End-user price							
D1	End-user price, excluding all taxes	$/liter	0.463	0.431	0.513	0.611	0.562	0.561
D2	End-user price, including all taxes	$/liter	0.495	0.463	0.545	0.643	0.594	0.593
E.	Price gap							
E1	Price gap, excluding all taxes (C1 − D1)	$/liter	0.073	−0.061	−0.045	−0.074	−0.111	−0.113
E2	Price gap, with fuel at counterfactual tax rates (C2 − D2)	$/liter	0.105	−0.049	−0.021	−0.041	−0.089	−0.078
F.	Consumption							
F1	Volume	Million liters	145.83	136.24	139.32	150.32	172.16	308.77
G.	Fiscal cost							
G1	Fiscal cost, excluding all taxes (E1 * F1)	$, millions	10.67	(8.29)	(6.28)	(11.06)	(19.13)	(34.77)
G2	Fiscal cost, with fuel at counterfactual tax rates (E2 * F1)	$, millions	15.37	(6.62)	(2.95)	(6.21)	(15.36)	(23.94)
G3	Fiscal expenditures (F1 * (E2 − E1))	$, millions	4.69	1.67	3.34	4.86	3.77	10.83

a. All prices and costs are expressed in current U.S. dollars.
b. CIF (cost, insurance, and freight) price + local distribution and retail margins.
c. VAT (value added tax) is zero, and excise tax is a unit tax.

Table A.27 Honduras—Kerosene

		Unit[a]	2008	2009	2010	2011	2012	2013
A.	Supply cost[b]	$/liter	0.913	0.523	0.667	0.896	0.939	0.912
B1.	Taxes, excluding transfers (includes VAT and excise taxes)[c]	$/liter	0.000	0.000	0.000	0.000	0.000	0.000
B2.	Taxes (counterfactual): standard VAT	$/liter	0.110	0.063	0.080	0.108	0.113	0.137
C.	Total reference benchmark price							
C1	Total reference price without taxes	$/liter	0.913	0.523	0.667	0.896	0.939	0.912
C2	Total reference price at counterfactual tax rates	$/liter	1.022	0.586	0.747	1.004	1.051	1.049
D.	End-user price							
D1	End-user price, excluding all taxes	$/liter	0.495	0.463	0.545	0.643	0.594	0.593
D2	End-user price, including all taxes	$/liter	0.495	0.463	0.545	0.643	0.594	0.593
E.	Price gap							
E1	Price gap, excluding all taxes (C1− D1)	$/liter	0.417	0.060	0.122	0.253	0.344	0.319
E2	Price gap, with fuel at counterfactual tax rates (C2 − D2)	$/liter	0.527	0.123	0.202	0.360	0.457	0.456
F.	Consumption							
F1	Volume	Million liters	112.54	116.59	117.01	118.24	119.21	97.61
G.	Fiscal cost							
G1	Fiscal cost, excluding all taxes (E1 * F1)	$, millions	46.95	7.00	14.32	29.90	41.06	31.13
G2	Fiscal cost, with fuel at counterfactual tax rates (E2 * F1)	$, millions	59.28	14.32	23.69	42.61	54.49	44.48
G3	Fiscal expenditures (F1 * (E2 − E1))	$, millions	12.32	7.32	9.37	12.71	13.43	13.35

a. All prices and costs are expressed in current U.S. dollars.
b. CIF (cost, insurance, and freight) price + local distribution and retail margins.
c. VAT (value added tax) and excise taxes are zero.

Table A.28 Mexico—Gasoline, Premium

		Unit[a]	2008	2009	2010	2011	2012	2013
A.	Supply cost[b]	$/liter	0.866	0.618	0.754	0.948	1.026	1.013
B1.	Taxes, excluding transfers (includes VAT and excise taxes)[c]	$/liter	−0.054	0.085	0.027	−0.113	−0.194	−0.067
B2.	Taxes (counterfactual): standard VAT	$/liter	0.130	0.099	0.121	0.152	0.164	0.162
C.	Total reference benchmark price							
C1	Total reference price without taxes	$/liter	0.866	0.618	0.754	0.948	1.026	1.013
C2	Total reference price at counterfactual tax rates	$/liter	0.995	0.717	0.875	1.099	1.191	1.175
D.	End-user price							
D1	End-user price, excluding all taxes	$/liter	0.860	0.613	0.749	0.942	1.021	1.007
D2	End-user price, including all taxes	$/liter	0.806	0.698	0.776	0.829	0.827	0.940
E.	Price gap							
E1	Price gap, excluding all taxes (C1 − D1)	$/liter	0.005	0.005	0.005	0.005	0.005	0.006
E2	Price gap, with fuel at counterfactual tax rates (C2 − D2)	$/liter	0.190	0.018	0.099	0.270	0.364	0.236
F.	Consumption							
F1	Volume	Million liters	4,578.99	3,370.66	3,034.98	3,147.58	5,075.39	6,896.02
G.	Fiscal cost							
G1	Fiscal cost, excluding all taxes (E1 * F1)	$, millions	24.82	15.36	15.36	17.05	27.47	42.58
G2	Fiscal cost, with fuel at counterfactual tax rates (E2 * F1)	$, millions	867.75	62.35	299.44	850.36	1,846.95	1,625.28
G3	Fiscal expenditures (F1 * (E2 − E1))	$, millions	842.93	46.98	284.09	833.31	1,819.48	1,582.69

a. All prices and costs are expressed in current U.S. dollars.
b. CIF (cost, insurance, and freight) price + local distribution and retail margins + other costs.
c. VAT (value added tax) and excise tax (IEPS) are unit taxes.

Table A.29 Mexico—Gasoline, Regular

		Unit[a]	2008	2009	2010	2011	2012	2013
A.	Supply cost[b]	$/liter	0.765	0.524	0.667	0.855	0.903	0.882
B1.	Taxes, excluding transfers (includes VAT and excise taxes)[c]	$/liter	−0.112	0.044	−0.011	−0.112	−0.123	0.016
B2.	Taxes (counterfactual): standard VAT	$/liter	0.115	0.084	0.107	0.137	0.144	0.141

table continues next page

Table A.29 Mexico—Gasoline, Regular *(continued)*

	Unitª	2008	2009	2010	2011	2012	2013
C.	**Total reference benchmark price**						
C1	Total reference price without taxes	$/liter	0.765	0.524	0.667	0.855	0.903	0.882
C2	Total reference price at counterfactual tax rates	$/liter	0.880	0.608	0.773	0.992	1.047	1.023
D.	**End-user price**						
D1	End-user price, excluding all taxes	$/liter	0.761	0.521	0.663	0.850	0.898	0.876
D2	End-user price, including all taxes	$/liter	0.649	0.564	0.651	0.739	0.775	0.892
E.	**Price gap**						
E1	Price gap, excluding all taxes (C1 – D1)	$/liter	0.004	0.004	0.004	0.005	0.005	0.006
E2	Price gap, with fuel at counterfactual tax rates (C2 – D2)	$/liter	0.231	0.044	0.122	0.253	0.273	0.131
F.	**Consumption**						
F1	Volume	Million liters	40,870.14	42,113.84	43,040.38	42,741.31	41,396.10	38,637.65
G.	**Fiscal cost**						
G1	Fiscal cost, excluding all taxes (E1 * F1)	$, millions	178.57	155.09	182.85	206.99	211.04	227.83
G2	Fiscal cost, with fuel at counterfactual tax rates (E2 * F1)	$, millions	9,431.38	1,846.50	5,267.70	10,831.77	11,287.20	5,068.63
G3	Fiscal expenditures (F1 * (E2 – E1))	$, millions	9,252.82	1,691.41	5,084.85	10,624.78	11,076.17	4,840.80

a. All prices and costs are expressed in current U.S. dollars.
b. CIF (cost, insurance, and freight) price + local distribution and retail margins + other costs.
c. VAT (value added tax) and excise tax (IEPS) are unit taxes.

Table A.30 Mexico—Diesel

	Unitª	2008	2009	2010	2011	2012	2013
A.	**Supply cost**[b]	$/liter	0.853	0.515	0.662	0.872	0.922	0.904
B1.	**Taxes, excluding transfers (includes VAT and excise taxes)**[c]	$/liter	−0.294	0.056	0.014	−0.108	−0.123	0.015
B2.	**Taxes (counterfactual): standard VAT**	$/liter	0.128	0.082	0.106	0.140	0.148	0.145
C.	**Total reference benchmark price**						
C1	Total reference price without taxes	$/liter	0.853	0.515	0.662	0.872	0.922	0.904
C2	Total reference price at counterfactual tax rates	$/liter	0.981	0.597	0.768	1.012	1.070	1.049
D.	**End-user price**						
D1	End-user price, excluding all taxes	$/liter	0.853	0.515	0.662	0.872	0.922	0.904

table continues next page

Table A.30 Mexico—Diesel *(continued)*

	Unit[a]	2008	2009	2010	2011	2012	2013
D2 End-user price, including all taxes	$/liter	0.560	0.570	0.676	0.764	0.799	0.919
E. Price gap							
E1 Price gap, excluding all taxes (C1 – D1)	$/liter	0.000	0.000	0.000	0.000	0.000	0.000
E2 Price gap, with fuel at counterfactual tax rates (C2 – D2)	$/liter	0.422	0.027	0.092	0.248	0.271	0.130
F. Consumption							
F1 Volume	Million liters	22,165.70	20,831.47	21,533.29	22,259.49	23,242.63	22,731.03
G. Fiscal cost							
G1 Fiscal cost, excluding all taxes (E1 * F1)	$, millions	0.00	0.00	0.00	(0.00)	(0.00)	0.00
G2 Fiscal cost, with fuel at counterfactual tax rates (E2 * F1)	$, millions	9,347.23	558.43	1,980.55	5,515.90	6,292.83	2,950.96
G3 Fiscal expenditures (F1 * (E2 – E1))	$, millions	9,347.23	558.43	1,980.55	5,515.90	6,292.83	2,950.96

a. All prices and costs are expressed in current U.S. dollars.
b. CIF (cost, insurance, and freight) price + local distribution and retail margins + other costs.
c. VAT (value added tax) and excise tax (IEPS) are unit taxes.

Table A.31 Mexico—Liquefied Petroleum Gas (LPG)

	Unit[a]	2008	2009	2010	2011	2012	2013
A. Supply cost[b]	$/liter	0.401	0.264	0.354	0.435	0.312	0.315
B1. Taxes, excluding transfers (includes VAT and excise taxes)[c]	$/liter	0.062	0.048	0.054	0.059	0.060	0.068
B2. Taxes (counterfactual): standard VAT	$/liter	0.060	0.042	0.057	0.070	0.050	0.050
C. Total reference benchmark price							
C1 Total reference price without taxes	$/liter	0.401	0.264	0.354	0.435	0.312	0.315
C2 Total reference price at counterfactual tax rates	$/liter	0.461	0.306	0.411	0.505	0.362	0.366
D. End-user price							
D1 End-user price, excluding all taxes	$/liter	0.415	0.323	0.362	0.391	0.402	0.457
D2 End-user price, including all taxes	$/liter	0.477	0.371	0.416	0.449	0.462	0.525
E. Price gap							
E1 Price gap, excluding all taxes (C1 – D1)	$/liter	−0.014	−0.059	−0.008	0.045	−0.090	−0.141
E2 Price gap, with fuel at counterfactual tax rates (C2 – D2)	$/liter	−0.016	−0.065	−0.006	0.056	−0.100	−0.160

table continues next page

Table A.31 Mexico—Liquefied Petroleum Gas (LPG) *(continued)*

	Unit[a]	2008	2009	2010	2011	2012	2013
F.	**Consumption**							
F1	Volume	Million liters	16,875.25	16,279.22	16,679.62	16,501.25	16,490.50	16,385.65
G.	**Fiscal cost**							
G1	Fiscal cost, excluding all taxes (E1 * F1)	$, millions	(233.47)	(956.68)	(132.05)	735.00	(1,483.91)	(2,317.80)
G2	Fiscal cost, with fuels at counterfactual tax rates (E2 * F1)	$, millions	(268.49)	(1,057.18)	(92.82)	917.05	(1,655.11)	(2,613.83)
G3	Fiscal expenditures (F1 * (E2 – E1))	$, millions	(35.02)	(100.50)	39.23	182.05	(171.19)	(296.02)

a. All prices and costs are expressed in current U.S. dollars.
b. CIF (cost, insurance, and freight) price + local distribution and retail margins + other costs.
c. VAT (value added tax) is a unit tax, and excise tax (IEPS) is zero.

Table A.32 Mexico—Kerosene

	Unit[a]	2008	2009	2010	2011	2012	2013
A.	**Supply cost[b]**	$/liter	0.776	0.442	0.579	0.800	0.819	0.788
B1.	**Taxes, excluding transfers (includes VAT and excise taxes)[c]**	$/liter	0.112	0.064	0.089	0.124	0.127	0.122
B2.	**Taxes (counterfactual): standard VAT**	$/liter	0.116	0.071	0.093	0.128	0.131	0.126
C.	**Total reference benchmark price**							
C1	Total reference price without taxes	$/liter	0.776	0.442	0.579	0.800	0.819	0.788
C2	Total reference price at counterfactual tax rates	$/liter	0.893	0.512	0.672	0.928	0.951	0.914
D.	**End-user price**							
D1	End-user price, excluding all taxes	$/liter	0.776	0.442	0.579	0.800	0.819	0.788
D2	End-user price, including all taxes	$/liter	0.889	0.506	0.668	0.924	0.946	0.910
E.	**Price gap**							
E1	Price gap, excluding all taxes (C1 – D1)	$/liter	0.000	0.000	0.000	0.000	0.000	0.000
E2	Price gap, with fuel at counterfactual tax rates (C2 – D2)	$/liter	0.004	0.007	0.003	0.004	0.005	0.004
F.	**Consumption**							
F1	Volume	Million liters	3,760.27	3,182.13	3,231.10	3,249.21	3,432.51	3,602.06
G.	**Fiscal cost**							
G1	Fiscal cost, excluding all taxes (E1 * F1)	$, millions	0.00	0.00	0.00	0.00	0.00	0.00
G2	Fiscal cost, with fuel at counterfactual tax rates (E2 * F1)	$, millions	16.06	21.79	10.29	14.30	15.47	15.61
G3	Fiscal expenditures (F1 * (E2 – E1))	$, millions	16.06	21.79	10.29	14.30	15.47	15.61

a. All prices and costs are expressed in current U.S. dollars.
b. CIF (cost, insurance, and freight) price + local distribution and retail margins + other costs.
c. VAT (value added tax) is a unit tax, and excise tax (IEPS) is zero.

Table A.33 Mexico—Fuel Oil

	Unit[a]	2008	2009	2010	2011	2012	2013
A.	Supply cost[b]	$/liter	0.439	0.308	0.419	0.550	0.611	0.558
B1.	Taxes, excluding transfers (includes VAT and excise taxes)[c]	$/liter	0.065	0.046	0.066	0.087	0.097	0.088
B2.	Taxes (counterfactual): standard VAT	$/liter	0.066	0.049	0.067	0.088	0.098	0.089
C.	Total reference benchmark price							
C1	Total reference price without taxes	$/liter	0.439	0.308	0.419	0.550	0.611	0.558
C2	Total reference price at counterfactual tax rates	$/liter	0.504	0.358	0.486	0.638	0.709	0.648
D.	End-user price							
D1	End-user price, excluding all taxes	$/liter	0.439	0.308	0.419	0.550	0.611	0.558
D2	End-user price, including all taxes	$/liter	0.504	0.354	0.485	0.637	0.708	0.647
E.	Price gap							
E1	Price gap, excluding all taxes (C1 − D1)	$/liter	0.000	0.000	0.000	0.000	0.000	0.000
E2	Price gap, with fuel at counterfactual tax rates (C2 − D2)	$/liter	0.001	0.004	0.001	0.001	0.001	0.001
F.	Consumption							
F1	Volume	Million liters	12,709.39	12,093.42	10,700.58	11,610.13	12,406.05	10,954.20
G.	Fiscal cost							
G1	Fiscal cost, excluding all taxes (E1 * F1)	$, millions	0.00	0.00	0.00	0.00	0.00	0.00
G2	Fiscal cost, with fuel at counterfactual tax rates (E2 * F1)	$, millions	9.48	43.61	7.62	10.86	12.89	10.40
G3	Fiscal expenditures (F1 * (E2 − E1))	$, millions	9.48	43.61	7.62	10.86	12.89	10.40

a. All prices and costs are expressed in current U.S. dollars.
b. CIF (cost, insurance, and freight) price + local distribution and retail margins + other costs.
c. VAT (value added tax) is a unit tax, and excise tax (IEPS) is zero.

Table A.34 Peru—Gasoline

	Unit[a]	2008	2009	2010	2011	2012	2013
A.	Supply cost[b]	$/liter	0.814	0.537	0.649	0.871	0.959	0.916
B1.	Taxes, excluding transfers (includes VAT and excise taxes)[c]	$/liter	0.362	0.283	0.313	0.348	0.376	0.362
B2.	Taxes (counterfactual): standard VAT	$/liter	0.155	0.102	0.123	0.157	0.173	0.165

table continues next page

Table A.34 Peru—Gasoline (continued)

		Unit[a]	2008	2009	2010	2011	2012	2013
C.	**Total reference benchmark price**							
C1	Total reference price without taxes	$/liter	0.814	0.537	0.649	0.871	0.959	0.916
C2	Total reference price at counterfactual tax rates	$/liter	0.969	0.639	0.772	1.027	1.132	1.081
D.	**End-user price**							
D1	End-user price, excluding all taxes	$/liter	0.786	0.526	0.687	0.866	0.948	0.907
D2	End-user price, including all taxes	$/liter	1.142	0.806	1.007	1.213	1.322	1.267
E.	**Price gap**							
E1	Price gap, excluding all taxes (C1 − D1)	$/liter	0.028	0.011	−0.038	0.005	0.011	0.009
E2	Price gap, with fuel at counterfactual tax rates (C2 − D2)	$/liter	−0.173	−0.168	−0.235	−0.185	−0.190	−0.186
F.	**Consumption**							
F1	Volume	Million liters	1,259.26	1,505.31	1,657.35	1,827.38	1,925.45	2,061.82
G.	**Fiscal cost**							
G1	Fiscal cost, excluding all taxes (E1 * F1)	$, millions	35.53	16.47	(63.39)	8.95	22.05	18.42
G2	Fiscal cost, with fuel at counterfactual tax rates (E2 * F1)	$, millions	(218.23)	(252.30)	(389.19)	(338.38)	(365.65)	(384.24)
G3	Fiscal expenditures (F1 * (E2 − E1))	$, millions	(253.76)	(268.77)	(325.80)	(347.33)	(387.71)	(402.66)

a. All prices and costs are expressed in current U.S. dollars.
b. CIF (cost, insurance, and freight) price + local distribution and retail margins.
c. VAT (value added tax, IGV) is 18 percent of the supply cost, and excise tax (ISC) is a mix of ad valorem (8 percent of the supply cost) and a unit tax.

Table A.35 Peru—Diesel

		Unit[a]	2008	2009	2010	2011	2012	2013
A.	**Supply cost**[b]	$/liter	0.894	0.610	0.707	0.925	0.951	0.926
B1.	**Taxes, excluding transfers (includes VAT and excise taxes)**[c]	$/liter	0.303	0.242	0.250	0.263	0.272	0.265
B2.	**Taxes (counterfactual): standard VAT**	$/liter	0.170	0.116	0.134	0.166	0.171	0.167
C.	**Total reference benchmark price**							
C1	Total reference price without taxes	$/liter	0.894	0.610	0.707	0.925	0.951	0.926
C2	Total reference price at counterfactual tax rates	$/liter	1.064	0.726	0.842	1.091	1.123	1.093

table continues next page

Table A.35 Peru—Diesel *(continued)*

	Unit[a]	2008	2009	2010	2011	2012	2013
D.	**End-user price**						
D1	End-user price, excluding all taxes	$/liter	0.725	0.587	0.738	0.925	1.056	1.021
D2	End-user price, including all taxes	$/liter	0.995	0.824	0.994	1.188	1.348	1.304
E.	**Price gap**						
E1	Price gap, excluding all taxes (C1 − D1)	$/liter	0.169	0.024	−0.031	0.000	−0.105	−0.095
E2	Price gap, with fuel at counterfactual tax rates (C2 − D2)	$/liter	0.069	−0.098	−0.152	−0.097	−0.225	−0.211
F.	**Consumption**						
F1	Volume	Million liters	4,271.03	4,531.01	4,842.05	5,423.52	5,665.50	5,899.36
G.	**Fiscal cost**						
G1	Fiscal cost, excluding all taxes (E1 * F1)	$, millions	723.16	106.54	(148.14)	(0.42)	(595.62)	(562.62)
G2	Fiscal cost, with fuel at counterfactual tax rates (E2 * F1)	$, millions	293.71	(445.47)	(737.55)	(525.75)	(1,275.85)	(1,246.22)
G3	Fiscal expenditures (F1 * (E2 − E1))	$, millions	(429.44)	(552.01)	(589.41)	(525.33)	(680.23)	(683.60)

a. All prices and costs are expressed in current U.S. dollars.
b. CIF (cost, insurance, and freight) price + local distribution and retail margins.
c. VAT (value added tax, IGV) is 18 percent of the supply cost, and excise tax (ISC) is a unit tax.

Table A.36 Peru—Liquefied Petroleum Gas (LPG)

		Unit[a]	2009	2010	2011	2012	2013
A.	**Supply cost[b]**	$/liter	0.545	0.647	0.734	0.641	0.688
B1.	**Taxes, excluding transfers (includes VAT and excise taxes)[c]**	$/liter	0.104	0.123	0.132	0.115	0.124
B2.	**Taxes (counterfactual): standard VAT**	$/liter	0.104	0.123	0.132	0.115	0.124
C.	**Total reference benchmark price**						
C1	Total reference price without taxes	$/liter	0.545	0.647	0.734	0.641	0.688
C2	Total reference price at counterfactual tax rates	$/liter	0.649	0.770	0.867	0.757	0.812
D.	**End-user price**						
D1	End-user price, excluding all taxes	$/liter	0.469	0.519	0.557	0.604	0.608
D2	End-user price, including all taxes	$/liter	0.573	0.642	0.689	0.720	0.732
E.	**Price gap**						
E1	Price gap, excluding all taxes (C1 − D1)	$/liter	0.076	0.128	0.177	0.037	0.080
E2	Price gap, with fuel at counterfactual tax rates (C2 − D2)	$/liter	0.076	0.128	0.177	0.037	0.080
F.	**Consumption**						
F1	Volume	Million liters	1,894.11	2,107.66	2,356.03	2,603.82	2,853.93

table continues next page

Table A.36 Peru—Liquefied Petroleum Gas (LPG) *(continued)*

	Unit[a]	2009	2010	2011	2012	2013
G.	**Fiscal cost**					
G1	Fiscal cost, excluding all taxes (E1 * F1)	$, millions	144.70	269.13	417.62	96.90	227.49
G2	Fiscal cost, with fuel at counterfactual tax rates (E2 * F1)	$, millions	144.70	269.13	417.62	96.90	227.49
G3	Fiscal expenditures (F1 * (E2 − E1))	$, millions	0.00	0.00	0.00	0.00	0.00

a. All prices and costs are expressed in current U.S. dollars.
b. CIF (cost, insurance, and freight) price + local distribution and retail margins.
c. VAT (value added tax, IGV) is 18 percent of the supply cost, and excise tax (ISC) is zero.

Table A.37 Peru—Natural Gas (Power)

		Unit[a]	2009	2010	2011	2012	2013
A.	**Supply cost[b]**	$/MPC	6.694	7.179	6.865	5.673	6.711
B1.	**Taxes, excluding transfers (includes VAT and excise taxes)[c]**	$/MPC	1.272	1.364	1.236	1.021	1.208
B2.	**Taxes (counterfactual): standard VAT**	$/MPC	1.272	1.364	1.236	1.021	1.208
C.	**Total reference benchmark price**						
C1	Total reference price without taxes	$/MPC	6.694	7.179	6.865	5.673	6.711
C2	Total reference price at counterfactual tax rates	$/MPC	7.966	8.543	8.101	6.694	7.919
D.	**End-user price**						
D1	End-user price, excluding all taxes	$/MPC	1.228	1.136	1.364	1.679	1.592
D2	End-user price, including all taxes	$/MPC	2.500	2.500	2.600	2.700	2.800
E.	**Price gap**						
E1	Price gap, excluding all taxes (C1 − D1)	$/MPC	5.466	6.043	5.501	3.994	5.119
E2	Price gap, with fuel at counterfactual tax rates (C2 − D2)	$/MPC	5.466	6.043	5.501	3.994	5.119
F.	**Consumption**						
F1	Volume	Million MPCs	86.07	94.90	107.75	114.87	114.65
G.	**Fiscal cost**						
G1	Fiscal cost, excluding all taxes (E1 * F1)	$, millions	470.43	573.49	592.72	458.73	586.88
G2	Fiscal cost, with fuel at counterfactual tax rates (E2 * F1)	$, millions	470.43	573.49	592.72	458.73	586.88
G3	Fiscal expenditures (F1 * (E2 − E1))	$, millions	0.00	0.00	0.00	0.00	0.00

a. All prices and costs are expressed in current U.S. dollars. MPC = thousand cubic feet.
b. CIF (cost, insurance, and freight) price + local distribution and retail margins.
c. VAT (value added tax) is 18 percent of the supply cost, and excise tax is zero.

Table A.38 Peru—Natural Gas (Households)

		Unit[a]	2009	2010	2011	2012	2013
A.	**Supply cost[b]**	$/MPC	6.694	7.179	6.865	5.673	6.711
B1.	**Taxes, excluding transfers (includes VAT and excise taxes)[c]**	$/MPC	1.272	1.364	1.236	1.021	1.208
B2.	**Taxes (counterfactual): standard VAT**	$/MPC	1.272	1.364	1.236	1.021	1.208
C.	**Total reference benchmark price**						
C1	Total reference price without taxes	$/MPC	6.694	7.179	6.865	5.673	6.711
C2	Total reference price at counterfactual tax rates	$/MPC	7.966	8.543	8.101	6.694	7.919

table continues next page

Table A.38 Peru—Natural Gas (Households) *(continued)*

		Unit[a]	2009	2010	2011	2012	2013
D.	**End-user price**						
D1	End-user price, excluding all taxes	$/MPC	2.228	2.236	2.564	2.879	2.992
D2	End-user price, including all taxes	$/MPC	3.500	3.600	3.800	3.900	4.200
E.	**Price gap**						
E1	Price gap, excluding all taxes (C1 − D1)	$/MPC	4.466	4.943	4.301	2.794	3.719
E2	Price gap, with fuel at counterfactual tax rates (C2 − D2)	$/MPC	4.466	4.943	4.301	2.794	3.719
F.	**Consumption**						
F1	Volume	Million MPCs	0.15	0.22	0.37	0.62	0.91
G.	**Fiscal cost**						
G1	Fiscal cost, excluding all taxes (E1 * F1)	$, millions	0.65	1.08	1.57	1.73	3.39
G2	Fiscal cost, with fuel at counterfactual tax rates (E2 * F1)	$, millions	0.65	1.08	1.57	1.73	3.39
G3	Fiscal expenditures (F1 * (E2 − E1))	$, millions	0.00	0.00	0.00	0.00	0.00

a. All prices and costs are expressed in current U.S dollars. MPC = thousand cubic feet.
b. CIF (cost, insurance, and freight) price + local distribution and retail margins.
c. VAT (value added tax) is 18 percent of the supply cost, and excise tax is zero.

Table A.39 Peru—Natural Gas (Industrial)

		Unit[a]	2009	2010	2011	2012	2013
A.	**Supply cost[b]**	$/MPC	6.694	7.179	6.865	5.673	6.711
B1.	**Taxes, excluding transfers (includes VAT and excise taxes)[c]**	$/MPC	1.272	1.364	1.236	1.021	1.208
B2.	**Taxes (counterfactual): standard VAT**	$/MPC	1.272	1.364	1.236	1.021	1.208
C.	**Total reference benchmark price**						
C1	Total reference price without taxes	$/MPC	6.694	7.179	6.865	5.673	6.711
C2	Total reference price at counterfactual tax rates	$/MPC	7.966	8.543	8.101	6.694	7.919
D.	**End-user price**						
D1	End-user price, excluding all taxes	$/MPC	2.328	2.336	2.664	3.079	3.092
D2	End-user price, including all taxes	$/MPC	3.600	3.700	3.900	4.100	4.300
E.	**Price gap**						
E1	Price gap, excluding all taxes (C1 − D1)	$/MPC	4.366	4.843	4.201	2.594	3.619
E2	Price gap, with fuel at counterfactual tax rates (C2 − D2)	$/MPC	4.366	4.843	4.201	2.594	3.619
F.	**Consumption**						
F1	Volume	Million MPCs	21.95	27.89	31.68	33.00	35.55
G.	**Fiscal cost**						
G1	Fiscal cost, excluding all taxes (E1 * F1)	$, millions	95.82	135.05	133.10	85.58	128.66
G2	Fiscal cost, with fuel at counterfactual tax rates (E2 * F1)	$, millions	95.82	135.05	133.10	85.58	128.66
G3	Fiscal expenditures (F1 * (E2 − E1))	$, millions	0.00	0.00	0.00	0.00	0.00

a. All prices and costs are expressed in current U.S. dollars. MPC = thousand cubic feet.
b. CIF (cost, insurance, and freight) price + local distribution and retail margins.
c. VAT (value added tax) is 18 percent of the supply cost, and excise tax is zero.

Table A.40 Peru—Natural Gas (Vehicles)

		Unit[a]	2009	2010	2011	2012	2013
A.	**Supply cost[b]**	$/MPC	6.694	7.179	6.865	5.673	6.711
B1.	**Taxes, excluding transfers (includes VAT and excise taxes)[c]**	$/MPC	1.272	1.364	1.236	1.021	1.208
B2.	**Taxes (counterfactual): standard VAT**	$/MPC	1.272	1.364	1.236	1.021	1.208
C.	**Total reference benchmark price**						
C1	Total reference price without taxes	$/MPC	6.694	7.179	6.865	5.673	6.711
C2	Total reference price at counterfactual tax rates	$/MPC	7.966	8.543	8.101	6.694	7.919
D.	**End-user price**						
D1	End-user price, excluding all taxes	$/MPC	0.428	0.336	0.564	0.779	3.092
D2	End-user price, including all taxes	$/MPC	1.700	1.700	1.800	1.800	4.300
E.	**Price gap**						
E1	Price gap, excluding all taxes (C1 − D1)	$/MPC	6.266	6.843	6.301	4.894	3.619
E2	Price gap, with fuel at counterfactual tax rates (C2 − D2)	$/MPC	6.266	6.843	6.301	4.894	3.619
F.	**Consumption**						
F1	Volume	Million MPCs	9.38	12.56	14.89	17.67	20.95
G.	**Fiscal cost**						
G1	Fiscal cost, excluding all taxes (E1 * F1)	$, millions	58.78	85.92	93.83	86.45	75.82
G2	Fiscal cost, with fuel at counterfactual tax rates (E2 * F1)	$, millions	58.78	85.92	93.83	86.45	75.82
G3	Fiscal expenditures (F1 * (E2 − E1))	$, millions	0.00	0.00	0.00	0.00	0.00

a. All prices and costs are expressed in current U.S. dollars. MPC = thousand cubic feet.
b. CIF (cost, insurance, and freight) price + local distribution and retail margins.
c. VAT (value added tax) is 18 percent of the supply cost, and excise tax is zero.

Electricity Fiscal Cost Tables

Table B.1 Bolivia—Residential

	Unit[a]	2009	2010	2011	2012
A. Supply cost[b]	$/kWh	0.106	0.116	0.140	0.157
B1. Taxes, excluding transfers (VAT)[c]	$/kWh	0.010	0.010	0.010	0.011
B2. Taxes (counterfactual): standard VAT	$/kWh	0.010	0.010	0.010	0.011
C. Total reference benchmark price					
C1 Total reference price without taxes	$/kWh	0.106	0.116	0.140	0.157
C2 Total reference price at counterfactual tax rates	$/kWh	0.116	0.126	0.150	0.168
D. End-user price					
D1 Applied tariff, excluding all taxes	$/kWh	0.077	0.077	0.081	0.082
D2 Applied tariff, including all taxes	$/kWh	0.086	0.086	0.091	0.092
E. Price gap					
E1 Price gap, excluding all taxes (C1 − D1)	$/kWh	0.030	0.039	0.059	0.076
E2 Price gap, with fuel at counterfactual tax rates (C2 − D2)	$/kWh	0.030	0.039	0.059	0.076
F. Consumption					
F1 Volume	1,000 GWh	1.75	1.91	2.11	2.23
G. Fiscal cost					
G1 Fiscal cost, excluding all taxes (E1 * F1)	$, millions	52.31	75.24	124.81	168.28
G2 Fiscal cost, with fuel at counterfactual tax rates (E2 * F1)	$, millions	52.31	75.24	124.81	168.28
G3 Fiscal expenditures (F1 * (E2 − E1))	$, millions	0.00	0.00	0.00	0.00

Source: National Committee for Load Dispatch (CNDC).
a. All prices and costs are expressed in current U.S. dollars. GWh = gigawatt-hour; kWh = kilowatt-hour.
b. Generation cost + transmission and distribution margins. Weighted average of each distributor's applied tariff.
c. VAT (value added tax) is 13 percent of the applied tariff.

Table B.2 Bolivia—General

		Unit[a]	2009	2010	2011	2012
A.	**Supply cost**[b]	$/kWh	0.106	0.116	0.140	0.157
B1.	**Taxes, excluding transfers (VAT)**[c]	$/kWh	0.014	0.014	0.015	0.015
B2.	**Taxes (counterfactual): standard VAT**	$/kWh	0.014	0.014	0.015	0.015
C.	**Total reference benchmark price**					
C1	Total reference price without taxes	$/kWh	0.106	0.116	0.140	0.157
C2	Total reference price at counterfactual tax rates	$/kWh	0.120	0.130	0.154	0.172
D.	**End-user price**					
D1	Applied tariff, excluding all taxes	$/kWh	0.106	0.107	0.113	0.117
D2	Applied tariff, including all taxes	$/kWh	0.120	0.121	0.128	0.132
E.	**Price gap**					
E1	Price gap, excluding all taxes (C1 − D1)	$/kWh	0.000	0.009	0.026	0.040
E2	Price gap, with fuel at counterfactual tax rates (C2 − D2)	$/kWh	0.000	0.009	0.026	0.040
F.	**Consumption**					
F1	Volume	1,000 GWh	0.87	0.94	1.03	1.08
G.	**Fiscal cost**					
G1	Fiscal cost, excluding all taxes (E1 * F1)	$, millions	0.36	8.17	27.17	43.50
G2	Fiscal cost, with fuel at counterfactual tax rates (E2 * F1)	$, millions	0.36	8.17	27.17	43.50
G3	Fiscal expenditures (F1 * (E2 − E1))	$, millions	0.00	0.00	0.00	0.00

Source: National Committee for Load Dispatch (CNDC).
a. All prices and costs are expressed in current U.S. dollars. GWh = gigawatt-hour; kWh = kilowatt-hour.
b. Generation cost + transmission and distribution margins. Weighted average of each distributor's applied tariff.
c. VAT (value added tax) is 13 percent of the applied tariff.

Table B.3 Bolivia—Industrial

		Unit[a]	2009	2010	2011	2012
A.	**Supply cost**[b]	$/kWh	0.106	0.116	0.140	0.157
B1.	**Taxes, excluding transfers (VAT)**[c]	$/kWh	0.007	0.007	0.007	0.008
B2.	**Taxes (counterfactual): standard VAT**	$/kWh	0.007	0.007	0.007	0.008
C.	**Total reference benchmark price**					
C1	Total reference price without taxes	$/kWh	0.106	0.116	0.140	0.157
C2	Total reference price at counterfactual tax rates	$/kWh	0.113	0.123	0.147	0.165
D.	**End-user price**					
D1	Applied tariff, excluding all taxes	$/kWh	0.053	0.053	0.057	0.058
D2	Applied tariff, including all taxes	$/kWh	0.060	0.060	0.065	0.065
E.	**Price gap**					
E1	Price gap, excluding all taxes (C1 − D1)	$/kWh	0.053	0.063	0.082	0.099
E2	Price gap, with fuel at counterfactual tax rates (C2 − D2)	$/kWh	0.053	0.063	0.082	0.099
F.	**Consumption**					
F1	Volume	1,000 GWh	1.19	1.29	1.42	1.57
G.	**Fiscal cost**					
G1	Fiscal cost, excluding all taxes (E1 * F1)	$, millions	63.18	80.75	117.00	155.74
G2	Fiscal cost, with fuel at counterfactual tax rates (E2 * F1)	$, millions	63.18	80.75	117.00	155.74
G3	Fiscal expenditures (F1 * (E2 − E1))	$, millions	0.00	0.00	0.00	(0.00)

Source: National Committee for Load Dispatch (CNDC).
a. All prices and costs are expressed in current U.S. dollars. GWh = gigawatt-hour; kWh = kilowatt-hour.
b. Generation cost + transmission and distribution margins. Weighted average of each distributor's applied tariff.
c. VAT (value added tax) is 13 percent of the applied tariff.

Table B.4 Bolivia—Mining

	Unit[a]	2009	2010	2011	2012
A. Supply cost[b]	$/kWh	0.106	0.116	0.140	0.157
B1. Taxes, excluding transfers (VAT)[c]	$/kWh	0.006	0.006	0.007	0.007
B2. Taxes (counterfactual): standard VAT	$/kWh	0.006	0.006	0.007	0.007
C. Total reference benchmark price					
C1 Total reference price without taxes	$/kWh	0.106	0.116	0.140	0.157
C2 Total reference price at counterfactual tax rates	$/kWh	0.112	0.122	0.146	0.164
D. End-user price					
D1 Applied tariff, excluding all taxes	$/kWh	0.045	0.046	0.050	0.053
D2 Applied tariff, including all taxes	$/kWh	0.051	0.052	0.057	0.060
E. Price gap					
E1 Price gap, excluding all taxes (C1 − D1)	$/kWh	0.061	0.070	0.089	0.104
E2 Price gap, with fuel at counterfactual tax rates (C2 − D2)	$/kWh	0.061	0.070	0.089	0.104
F. Consumption					
F1 Volume	1,000 GWh	0.16	0.20	0.11	0.24
G. Fiscal cost					
G1 Fiscal cost, excluding all taxes (E1 * F1)	$, millions	9.95	14.05	10.17	25.32
G2 Fiscal cost, with fuel at counterfactual tax rates (E2 * F1)	$, millions	9.95	14.05	10.17	25.32
G3 Fiscal expenditures (F1 * (E2 − E1))	$, millions	0.00	0.00	0.00	0.00

Source: National Committee for Load Dispatch (CNDC).
a. All prices and costs are expressed in current U.S. dollars. GWh = gigawatt-hour; kWh = kilowatt-hour.
b. Generation cost + transmission and distribution margins. Weighted average of each distributor's applied tariff.
c. VAT (value added tax) is 13 percent of the applied tariff.

Table B.5 Bolivia—Public Lighting

	Unit[a]	2009	2010	2011	2012
A. Supply cost[b]	$/kWh	0.106	0.116	0.140	0.157
B1. Taxes, excluding transfers (VAT)[c]	$/kWh	0.012	0.012	0.012	0.013
B2. Taxes (counterfactual): standard VAT	$/kWh	0.012	0.012	0.012	0.013
C. Total reference benchmark price					
C1 Total reference price without taxes	$/kWh	0.106	0.116	0.140	0.157
C2 Total reference price at counterfactual tax rates	$/kWh	0.118	0.127	0.152	0.170
D. End-user price					
D1 Applied tariff, excluding all taxes	$/kWh	0.093	0.089	0.094	0.099
D2 Applied tariff, including all taxes	$/kWh	0.105	0.100	0.107	0.112
E. Price gap					
E1 Price gap, excluding all taxes (C1 − D1)	$/kWh	0.014	0.027	0.045	0.058
E2 Price gap, with fuel at counterfactual tax rates (C2 − D2)	$/kWh	0.014	0.027	0.045	0.058
F. Consumption					
F1 Volume	1,000 GWh	0.22	0.24	0.25	0.27
G. Fiscal cost					
G1 Fiscal cost, excluding all taxes (E1 * F1)	$, millions	3.00	6.38	11.50	15.61
G2 Fiscal cost, with fuel at counterfactual tax rates (E2 * F1)	$, millions	3.00	6.38	11.50	15.61
G3 Fiscal expenditures (F1 * (E2 − E1))	$, millions	0.00	0.00	0.00	(0.00)

Source: National Committee for Load Dispatch (CNDC).
a. All prices and costs are expressed in current U.S. dollars. GWh = gigawatt-hour; kWh = kilowatt-hour.
b. Generation cost + transmission and distribution margins. Weighted average of each distributor's applied tariff.
c. VAT (value added tax) is 13 percent of the applied tariff.

Table B.6 Bolivia—Other

		Unit[a]	2009	2010	2011	2012
A.	Supply cost[b]	$/kWh	0.106	0.116	0.140	0.157
B1.	Taxes, excluding transfers (VAT)[c]	$/kWh	0.005	0.005	0.006	0.006
B2.	Taxes (counterfactual): standard VAT	$/kWh	0.005	0.005	0.006	0.006
C.	Total reference benchmark price					
C1	Total reference price without taxes	$/kWh	0.106	0.116	0.140	0.157
C2	Total reference price at counterfactual tax rates	$/kWh	0.111	0.121	0.146	0.164
D.	End-user price					
D1	Applied tariff, excluding all taxes	$/kWh	0.038	0.042	0.045	0.049
D2	Applied tariff, including all taxes	$/kWh	0.043	0.047	0.051	0.055
E.	Price gap					
E1	Price gap, excluding all taxes (C1 − D1)	$/kWh	0.068	0.074	0.094	0.108
E2	Price gap, with fuel at counterfactual tax rates (C2 − D2)	$/kWh	0.068	0.074	0.094	0.108
F.	Consumption					
F1	Volume	1,000 GWh	0.16	0.18	0.20	0.25
G.	Fiscal cost					
G1	Fiscal cost, excluding all taxes (E1 * F1)	$, millions	10.95	13.55	18.49	26.64
G2	Fiscal cost, with fuel at counterfactual tax rates (E2 * F1)	$, millions	10.95	13.55	18.49	26.64
G3	Fiscal expenditures (F1 * (E2 − E1))	$, millions	0.00	0.00	0.00	0.00

Source: National Committee for Load Dispatch (CNDC).
a. All prices and costs are expressed in current U.S. dollars. GWh = gigawatt-hour; kWh = kilowatt-hour.
b. Generation cost + transmission and distribution margins. Weighted average of each distributor's applied tariff.
c. VAT (value added tax) is 13 percent of the applied tariff.

Table B.7 Brazil—Commercial

		Unit[a]	2008	2009	2010	2011	2012	2013
A.	Supply cost[b]	$/kWh	0.136	0.075	0.109	0.093	0.147	0.160
B1.	Taxes, excluding transfers (VAT)[c]	$/kWh	0.023	0.022	0.026	0.029	0.025	0.019
B2.	Taxes (counterfactual): standard VAT	$/kWh	0.023	0.022	0.026	0.029	0.025	0.019
C.	Total reference benchmark price							
C1	Total reference price without taxes	$/kWh	0.136	0.075	0.109	0.093	0.147	0.160
C2	Total reference price at counterfactual tax rates	$/kWh	0.159	0.097	0.135	0.121	0.172	0.179
D.	End-user price							
D1	Applied tariff, excluding all taxes	$/kWh	0.137	0.129	0.154	0.168	0.146	0.112
D2	Applied tariff, including all taxes	$/kWh	0.160	0.151	0.180	0.197	0.171	0.131
E.	Price gap							
E1	Price gap, excluding all taxes (C1 − D1)	$/kWh	−0.001	−0.054	−0.045	−0.075	0.001	0.049
E2	Price gap, with fuel at counterfactual tax rates (C2 − D2)	$/kWh	−0.001	−0.054	−0.045	−0.075	0.001	0.049
F.	Consumption							
F1	Volume	1,000 GWh	60.17	63.63	67.03	70.81	75.30	77.81

table continues next page

Table B.7 Brazil—Commercial *(continued)*

		Unit[a]	2008	2009	2010	2011	2012	2013
G.	**Fiscal cost**							
G1	Fiscal cost, excluding all taxes (E1 * F1)	$, millions	(69.60)	(3,455.98)	(3,014.19)	(5,345.04)	85.84	3,779.25
G2	Fiscal cost, with fuel at counterfactual tax rates (E2 * F1)	$, millions	(69.60)	(3,455.98)	(3,014.19)	(5,345.04)	85.84	3,779.25
G3	Fiscal expenditures (F1 * (E2 − E1))	$, millions	0.00	0.00	0.00	0.00	0.00	0.00

Source: Electric Energy Commercialization Chamber (CCEE).
a. All prices and costs are expressed in current U.S. dollars. GWh = gigawatt-hour; kWh = kilowatt-hour.
b. Generation cost (Balance Liquidation Value) + transmission and distribution margins (weighted average by company).
c. VAT (value added tax, ICMS) is an ad valorem state tax.

Table B.8 Brazil—Self-Consumption

		Unit[a]	2008	2009	2010	2011	2012	2013
A.	**Supply cost[b]**	$/kWh	0.151	0.296	0.370	0.340	0.396	0.347
B1.	**Taxes, excluding transfers (VAT)[c]**	$/kWh	0.025	0.023	0.029	0.030	0.027	0.021
B2.	**Taxes (counterfactual): standard VAT**	$/kWh	0.025	0.023	0.029	0.030	0.027	0.021
C.	**Total reference benchmark price**							
C1	Total reference price without taxes	$/kWh	0.151	0.296	0.370	0.340	0.396	0.347
C2	Total reference price at counterfactual tax rates	$/kWh	0.176	0.319	0.399	0.371	0.423	0.368
D.	**End-user price**							
D1	Applied tariff, excluding all taxes	$/kWh	0.146	0.138	0.168	0.178	0.158	0.123
D2	Applied tariff, including all taxes	$/kWh	0.171	0.161	0.197	0.208	0.185	0.144
E.	**Price gap**							
E1	Price gap, excluding all taxes (C1 − D1)	$/kWh	0.005	0.158	0.202	0.162	0.238	0.224
E2	Price gap, with fuel at counterfactual tax rates (C2 − D2)	$/kWh	0.005	0.158	0.202	0.162	0.238	0.224
F.	**Consumption**							
F1	Volume	1,000 GWh	0.44	0.47	0.51	0.45	0.50	0.53
G.	**Fiscal cost**							
G1	Fiscal cost, excluding all taxes (E1 * F1)	$, millions	2.08	74.22	103.65	73.44	119.02	118.25
G2	Fiscal cost, with fuel at counterfactual tax rates (E2 * F1)	$, millions	2.08	74.22	103.65	73.44	119.02	118.25
G3	Fiscal expenditures (F1 * (E2 − E1))	$, millions	0.00	0.00	0.00	0.00	0.00	0.00

Source: Electric Energy Commercialization Chamber (CCEE).
a. All prices and costs are expressed in current U.S. dollars. GWh = gigawatt-hour; kWh = kilowatt-hour.
b. Generation cost (Balance Liquidation Value) + transmission and distribution margins (weighted average by company).
c. VAT (value added tax, ICMS) is an ad valorem state tax.

Table B.9 Brazil—Public Lighting

		Unit[a]	2008	2009	2010	2011	2012	2013
A.	**Supply cost[b]**	$/kWh	0.137	0.073	0.095	0.077	0.135	0.155
B1.	**Taxes, excluding transfers (VAT)[c]**	$/kWh	0.013	0.013	0.015	0.017	0.015	0.012
B2.	**Taxes (counterfactual): standard VAT**	$/kWh	0.013	0.013	0.015	0.017	0.015	0.012
C.	**Total reference benchmark price**							
C1	Total reference price without taxes	$/kWh	0.137	0.073	0.095	0.077	0.135	0.155

table continues next page

Table B.9 Brazil—Public Lighting *(continued)*

		Unit[a]	2008	2009	2010	2011	2012	2013
C2	Total reference price at counterfactual tax rates	$/kWh	0.150	0.086	0.110	0.094	0.150	0.167
D.	**End-user price**							
D1	Applied tariff, excluding all taxes	$/kWh	0.077	0.074	0.089	0.098	0.087	0.068
D2	Applied tariff, including all taxes	$/kWh	0.090	0.086	0.105	0.114	0.101	0.080
E.	**Price gap**							
E1	Price gap, excluding all taxes (C1 – D1)	$/kWh	0.060	−0.001	0.006	−0.020	0.049	0.087
E2	Price gap, with fuel at counterfactual tax rates (C2 – D2)	$/kWh	0.060	−0.001	0.006	−0.020	0.049	0.087
F.	**Consumption**							
F1	Volume	1,000 GWh	11.44	11.81	12.12	12.68	13.03	13.62
G.	**Fiscal cost**							
G1	Fiscal cost, excluding all taxes (E1 * F1)	$, millions	684.65	(8.66)	71.36	(256.12)	632.53	1,181.39
G2	Fiscal cost, with fuel at counterfactual tax rates (E2 * F1)	$, millions	684.65	(8.66)	71.36	(256.12)	632.53	1,181.39
G3	Fiscal expenditures (F1 * (E2 – E1))	$, millions	0.00	0.00	0.00	0.00	0.00	0.00

Source: Electric Energy Commercialization Chamber (CCEE).
a. All prices and costs are expressed in current U.S. dollars. GWh = gigawatt-hour; kWh = kilowatt-hour.
b. Generation cost (Balance Liquidation Value) + transmission and distribution margins (weighted average by company).
c. VAT (value added tax, ICMS) is an ad valorem state tax.

Table B.10 Brazil—Industrial

		Unit[a]	2008	2009	2010	2011	2012	2013
A.	**Supply cost**[b]	$/kWh	0.105	0.047	0.079	0.059	0.119	0.141
B1.	**Taxes, excluding transfers (VAT)**[c]	$/kWh	0.019	0.018	0.022	0.024	0.021	0.016
B2.	**Taxes (counterfactual): standard VAT**	$/kWh	0.019	0.018	0.022	0.024	0.021	0.016
C.	**Total reference benchmark price**							
C1	Total reference price without taxes	$/kWh	0.105	0.047	0.079	0.059	0.119	0.141
C2	Total reference price at counterfactual tax rates	$/kWh	0.125	0.066	0.101	0.084	0.140	0.157
D.	**End-user price**							
D1	Applied tariff, excluding all taxes	$/kWh	0.113	0.109	0.129	0.142	0.124	0.096
D2	Applied tariff, including all taxes	$/kWh	0.133	0.127	0.151	0.166	0.145	0.113
E.	**Price gap**							
E1	Price gap, excluding all taxes (C1 – D1)	$/kWh	−0.008	−0.061	−0.050	−0.082	−0.005	0.044
E2	Price gap, with fuel at counterfactual tax rates (C2 – D2)	$/kWh	−0.008	−0.061	−0.050	−0.082	−0.005	0.044

table continues next page

Table B.10 Brazil—Industrial (continued)

		Unit[a]	2008	2009	2010	2011	2012	2013
F.	**Consumption**							
F1	Volume	1,000 GWh	70.34	68.61	71.07	69.04	64.54	61.78
G.	**Fiscal cost**							
G1	Fiscal cost, excluding all taxes (E1 * F1)	$, millions	(569.03)	(4,217.21)	(3,561.64)	(5,678.75)	(317.24)	2,742.19
G2	Fiscal cost, with fuel at counterfactual tax rates (E2 * F1)	$, millions	(569.03)	(4,217.21)	(3,561.64)	(5,678.75)	(317.24)	2,742.19
G3	Fiscal expenditures (F1 * (E2 − E1))	$, millions	0.00	0.00	0.00	0.00	0.00	0.00

Source: Electric Energy Commercialization Chamber (CCEE).

a. All prices and costs are expressed in current U.S. dollars. GWh = gigawatt-hour; kWh = kilowatt-hour.

b. Generation cost (Balance Liquidation Value) + transmission and distribution margins (weighted average by company).

c. VAT (value added tax, ICMS) is an ad valorem state tax.

Table B.11 Brazil—Public

		Unit[a]	2008	2009	2010	2011	2012	2013
A.	**Supply cost**[b]	$/kWh	0.137	0.077	0.113	0.097	0.149	0.163
B1.	**Taxes, excluding transfers (VAT)**[c]	$/kWh	0.025	0.024	0.028	0.031	0.027	0.020
B2.	**Taxes (counterfactual): standard VAT**	$/kWh	0.025	0.024	0.028	0.031	0.027	0.020
C.	**Total reference benchmark price**							
C1	Total reference price without taxes	$/kWh	0.137	0.077	0.113	0.097	0.149	0.163
C2	Total reference price at counterfactual tax rates	$/kWh	0.162	0.100	0.141	0.127	0.176	0.183
D.	**End-user price**							
D1	Applied tariff, excluding all taxes	$/kWh	0.149	0.138	0.166	0.182	0.157	0.120
D2	Applied tariff, including all taxes	$/kWh	0.174	0.162	0.194	0.213	0.184	0.140
E.	**Price gap**							
E1	Price gap, excluding all taxes (C1 − D1)	$/kWh	−0.012	−0.062	−0.053	−0.085	−0.008	0.043
E2	Price gap, with fuel at counterfactual tax rates (C2 − D2)	$/kWh	−0.012	−0.062	−0.053	−0.085	−0.008	0.043
F.	**Consumption**							
F1	Volume	1,000 GWh	11.51	12.13	12.84	13.29	14.11	14.66
G.	**Fiscal cost**							
G1	Fiscal cost, excluding all taxes (E1 * F1)	$, millions	(132.93)	(746.18)	(679.93)	(1,131.69)	(112.25)	627.37
G2	Fiscal cost, with fuel at counterfactual tax rates (E2 * F1)	$, millions	(132.93)	(746.18)	(679.93)	(1,131.69)	(112.25)	627.37
G3	Fiscal expenditures (F1 * (E2 − E1))	$, millions	0.00	0.00	0.00	0.00	0.00	0.00

Source: Electric Energy Commercialization Chamber (CCEE).

a. All prices and costs are expressed in current U.S. dollars. GWh = gigawatt-hour; kWh = kilowatt-hour.

b. Generation cost (Balance Liquidation Value) + transmission and distribution margins (weighted average by company).

c. VAT (value added tax, ICMS) is an ad valorem state tax.

Table B.12 Brazil—Residential

	Unit[a]	2008	2009	2010	2011	2012	2013
A.	**Supply cost**[b]	$/kWh	0.158	0.099	0.138	0.124	0.174	0.177
B1.	**Taxes, excluding transfers (VAT)**[c]	$/kWh	0.024	0.023	0.027	0.030	0.026	0.020
B2.	**Taxes (counterfactual): standard VAT**	$/kWh	0.024	0.023	0.027	0.030	0.026	0.020
C.	**Total reference benchmark price**							
C1	Total reference price without taxes	$/kWh	0.158	0.099	0.138	0.124	0.174	0.177
C2	Total reference price at counterfactual tax rates	$/kWh	0.182	0.122	0.165	0.155	0.200	0.197
D.	**End-user price**							
D1	Applied tariff, excluding all taxes	$/kWh	0.139	0.135	0.162	0.177	0.156	0.116
D2	Applied tariff, including all taxes	$/kWh	0.163	0.158	0.189	0.207	0.182	0.135
E.	**Price gap**							
E1	Price gap, excluding all taxes (C1 − D1)	$/kWh	0.019	−0.036	−0.024	−0.053	0.018	0.061
E2	Price gap, with fuel at counterfactual tax rates (C2 − D2)	$/kWh	0.019	−0.036	−0.024	−0.053	0.018	0.061
F.	**Consumption**							
F1	Volume	1,000 GWh	95.81	101.97	108.54	113.27	119.03	126.52
G.	**Fiscal cost**							
G1	Fiscal cost, excluding all taxes (E1 * F1)	$, millions	1,851.93	(3,669.42)	(2,593.53)	(5,989.36)	2,150.51	7,761.26
G2	Fiscal cost, with fuel at counterfactual tax rates (E2 * F1)	$, millions	1,851.93	(3,669.42)	(2,593.53)	(5,989.36)	2,150.51	7,761.26
G3	Fiscal expenditures (F1 * (E2 − E1))	$, millions	0.00	0.00	0.00	0.00	0.00	0.00

Source: Electric Energy Commercialization Chamber (CCEE).
a. All prices and costs are expressed in current U.S. dollars. GWh = gigawatt-hour; kWh = kilowatt-hour.
b. Generation cost (Balance Liquidation Value) + transmission and distribution margins (weighted average by company).
c. VAT (value added tax, ICMS) is an ad valorem state tax.

Table B.13 Brazil—Rural

	Unit[a]	2008	2009	2010	2011	2012	2013
A.	**Supply cost**[b]	$/kWh	0.118	0.077	0.096	0.078	0.134	0.154
B1.	**Taxes, excluding transfers (VAT)**[c]	$/kWh	0.013	0.017	0.017	0.019	0.016	0.013
B2.	**Taxes (counterfactual): standard VAT**	$/kWh	0.013	0.017	0.017	0.019	0.016	0.013
C.	**Total reference benchmark price**							
C1	Total reference price without taxes	$/kWh	0.118	0.077	0.096	0.078	0.134	0.154
C2	Total reference price at counterfactual tax rates	$/kWh	0.131	0.094	0.113	0.097	0.151	0.167

table continues next page

Table B.13 Brazil—Rural *(continued)*

	Unit[a]	2008	2009	2010	2011	2012	2013
D.	**End-user price**						
D1	Applied tariff, excluding all taxes	$/kWh	0.077	0.101	0.102	0.111	0.097	0.077
D2	Applied tariff, including all taxes	$/kWh	0.090	0.118	0.119	0.130	0.113	0.090
E.	**Price gap**						
E1	Price gap, excluding all taxes (C1 − D1)	$/kWh	0.041	−0.024	−0.006	−0.033	0.037	0.077
E2	Price gap, with fuel at counterfactual tax rates (C2 − D2)	$/kWh	0.041	−0.024	−0.006	−0.033	0.037	0.077
F.	**Consumption**						
F1	Volume	1,000 GWh	14.19	13.79	14.17	14.48	15.58	16.15
G.	**Fiscal cost**						
G1	Fiscal cost, excluding all taxes (E1 * F1)	$, millions	579.64	(327.68)	(83.07)	(481.02)	580.89	1,247.85
G2	Fiscal cost, with fuel at counterfactual tax rates (E2 * F1)	$, millions	579.64	(327.68)	(83.07)	(481.02)	580.89	1,247.85
G3	Fiscal expenditures (F1 * (E2 − E1))	$, millions	0.00	0.00	0.00	0.00	0.00	0.00

Source: Electric Energy Commercialization Chamber (CCEE).
a. All prices and costs are expressed in current U.S. dollars. GWh = gigawatt-hour; kWh = kilowatt-hour.
b. Generation cost (Balance Liquidation Value) + transmission and distribution margins (weighted average by company).
c. VAT (value added tax, ICMS) is an ad valorem state tax.

Table B.14 Brazil—Public Service

	Unit[a]	2008	2009	2010	2011	2012	2013
A.	**Supply cost[b]**	$/kWh	0.104	0.047	0.077	0.056	0.117	0.140
B1.	**Taxes, excluding transfers (VAT)[c]**	$/kWh	0.017	0.016	0.020	0.021	0.019	0.014
B2.	**Taxes (counterfactual): standard VAT**	$/kWh	0.017	0.016	0.020	0.021	0.019	0.014
C.	**Total reference benchmark price**						
C1	Total reference price without taxes	$/kWh	0.104	0.047	0.077	0.056	0.117	0.140
C2	Total reference price at counterfactual tax rates	$/kWh	0.121	0.064	0.096	0.078	0.136	0.154
D.	**End-user price**						
D1	Applied tariff, excluding all taxes	$/kWh	0.100	0.096	0.115	0.126	0.110	0.084
D2	Applied tariff, including all taxes	$/kWh	0.117	0.112	0.135	0.147	0.128	0.098
E.	**Price gap**						
E1	Price gap, excluding all taxes (C1 − D1)	$/kWh	0.004	−0.048	−0.038	−0.069	0.007	0.056
E2	Price gap, with fuel at counterfactual tax rates (C2 − D2)	$/kWh	0.004	−0.048	−0.038	−0.069	0.007	0.056
F.	**Consumption**						
F1	Volume	1,000 GWh	10.80	10.96	11.13	11.36	11.75	12.05
G.	**Fiscal cost**						
G1	Fiscal cost, excluding all taxes (E1 * F1)	$, millions	41.40	(528.40)	(423.69)	(787.69)	86.59	673.82
G2	Fiscal cost, with fuel at counterfactual tax rates (E2 * F1)	$, millions	41.40	(528.40)	(423.69)	(787.69)	86.59	673.82
G3	Fiscal expenditures (F1 * (E2 − E1))	$, millions	0.00	0.00	(0.00)	0.00	0.00	0.00

Source: Electric Energy Commercialization Chamber (CCEE).
a. All prices and costs are expressed in current U.S. dollars. GWh = gigawatt-hour; kWh = kilowatt-hour.
b. Generation cost (Balance Liquidation Value) + transmission and distribution margins (weighted average by company).
c. VAT (value added tax, ICMS) is an ad valorem state tax.

Table B.15 Colombia—Residential (Estrato 1)

		Unit[a]	2008	2009	2010	2011	2012	2013
A.	Supply cost[b]	$/kWh	0.144	0.139	0.170	0.194	0.193	0.186
B1.	Taxes, excluding transfers (VAT)[c]	$/kWh	0.000	0.000	0.000	0.000	0.000	0.000
B2.	Taxes (counterfactual): standard VAT	$/kWh	0.010	0.009	0.012	0.013	0.014	0.014
C.	Total reference benchmark price							
C1	Total reference price without taxes	$/kWh	0.144	0.139	0.170	0.194	0.193	0.186
C2	Total reference price at counterfactual tax rates	$/kWh	0.154	0.148	0.182	0.207	0.206	0.200
D.	End-user price							
D1	Applied tariff, excluding all taxes	$/kWh	0.061	0.055	0.075	0.080	0.085	0.085
D2	Applied tariff, including all taxes	$/kWh	0.061	0.055	0.075	0.080	0.085	0.085
E.	Price gap							
E1	Price gap, excluding all taxes (C1 − D1)	$/kWh	0.083	0.085	0.095	0.114	0.107	0.101
E2	Price gap, with fuel at counterfactual tax rates (C2 − D2)	$/kWh	0.092	0.093	0.107	0.127	0.121	0.115
F.	Consumption							
F1	Volume	1,000 GWh	4.19	4.44	4.68	4.74	4.54	5.20
G.	Fiscal cost							
G1	Fiscal cost, excluding all taxes (E1 * F1)	$, millions	346.00	375.42	445.87	541.52	487.01	526.93
G2	Fiscal cost, with fuel at counterfactual tax rates (E2 * F1)	$, millions	387.13	414.21	502.04	602.31	548.98	597.74
G3	Fiscal expenditures (F1 * (E2 − E1))	$, millions	41.13	38.79	56.17	60.79	61.97	70.81

Source: Codensa (distributor).
a. All prices and costs are expressed in current U.S. dollars. GWh = gigawatt-hour; kWh = kilowatt-hour.
b. Generation cost (average cost of electricity) + transmission and distribution margins (tariff publications) + other costs.
c. VAT (value added tax) is 16 percent of the applied tariff only for fuel oil used in electricity.

Table B.16 Colombia—Residential (Estrato 2)

		Unit[a]	2008	2009	2010	2011	2012	2013
A.	Supply cost[b]	$/kWh	0.144	0.139	0.170	0.194	0.193	0.186
B1.	Taxes, excluding transfers (VAT)[c]	$/kWh	0.000	0.000	0.000	0.000	0.000	0.000
B2.	Taxes (counterfactual): standard VAT	$/kWh	0.012	0.011	0.015	0.016	0.017	0.017
C.	Total reference benchmark price							
C1	Total reference price without taxes	$/kWh	0.144	0.139	0.170	0.194	0.193	0.186
C2	Total reference price at counterfactual tax rates	$/kWh	0.156	0.150	0.185	0.210	0.210	0.203
D.	End-user price							
D1	Applied tariff, excluding all taxes	$/kWh	0.074	0.068	0.094	0.100	0.107	0.106
D2	Applied tariff, including all taxes	$/kWh	0.074	0.068	0.094	0.100	0.107	0.106
E.	Price gap							
E1	Price gap, excluding all taxes (C1 − D1)	$/kWh	0.070	0.071	0.077	0.094	0.086	0.080
E2	Price gap, with fuel at counterfactual tax rates (C2 − D2)	$/kWh	0.082	0.082	0.092	0.110	0.103	0.097
F.	Consumption							
F1	Volume	1,000 GWh	6.30	6.45	6.60	6.45	6.31	7.03

table continues next page

Table B.16 Colombia—Residential (Estrato 2) *(continued)*

		Unit[a]	2008	2009	2010	2011	2012	2013
G.	**Fiscal cost**							
G1	Fiscal cost, excluding all taxes (E1 * F1)	$, millions	442.48	457.88	504.53	606.83	542.71	562.94
G2	Fiscal cost, with fuel at counterfactual tax rates (E2 * F1)	$, millions	516.62	528.40	603.45	710.10	650.46	682.64
G3	Fiscal expenditures (F1 * (E2 – E1))	$, millions	74.13	70.52	98.92	103.27	107.75	119.70

Source: Codensa (distributor).

a. All prices and costs are expressed in current U.S. dollars. GWh = gigawatt-hour; kWh = kilowatt-hour.

b. Generation cost (average cost of electricity) + transmission and distribution margins (tariff publications) + other costs.

c. VAT (value added tax) is 16 percent of the applied tariff only for fuel oil used in electricity.

Table B.17 Colombia—Residential (Estrato 3)

		Unit[a]	2008	2009	2010	2011	2012	2013
A.	**Supply cost[b]**	$/kWh	0.144	0.139	0.170	0.194	0.193	0.186
B1.	Taxes, excluding transfers (VAT)[c]	$/kWh	0.000	0.000	0.000	0.000	0.000	0.000
B2.	Taxes (counterfactual): standard VAT	$/kWh	0.020	0.019	0.025	0.027	0.027	0.026
C.	**Total reference benchmark price**							
C1	Total reference price without taxes	$/kWh	0.144	0.139	0.170	0.194	0.193	0.186
C2	Total reference price at counterfactual tax rates	$/kWh	0.163	0.158	0.196	0.221	0.219	0.212
D.	**End-user price**							
D1	Applied tariff, excluding all taxes	$/kWh	0.122	0.116	0.159	0.167	0.166	0.160
D2	Applied tariff, including all taxes	$/kWh	0.122	0.116	0.159	0.167	0.166	0.160
E.	**Price gap**							
E1	Price gap, excluding all taxes (C1 – D1)	$/kWh	0.022	0.023	0.011	0.027	0.026	0.026
E2	Price gap, with fuel at counterfactual tax rates (C2 – D2)	$/kWh	0.041	0.042	0.036	0.054	0.053	0.052
F.	**Consumption**							
F1	Volume	1,000 GWh	4.51	4.54	4.52	4.28	4.20	4.71
G.	**Fiscal cost**							
G1	Fiscal cost, excluding all taxes (E1 * F1)	$, millions	97.38	105.03	49.19	115.16	110.32	124.87
G2	Fiscal cost, with fuel at counterfactual tax rates (E2 * F1)	$, millions	185.68	189.33	164.44	229.62	222.08	245.52
G3	Fiscal expenditures (F1 * (E2 – E1))	$, millions	88.29	84.30	115.25	114.47	111.76	120.64

Source: Codensa (distributor).

a. All prices and costs are expressed in current U.S. dollars. GWh = gigawatt-hour; kWh = kilowatt-hour.

b. Generation cost (average cost of electricity) + transmission and distribution margins (tariff publications) + other costs.

c. VAT (value added tax) is 16 percent of the applied tariff only for fuel oil used in electricity.

Table B.18 Colombia—Residential (Estrato 4)

		Unit[a]	2008	2009	2010	2011	2012	2013
A.	**Supply cost[b]**	$/kWh	0.144	0.139	0.170	0.194	0.193	0.186
B1.	Taxes, excluding transfers (VAT)[c]	$/kWh	0.000	0.000	0.000	0.000	0.000	0.000
B2.	Taxes (counterfactual): standard VAT	$/kWh	0.023	0.022	0.030	0.031	0.031	0.030

table continues next page

Table B.18 Colombia—Residential (Estrato 4) *(continued)*

		Unit[a]	2008	2009	2010	2011	2012	2013
C.	**Total reference benchmark price**							
C1	Total reference price without taxes	$/kWh	0.144	0.139	0.170	0.194	0.193	0.186
C2	Total reference price at counterfactual tax rates	$/kWh	0.167	0.161	0.200	0.226	0.224	0.217
D.	**End-user price**							
D1	Applied tariff, excluding all taxes	$/kWh	0.144	0.137	0.187	0.197	0.196	0.188
D2	Applied tariff, including all taxes	$/kWh	0.144	0.137	0.187	0.197	0.196	0.188
E.	**Price gap**							
E1	Price gap, excluding all taxes (C1 − D1)	$/kWh	0.000	0.003	−0.017	−0.003	−0.003	−0.002
E2	Price gap, with fuel at counterfactual tax rates (C2 − D2)	$/kWh	0.023	0.025	0.013	0.029	0.028	0.028
F.	**Consumption**							
F1	Volume	1,000 GWh	1.54	1.59	1.62	1.56	1.55	1.78
G.	**Fiscal cost**							
G1	Fiscal cost, excluding all taxes (E1 * F1)	$, millions	0.00	4.21	(27.96)	(4.04)	(4.79)	(3.10)
G2	Fiscal cost, with fuel at counterfactual tax rates (E2 * F1)	$, millions	35.53	38.86	20.69	45.05	43.88	50.56
G3	Fiscal expenditures (F1 * (E2 − E1))	$, millions	35.53	34.65	48.66	49.10	48.67	53.65

Source: Codensa (distributor).

a. All prices and costs are expressed in current U.S. dollars. GWh = gigawatt-hour; kWh = kilowatt-hour.

b. Generation cost (average cost of electricity) + transmission and distribution margins (tariff publications) + other costs.

c. VAT (value added tax) is 16 percent of the applied tariff only for fuel oil used in electricity.

Table B.19 Colombia—Residential (Estrato 5)

		Unit[a]	2008	2009	2010	2011	2012	2013
A.	**Supply cost[b]**	$/kWh	0.144	0.139	0.170	0.194	0.193	0.186
B1.	**Taxes, excluding transfers (VAT)[c]**	$/kWh	0.000	0.000	0.000	0.000	0.000	0.000
B2.	**Taxes (counterfactual): standard VAT**	$/kWh	0.028	0.026	0.036	0.038	0.038	0.036
C.	**Total reference benchmark price**							
C1	Total reference price without taxes	$/kWh	0.144	0.139	0.170	0.194	0.193	0.186
C2	Total reference price at counterfactual tax rates	$/kWh	0.171	0.166	0.206	0.232	0.230	0.223
D.	**End-user price**							
D1	Applied tariff, excluding all taxes	$/kWh	0.173	0.164	0.225	0.236	0.235	0.226
D2	Applied tariff, including all taxes	$/kWh	0.173	0.164	0.225	0.236	0.235	0.226
E.	**Price gap**							
E1	Price gap, excluding all taxes (C1 − D1)	$/kWh	−0.029	−0.025	−0.055	−0.042	−0.042	−0.039
E2	Price gap, with fuel at counterfactual tax rates (C2 − D2)	$/kWh	−0.001	0.002	−0.019	−0.004	−0.005	−0.003
F.	**Consumption**							
F1	Volume	1,000 GWh	0.86	0.87	0.86	0.81	0.81	0.89

table continues next page

Table B.19 Colombia—Residential (Estrato 5) *(continued)*

	Unit[a]	2008	2009	2010	2011	2012	2013
G.	**Fiscal cost**							
G1	Fiscal cost, excluding all taxes (E1 * F1)	$, millions	(24.59)	(21.34)	(47.07)	(33.95)	(34.25)	(35.12)
G2	Fiscal cost, with fuel at counterfactual tax rates (E2 * F1)	$, millions	(0.98)	1.35	(16.12)	(3.37)	(3.77)	(2.89)
G3	Fiscal expenditures (F1 * (E2 − E1))	$, millions	23.61	22.69	30.96	30.57	30.48	32.23

Source: Codensa (distributor).
a. All prices and costs are expressed in current U.S. dollars. GWh = gigawatt-hour; kWh = kilowatt-hour.
b. Generation cost (average cost of electricity) + transmission and distribution margins (tariff publications) + other costs.
c. VAT (value added tax) is 16 percent of the applied tariff only for fuel oil used in electricity.

Table B.20 Colombia—Residential (Estrato 6)

		Unit[a]	2008	2009	2010	2011	2012	2013
A.	**Supply cost[b]**	$/kWh	0.144	0.139	0.170	0.194	0.193	0.186
B1.	**Taxes, excluding transfers (VAT)[c]**	$/kWh	0.000	0.000	0.000	0.000	0.000	0.000
B2.	**Taxes (counterfactual): standard VAT**	$/kWh	0.028	0.026	0.036	0.038	0.038	0.036
C.	**Total reference benchmark price**							
C1	Total reference price without taxes	$/kWh	0.144	0.139	0.170	0.194	0.193	0.186
C2	Total reference price at counterfactual tax rates	$/kWh	0.171	0.166	0.206	0.232	0.230	0.223
D.	**End-user price**							
D1	Applied tariff, excluding all taxes	$/kWh	0.173	0.164	0.225	0.236	0.235	0.226
D2	Applied tariff, including all taxes	$/kWh	0.173	0.164	0.225	0.236	0.235	0.226
E.	**Price gap**							
E1	Price gap, excluding all taxes (C1 − D1)	$/kWh	−0.029	−0.025	−0.055	−0.042	−0.042	−0.039
E2	Price gap, with fuel at counterfactual tax rates (C2 − D2)	$/kWh	−0.001	0.002	−0.019	−0.004	−0.005	−0.003
F.	**Consumption**							
F1	Volume	1,000 GWh	0.77	0.80	0.79	0.73	0.74	0.82
G.	**Fiscal cost**							
G1	Fiscal cost, excluding all taxes (E1 * F1)	$, millions	(22.06)	(19.74)	(43.30)	(30.59)	(31.25)	(32.25)
G2	Fiscal cost, with fuel at counterfactual tax rates (E2 * F1)	$, millions	(0.88)	1.25	(14.82)	(3.04)	(3.44)	(2.66)
G3	Fiscal expenditures (F1 * (E2 − E1))	$, millions	21.18	20.99	28.47	27.55	27.81	29.59

Source: Codensa (distributor).
a. All prices and costs are expressed in current U.S. dollars. GWh = gigawatt-hour; kWh = kilowatt-hour.
b. Generation cost (average cost of electricity) + transmission and distribution margins (tariff publications) + other costs.
c. VAT (value added tax) is 16 percent of the applied tariff only for fuel oil used in electricity.

Table B.21 Colombia—Public Sector

		Unit[a]	2008	2009	2010	2011	2012	2013
A.	**Supply cost[b]**	$/kWh	0.144	0.139	0.170	0.194	0.193	0.186
B1.	**Taxes, excluding transfers (VAT)[c]**	$/kWh	0.000	0.000	0.000	0.000	0.000	0.000
B2.	**Taxes (counterfactual): standard VAT**	$/kWh	0.023	0.022	0.030	0.031	0.031	0.030
C.	**Total reference benchmark price**							
C1	Total reference price without taxes	$/kWh	0.144	0.139	0.170	0.194	0.193	0.186
C2	Total reference price at counterfactual tax rates	$/kWh	0.167	0.161	0.200	0.226	0.224	0.217

table continues next page

Table B.21 Colombia—Public Sector *(continued)*

		Unit[a]	2008	2009	2010	2011	2012	2013
D.	**End-user price**							
D1	Applied tariff, excluding all taxes	$/kWh	0.144	0.137	0.187	0.197	0.196	0.188
D2	Applied tariff, including all taxes	$/kWh	0.144	0.137	0.187	0.197	0.196	0.188
E.	**Price gap**							
E1	Price gap, excluding all taxes (C1 − D1)	$/kWh	0.000	0.003	−0.017	−0.003	−0.003	−0.002
E2	Price gap, with fuel at counterfactual tax rates (C2 − D2)	$/kWh	0.023	0.025	0.013	0.029	0.028	0.028
F.	**Consumption**							
F1	Volume	1,000 GWh	1.51	1.71	1.67	1.46	1.42	1.72
G.	**Fiscal cost**							
G1	Fiscal cost, excluding all taxes (E1 * F1)	$, millions	0.00	4.53	(28.84)	(3.78)	(4.37)	(2.98)
G2	Fiscal cost, with fuel at counterfactual tax rates (E2 * F1)	$, millions	34.70	41.85	21.34	42.16	39.98	48.69
G3	Fiscal expenditures (F1 * (E2 − E1))	$, millions	34.70	37.31	50.19	45.95	44.35	51.67

Source: Codensa (distributor).
a. All prices and costs are expressed in current U.S. dollars. GWh = gigawatt-hour; kWh = kilowatt-hour.
b. Generation cost (average cost of electricity) + transmission and distribution margins (tariff publications) + other costs.
c. VAT (value added tax) is 16 percent of the applied tariff only for fuel oil used in electricity.

Table B.22 Colombia—Commercial

		Unit[a]	2008	2009	2010	2011	2012	2013
A.	**Supply cost**[b]	$/kWh	0.144	0.139	0.170	0.194	0.193	0.186
B1.	**Taxes, excluding transfers (VAT)**[c]	$/kWh	0.000	0.000	0.000	0.000	0.000	0.000
B2.	**Taxes (counterfactual): standard VAT**	$/kWh	0.028	0.026	0.036	0.038	0.038	0.036
C.	**Total reference benchmark price**							
C1	Total reference price without taxes	$/kWh	0.144	0.139	0.170	0.194	0.193	0.186
C2	Total reference price at counterfactual tax rates	$/kWh	0.171	0.166	0.206	0.232	0.230	0.223
D.	**End-user price**							
D1	Applied tariff, excluding all taxes	$/kWh	0.173	0.164	0.225	0.236	0.235	0.226
D2	Applied tariff, including all taxes	$/kWh	0.173	0.164	0.225	0.236	0.235	0.226
E.	**Price gap**							
E1	Price gap, excluding all taxes (C1 − D1)	$/kWh	−0.029	−0.025	−0.055	−0.042	−0.042	−0.039
E2	Price gap, with fuel at counterfactual tax rates (C2 − D2)	$/kWh	−0.001	0.002	−0.019	−0.004	−0.005	−0.003
F.	**Consumption**							
F1	Volume	1,000 GWh	8.65	9.18	9.70	8.77	8.00	10.17
G.	**Fiscal cost**							
G1	Fiscal cost, excluding all taxes (E1 * F1)	$, millions	(248.73)	(226.37)	(530.69)	(367.95)	(337.75)	(400.36)
G2	Fiscal cost, with fuel at counterfactual tax rates (E2 * F1)	$, millions	(9.95)	14.34	(181.69)	(36.56)	(37.18)	(32.98)
G3	Fiscal expenditures (F1 * (E2 − E1))	$, millions	238.78	240.70	348.99	331.40	300.56	367.38

Source: Codensa (distributor).
a. All prices and costs are expressed in current U.S. dollars. GWh = gigawatt-hour; kWh = kilowatt-hour.
b. Generation cost (average cost of electricity) + transmission and distribution margins (tariff publications) + other costs.
c. VAT (value added tax) is 16 percent of the applied tariff only for fuel oil used in electricity.

Table B.23 Colombia—Industrial

		Unit[a]	2008	2009	2010	2011	2012	2013
A.	Supply cost[b]	$/kWh	0.144	0.139	0.170	0.194	0.193	0.186
B1.	Taxes, excluding transfers (VAT)[c]	$/kWh	0.000	0.000	0.000	0.000	0.000	0.000
B2.	Taxes (counterfactual): standard VAT	$/kWh	0.023	0.022	0.030	0.031	0.031	0.030
C.	Total reference benchmark price							
C1	Total reference price without taxes	$/kWh	0.144	0.139	0.170	0.194	0.193	0.186
C2	Total reference price at counterfactual tax rates	$/kWh	0.167	0.161	0.200	0.226	0.224	0.217
D.	End-user price							
D1	Applied tariff, excluding all taxes	$/kWh	0.144	0.137	0.187	0.197	0.196	0.188
D2	Applied tariff, including all taxes	$/kWh	0.144	0.137	0.187	0.197	0.196	0.188
E.	Price gap							
E1	Price gap, excluding all taxes (C1 − D1)	$/kWh	0.000	0.003	−0.017	−0.003	−0.003	−0.002
E2	Price gap, with fuel at counterfactual tax rates (C2 − D2)	$/kWh	0.023	0.025	0.013	0.029	0.028	0.028
F.	Consumption							
F1	Volume	1,000 GWh	14.67	14.10	13.79	11.76	10.44	13.56
G.	Fiscal cost							
G1	Fiscal cost, excluding all taxes (E1 * F1)	$, millions	0.00	37.44	(237.67)	(30.52)	(32.18)	(23.56)
G2	Fiscal cost, with fuel at counterfactual tax rates (E2 * F1)	$, millions	337.56	345.71	175.85	339.97	294.68	384.66
G3	Fiscal expenditures (F1 * (E2 − E1))	$, millions	337.56	308.27	413.52	370.49	326.86	408.22

Source: Codensa (distributor).
a. All prices and costs are expressed in current U.S. dollars. GWh = gigawatt-hour; kWh = kilowatt-hour.
b. Generation cost (average cost of electricity) + transmission and distribution margins (tariff publications) + other costs.
c. VAT (value added tax) is 16 percent of the applied tariff only for fuel oil used in electricity.

Table B.24 Colombia—Low Tension

		Unit[a]	2008	2009	2010	2011	2012	2013
A.	Supply cost[b]	$/kWh	0.144	0.139	0.170	0.194	0.193	0.186
B1.	Taxes, excluding transfers (VAT)[c]	$/kWh	0.000	0.000	0.000	0.000	0.000	0.000
B2.	Taxes (counterfactual): standard VAT	$/kWh	0.023	0.022	0.030	0.031	0.031	0.030
C.	Total reference benchmark price							
C1	Total reference price without taxes	$/kWh	0.144	0.139	0.170	0.194	0.193	0.186
C2	Total reference price at counterfactual tax rates	$/kWh	0.167	0.161	0.200	0.226	0.224	0.217
D.	End-user price							
D1	Applied tariff, excluding all taxes	$/kWh	0.144	0.137	0.187	0.197	0.196	0.188
D2	Applied tariff, including all taxes	$/kWh	0.144	0.137	0.187	0.197	0.196	0.188
E.	Price gap							
E1	Price gap, excluding all taxes (C1 − D1)	$/kWh	0.000	0.003	−0.017	−0.003	−0.003	−0.002
E2	Price gap, with fuel at counterfactual tax rates (C2 − D2)	$/kWh	0.023	0.025	0.013	0.029	0.028	0.028
F.	Consumption							
F1	Volume	1,000 GWh	26.67	27.12	25.93	28.26	25.91	30.28

table continues next page

Table B.24 Colombia—Low Tension *(continued)*

	Unit[a]	2008	2009	2010	2011	2012	2013
G.	**Fiscal cost**							
G1	Fiscal cost, excluding all taxes (E1 * F1)	$, millions	0.00	72.01	(446.97)	(73.30)	(79.90)	(52.62)
G2	Fiscal cost, with fuel at counterfactual tax rates (E2 * F1)	$, millions	613.63	664.87	330.72	816.64	731.71	859.20
G3	Fiscal expenditures (F1 * (E2 – E1))	$, millions	613.63	592.86	777.69	889.94	811.61	911.83

Source: Codensa (distributor).
a. All prices and costs are expressed in current U.S. dollars. GWh = gigawatt-hour; kWh = kilowatt-hour.
b. Generation cost (average cost of electricity) + transmission and distribution margins (tariff publications) + other costs.
c. VAT (value added tax) is 16 percent of the applied tariff only for fuel oil used in electricity.

Table B.25 Colombia—Medium Tension

		Unit[a]	2008	2009	2010	2011	2012	2013
A.	**Supply cost[b]**	$/kWh	0.144	0.139	0.170	0.194	0.193	0.186
B1.	**Taxes, excluding transfers (VAT)[c]**	$/kWh	0.000	0.000	0.000	0.000	0.000	0.000
B2.	**Taxes (counterfactual): standard VAT**	$/kWh	0.019	0.018	0.021	0.025	0.024	0.023
C.	**Total reference benchmark price**							
C1	Total reference price without taxes	$/kWh	0.144	0.139	0.170	0.194	0.193	0.186
C2	Total reference price at counterfactual tax rates	$/kWh	0.163	0.158	0.191	0.219	0.217	0.210
D.	**End-user price**							
D1	Applied tariff, excluding all taxes	$/kWh	0.119	0.114	0.131	0.154	0.152	0.146
D2	Applied tariff, including all taxes	$/kWh	0.119	0.114	0.131	0.154	0.152	0.146
E.	**Price gap**							
E1	Price gap, excluding all taxes (C1 – D1)	$/kWh	0.025	0.025	0.039	0.040	0.041	0.040
E2	Price gap, with fuel at counterfactual tax rates (C2 – D2)	$/kWh	0.044	0.043	0.060	0.065	0.065	0.064
F.	**Consumption**							
F1	Volume	1,000 GWh	7.15	7.72	7.57	7.24	7.38	9.78
G.	**Fiscal cost**							
G1	Fiscal cost, excluding all taxes (E1 * F1)	$, millions	178.52	193.64	293.71	289.24	299.24	392.06
G2	Fiscal cost, with fuel at counterfactual tax rates (E2 * F1)	$, millions	314.49	334.63	453.03	468.00	478.84	621.11
G3	Fiscal expenditures (F1 * (E2 – E1))	$, millions	135.97	140.99	159.32	178.76	179.59	229.05

Source: Codensa (distributor).
a. All prices and costs are expressed in current U.S. dollars. GWh = gigawatt-hour; kWh = kilowatt-hour.
b. Generation cost (average cost of electricity) + transmission and distribution margins (tariff publications) + other costs.
c. VAT (value added tax) is 16 percent of the applied tariff only for fuel oil used in electricity.

Table B.26 Colombia—High Tension

		Unit[a]	2008	2009	2010	2011	2012	2013
A.	**Supply cost[b]**	$/kWh	0.144	0.139	0.170	0.194	0.193	0.186
B1.	**Taxes, excluding transfers (VAT)[c]**	$/kWh	0.000	0.000	0.000	0.000	0.000	0.000
B2.	**Taxes (counterfactual): standard VAT**	$/kWh	0.017	0.016	0.020	0.023	0.023	0.022
C.	**Total reference benchmark price**							
C1	Total reference price without taxes	$/kWh	0.144	0.139	0.170	0.194	0.193	0.186

table continues next page

Table B.26 Colombia—High Tension (continued)

		Unit[a]	2008	2009	2010	2011	2012	2013
C2	Total reference price at counterfactual tax rates	$/kWh	0.161	0.156	0.190	0.217	0.216	0.208
D.	End-user price							
D1	Applied tariff, excluding all taxes	$/kWh	0.109	0.103	0.123	0.145	0.143	0.137
D2	Applied tariff, including all taxes	$/kWh	0.109	0.103	0.123	0.145	0.143	0.137
E.	Price gap							
E1	Price gap, excluding all taxes (C1 − D1)	$/kWh	0.035	0.037	0.047	0.049	0.050	0.050
E2	Price gap, with fuel at counterfactual tax rates (C2 − D2)	$/kWh	0.052	0.053	0.067	0.073	0.072	0.072
F.	Consumption							
F1	Volume	1,000 GWh	6.68	6.42	5.88	5.13	4.83	6.14
G.	Fiscal cost							
G1	Fiscal cost, excluding all taxes (E1 * F1)	$, millions	231.14	234.60	278.85	253.70	239.29	305.06
G2	Fiscal cost, with fuel at counterfactual tax rates (E2 * F1)	$, millions	347.92	340.18	394.40	372.67	349.80	439.39
G3	Fiscal expenditures (F1 * (E2 − E1))	$, millions	116.78	105.58	115.55	118.96	110.52	134.32

Source: Codensa (distributor).

a. All prices and costs are expressed in current U.S. dollars. GWh = gigawatt-hour; kWh = kilowatt-hour.

b. Generation cost (average cost of electricity) + transmission and distribution margins (tariff publications) + other costs.

c. VAT (value added tax) is 16 percent of the applied tariff only for fuel oil used in electricity.

Table B.27 Colombia—Regional Subtransmission

		Unit[a]	2008	2009	2010	2011	2012	2013
A.	**Supply cost**[b]	$/kWh	0.144	0.139	0.170	0.194	0.193	0.186
B1.	**Taxes, excluding transfers (VAT)**[c]	$/kWh	0.000	0.000	0.000	0.000	0.000	0.000
B2.	**Taxes (counterfactual): standard VAT**	$/kWh	0.014	0.014	0.017	0.020	0.019	0.019
C.	**Total reference benchmark price**							
C1	Total reference price without taxes	$/kWh	0.144	0.139	0.170	0.194	0.193	0.186
C2	Total reference price at counterfactual tax rates	$/kWh	0.158	0.153	0.187	0.214	0.212	0.205
D.	**End-user price**							
D1	Applied tariff, excluding all taxes	$/kWh	0.090	0.088	0.106	0.123	0.119	0.116
D2	Applied tariff, including all taxes	$/kWh	0.090	0.088	0.106	0.123	0.119	0.116
E.	**Price gap**							
E1	Price gap, excluding all taxes (C1 − D1)	$/kWh	0.054	0.051	0.065	0.071	0.074	0.070
E2	Price gap, with fuel at counterfactual tax rates (C2 − D2)	$/kWh	0.069	0.065	0.081	0.091	0.093	0.089
F.	**Consumption**							
F1	Volume	1,000 GWh	4.29	4.30	4.07	2.44	2.17	3.17
G.	**Fiscal cost**							
G1	Fiscal cost, excluding all taxes (E1 * F1)	$, millions	232.31	219.89	262.60	172.48	160.44	222.53
G2	Fiscal cost, with fuel at counterfactual tax rates (E2 * F1)	$, millions	293.77	280.44	331.39	220.59	201.76	281.55
G3	Fiscal expenditures (F1 * (E2 − E1))	$, millions	61.46	60.55	68.79	48.11	41.32	59.02

Source: Codensa (distributor).

a. All prices and costs are expressed in current U.S. dollars. GWh = gigawatt-hour; kWh = kilowatt-hour.

b. Generation cost (average cost of electricity) + transmission and distribution margins (tariff publications) + other costs.

c. VAT (value added tax) is 16 percent of the applied tariff only for fuel oil used in electricity.

Energy Pricing Policies for Inclusive Growth in Latin America and the Caribbean
http://dx.doi.org/10.1596/978-1-4648-1111-1

Table B.28 Dominican Republic—Residential (Less Than 200 kWh)

		Unit[a]	2008	2009	2010	2011	2012	2013
A.	Supply cost[b]	$/kWh	0.250	0.200	0.222	0.267	0.279	0.260
B1.	Taxes, excluding transfers (VAT)[c]	$/kWh	0.000	0.000	0.000	0.000	0.000	0.000
B2.	Taxes (counterfactual): standard VAT	$/kWh	0.014	0.015	0.016	0.018	0.020	0.019
C.	Total reference benchmark price							
C1	Total reference price without taxes	$/kWh	0.250	0.200	0.222	0.267	0.279	0.260
C2	Total reference price at counterfactual tax rates	$/kWh	0.264	0.215	0.238	0.285	0.299	0.279
D.	End-user price							
D1	Applied tariff, excluding all taxes	$/kWh	0.089	0.095	0.100	0.112	0.113	0.106
D2	Applied tariff, including all taxes	$/kWh	0.089	0.095	0.100	0.112	0.113	0.106
E.	Price gap							
E1	Price gap, excluding all taxes (C1 − D1)	$/kWh	0.160	0.104	0.122	0.155	0.166	0.154
E2	Price gap, with fuel at counterfactual tax rates (C2 − D2)	$/kWh	0.175	0.120	0.138	0.173	0.187	0.173
F.	Consumption							
F1	Volume	1,000 GWh	1.47	1.66	1.73	1.73	2.05	2.20
G.	Fiscal cost							
G1	Fiscal cost, excluding all taxes (E1 * F1)	$, millions	235.76	173.44	210.50	268.15	341.45	338.98
G2	Fiscal cost, with fuel at counterfactual tax rates (E2 * F1)	$, millions	256.82	198.77	238.17	299.37	383.18	381.11
G3	Fiscal expenditures (F1 * (E2 − E1))	$, millions	21.05	25.32	27.67	31.22	41.73	42.13

Source: Coordinating Body of Interconnected System (OC).
a. All prices and costs are expressed in current U.S. dollars. GWh = gigawatt-hour; kWh = kilowatt-hour.
b. Generation cost + transmission and distribution margins: indexed tariff.
c. Electricity sector is exempted from VAT (value added tax).

Table B.29 Dominican Republic—Residential (between 200 kWh and 300 kWh)

		Unit[a]	2008	2009	2010	2011	2012	2013
A.	Supply cost[b]	$/kWh	0.250	0.200	0.222	0.267	0.279	0.260
B1.	Taxes, excluding transfers (VAT)[c]	$/kWh	0.000	0.000	0.000	0.000	0.000	0.000
B2.	Taxes (counterfactual): standard VAT	$/kWh	0.022	0.024	0.031	0.028	0.032	0.030
C.	Total reference benchmark price							
C1	Total reference price without taxes	$/kWh	0.250	0.200	0.222	0.267	0.279	0.260
C2	Total reference price at counterfactual tax rates	$/kWh	0.271	0.223	0.253	0.295	0.311	0.290
D.	End-user price							
D1	Applied tariff, excluding all taxes	$/kWh	0.135	0.148	0.195	0.177	0.177	0.167
D2	Applied tariff, including all taxes	$/kWh	0.135	0.148	0.195	0.177	0.177	0.167
E.	Price gap							
E1	Price gap, excluding all taxes (C1 − D1)	$/kWh	0.115	0.052	0.027	0.091	0.102	0.093
E2	Price gap, with fuel at counterfactual tax rates (C2 − D2)	$/kWh	0.136	0.076	0.058	0.119	0.134	0.123
F.	Consumption							
F1	Volume	1,000 GWh	0.22	0.24	0.26	0.25	0.29	0.32

table continues next page

Table B.29 Dominican Republic—Residential (between 200 kWh and 300 kWh) *(continued)*

		Unit[a]	2008	2009	2010	2011	2012	2013
G.	**Fiscal cost**							
G1	Fiscal cost, excluding all taxes (E1 * F1)	$, millions	25.08	12.64	6.94	22.53	30.05	29.51
G2	Fiscal cost, with fuel at counterfactual tax rates (E2 * F1)	$, millions	29.80	18.37	14.92	29.56	39.45	39.00
G3	Fiscal expenditures (F1 * (E2 − E1))	$, millions	4.72	5.73	7.98	7.03	9.40	9.49

Source: Coordinating Body of Interconnected System (OC).
a. All prices and costs are expressed in current U.S. dollars. GWh = gigawatt-hour; kWh = kilowatt-hour.
b. Generation cost + transmission and distribution margins: indexed tariff.
c. Electricity sector is exempted from VAT (value added tax).

Table B.30 Dominican Republic—Residential (between 300 kWh and 700 kWh)

		Unit[a]	2008	2009	2010	2011	2012	2013
A.	**Supply cost[b]**	$/kWh	0.308	0.246	0.273	0.329	0.344	0.320
B1.	**Taxes, excluding transfers (VAT)[c]**	$/kWh	0.000	0.000	0.000	0.000	0.000	0.000
B2.	**Taxes (counterfactual): standard VAT**	$/kWh	0.032	0.036	0.039	0.044	0.050	0.047
C.	**Total reference benchmark price**							
C1	Total reference price without taxes	$/kWh	0.308	0.246	0.273	0.329	0.344	0.320
C2	Total reference price at counterfactual tax rates	$/kWh	0.340	0.282	0.312	0.373	0.394	0.367
D.	**End-user price**							
D1	Applied tariff, excluding all taxes	$/kWh	0.201	0.226	0.245	0.275	0.276	0.260
D2	Applied tariff, including all taxes	$/kWh	0.201	0.226	0.245	0.275	0.276	0.260
E.	**Price gap**							
E1	Price gap, excluding all taxes (C1 − D1)	$/kWh	0.107	0.020	0.028	0.054	0.068	0.061
E2	Price gap, with fuel at counterfactual tax rates (C2 − D2)	$/kWh	0.139	0.056	0.067	0.098	0.118	0.107
F.	**Consumption**							
F1	Volume	1,000 GWh	0.19	0.21	0.22	0.20	0.24	0.26
G.	**Fiscal cost**							
G1	Fiscal cost, excluding all taxes (E1 * F1)	$, millions	20.10	4.23	6.20	10.96	16.43	15.72
G2	Fiscal cost, with fuel at counterfactual tax rates (E2 * F1)	$, millions	26.14	11.91	14.84	19.97	28.45	27.86
G3	Fiscal expenditures (F1 * (E2 − E1))	$, millions	6.04	7.68	8.64	9.01	12.03	12.14

Source: Coordinating Body of Interconnected System (OC).
a. All prices and costs are expressed in current U.S. dollars. GWh = gigawatt-hour; kWh = kilowatt-hour.
b. Generation cost + transmission and distribution margins: indexed tariff.
c. Electricity sector is exempted from VAT (value added tax).

Table B.31 Dominican Republic—Residential (between 700 kWh and 1,000 kWh)

		Unit[a]	2008	2009	2010	2011	2012	2013
A.	**Supply cost[b]**	$/kWh	0.308	0.246	0.273	0.329	0.344	0.320
B1.	**Taxes, excluding transfers (VAT)[c]**	$/kWh	0.000	0.000	0.000	0.000	0.000	0.000
B2.	**Taxes (counterfactual): standard VAT**	$/kWh	0.039	0.040	0.040	0.045	0.051	0.048
C.	**Total reference benchmark price**							
C1	Total reference price without taxes	$/kWh	0.308	0.246	0.273	0.329	0.344	0.320
C2	Total reference price at counterfactual tax rates	$/kWh	0.347	0.285	0.313	0.374	0.395	0.368

table continues next page

Energy Pricing Policies for Inclusive Growth in Latin America and the Caribbean
http://dx.doi.org/10.1596/978-1-4648-1111-1

Table B.31 Dominican Republic—Residential (between 700 kWh and 1,000 kWh) *(continued)*

		Unit[a]	2008	2009	2010	2011	2012	2013
D.	**End-user price**							
D1	Applied tariff, excluding all taxes	$/kWh	0.246	0.247	0.251	0.281	0.282	0.266
D2	Applied tariff, including all taxes	$/kWh	0.246	0.247	0.251	0.281	0.282	0.266
E.	**Price gap**							
E1	Price gap, excluding all taxes (C1 − D1)	$/kWh	0.062	−0.001	0.023	0.048	0.062	0.055
E2	Price gap, with fuel at counterfactual tax rates (C2 − D2)	$/kWh	0.101	0.038	0.063	0.093	0.113	0.103
F.	**Consumption**							
F1	Volume	1,000 GWh	0.47	0.49	0.45	0.40	0.47	0.50
G.	**Fiscal cost**							
G1	Fiscal cost, excluding all taxes (E1 * F1)	$, millions	28.84	(0.68)	10.10	18.81	28.90	27.50
G2	Fiscal cost, with fuel at counterfactual tax rates (E2 * F1)	$, millions	47.15	18.89	28.08	36.61	52.67	51.50
G3	Fiscal expenditures (F1 * (E2 − E1))	$, millions	18.31	19.57	17.98	17.80	23.77	24.00

Source: Coordinating Body of Interconnected System (OC).
a. All prices and costs are expressed in current U.S. dollars. GWh = gigawatt-hour; kWh = kilowatt-hour.
b. Generation cost + transmission and distribution margins: indexed tariff.
c. Electricity sector is exempted from VAT (value added tax).

Table B.32 Dominican Republic—Nonresidential (Less Than 200 kWh)

		Unit[a]	2008	2009	2010	2011	2012	2013
A.	**Supply cost[b]**	$/kWh	0.250	0.200	0.222	0.267	0.279	0.260
B1.	**Taxes, excluding transfers (VAT)[c]**	$/kWh	0.000	0.000	0.000	0.000	0.000	0.000
B2.	**Taxes (counterfactual): standard VAT**	$/kWh	0.020	0.021	0.022	0.024	0.027	0.026
C.	**Total reference benchmark price**							
C1	Total reference price without taxes	$/kWh	0.250	0.200	0.222	0.267	0.279	0.260
C2	Total reference price at counterfactual tax rates	$/kWh	0.269	0.220	0.243	0.291	0.306	0.286
D.	**End-user price**							
D1	Applied tariff, excluding all taxes	$/kWh	0.123	0.129	0.135	0.151	0.152	0.143
D2	Applied tariff, including all taxes	$/kWh	0.123	0.129	0.135	0.151	0.152	0.143
E.	**Price gap**							
E1	Price gap, excluding all taxes (C1 − D1)	$/kWh	0.127	0.070	0.087	0.116	0.127	0.117
E2	Price gap, with fuel at counterfactual tax rates (C2 − D2)	$/kWh	0.147	0.091	0.108	0.140	0.155	0.143
F.	**Consumption**							
F1	Volume	1,000 GWh	0.16	0.18	0.19	0.20	0.23	0.25
G.	**Fiscal cost**							
G1	Fiscal cost, excluding all taxes (E1 * F1)	$, millions	19.87	12.80	16.51	22.92	29.90	29.53
G2	Fiscal cost, with fuel at counterfactual tax rates (E2 * F1)	$, millions	22.95	16.57	20.61	27.73	36.31	36.00
G3	Fiscal expenditures (F1 * (E2 − E1))	$, millions	3.07	3.77	4.10	4.80	6.41	6.48

Source: Coordinating Body of Interconnected System (OC).
a. All prices and costs are expressed in current U.S. dollars. GWh = gigawatt-hour; kWh = kilowatt-hour.
b. Generation cost + transmission and distribution margins: indexed tariff.
c. Electricity sector is exempted from VAT (value added tax).

Table B.33 Dominican Republic—Nonresidential (between 200 kWh and 300 kWh)

		Unit[a]	2008	2009	2010	2011	2012	2013
A.	Supply cost[b]	$/kWh	0.250	0.200	0.222	0.267	0.284	0.260
B1.	Taxes, excluding transfers (VAT)[c]	$/kWh	0.000	0.000	0.000	0.000	0.000	0.000
B2.	Taxes (counterfactual): standard VAT	$/kWh	0.024	0.028	0.031	0.035	0.051	0.047
C.	Total reference benchmark price							
C1	Total reference price without taxes	$/kWh	0.250	0.200	0.222	0.267	0.284	0.260
C2	Total reference price at counterfactual tax rates	$/kWh	0.273	0.228	0.253	0.302	0.336	0.307
D.	End-user price							
D1	Applied tariff, excluding all taxes	$/kWh	0.147	0.174	0.195	0.218	0.284	0.260
D2	Applied tariff, including all taxes	$/kWh	0.147	0.174	0.195	0.218	0.284	0.260
E.	Price gap							
E1	Price gap, excluding all taxes (C1 − D1)	$/kWh	0.103	0.025	0.027	0.049	0.000	0.000
E2	Price gap, with fuel at counterfactual tax rates (C2 − D2)	$/kWh	0.126	0.053	0.058	0.084	0.051	0.047
F.	Consumption							
F1	Volume	1,000 GWh	0.04	0.05	0.05	0.05	0.06	0.06
G.	Fiscal cost							
G1	Fiscal cost, excluding all taxes (E1 * F1)	$, millions	4.08	1.16	1.29	2.35	0.00	0.00
G2	Fiscal cost, with fuel at counterfactual tax rates (E2 * F1)	$, millions	5.01	2.46	2.77	4.03	2.92	2.87
G3	Fiscal expenditures (F1 * (E2 − E1))	$, millions	0.93	1.29	1.48	1.69	2.92	2.87

Source: Coordinating Body of Interconnected System (OC).
a. All prices and costs are expressed in current U.S. dollars. GWh = gigawatt-hour; kWh = kilowatt-hour.
b. Generation cost + transmission and distribution margins: indexed tariff.
c. Electricity sector is exempted from VAT (value added tax).

Table B.34 Dominican Republic—Nonresidential (between 300 kWh and 700 kWh)

		Unit[a]	2008	2009	2010	2011	2012	2013
A.	Supply cost[b]	$/kWh	0.308	0.246	0.273	0.329	0.343	0.320
B1.	Taxes, excluding transfers (VAT)[c]	$/kWh	0.000	0.000	0.000	0.000	0.000	0.000
B2.	Taxes (counterfactual): standard VAT	$/kWh	0.036	0.039	0.041	0.046	0.052	0.049
C.	Total reference benchmark price							
C1	Total reference price without taxes	$/kWh	0.308	0.246	0.273	0.329	0.343	0.320
C2	Total reference price at counterfactual tax rates	$/kWh	0.344	0.285	0.314	0.375	0.394	0.369
D.	End-user price							
D1	Applied tariff, excluding all taxes	$/kWh	0.227	0.242	0.255	0.286	0.287	0.270
D2	Applied tariff, including all taxes	$/kWh	0.227	0.242	0.255	0.286	0.287	0.270
E.	Price gap							
E1	Price gap, excluding all taxes (C1 − D1)	$/kWh	0.081	0.004	0.018	0.043	0.055	0.050
E2	Price gap, with fuel at counterfactual tax rates (C2 − D2)	$/kWh	0.117	0.043	0.059	0.088	0.107	0.099
F.	Consumption							
F1	Volume	1,000 GWh	0.05	0.06	0.06	0.06	0.07	0.07

table continues next page

Table B.34 Dominican Republic—Nonresidential (between 300 kWh and 700 kWh) *(continued)*

		Unit[a]	2008	2009	2010	2011	2012	2013
G.	**Fiscal cost**							
G1	Fiscal cost, excluding all taxes (E1 * F1)	$, millions	3.87	0.22	1.07	2.49	3.83	3.70
G2	Fiscal cost, with fuel at counterfactual tax rates (E2 * F1)	$, millions	5.60	2.41	3.45	5.16	7.40	7.31
G3	Fiscal expenditures (F1 * (E2 − E1))	$, millions	1.73	2.19	2.39	2.67	3.57	3.60

Source: Coordinating Body of Interconnected System (OC).
a. All prices and costs are expressed in current U.S. dollars. GWh = gigawatt-hour; kWh = kilowatt-hour.
b. Generation cost + transmission and distribution margins: indexed tariff.
c. Electricity sector is exempted from VAT (value added tax).

Table B.35 Dominican Republic—Nonresidential (between 700 kWh and 1,000 kWh)

		Unit[a]	2008	2009	2010	2011	2012	2013
A.	**Supply cost[b]**	$/kWh	0.308	0.246	0.273	0.329	0.343	0.320
B1.	**Taxes, excluding transfers (VAT)[c]**	$/kWh	0.000	0.000	0.000	0.000	0.000	0.000
B2.	**Taxes (counterfactual): standard VAT**	$/kWh	0.041	0.041	0.041	0.046	0.052	0.049
C.	**Total reference benchmark price**							
C1	Total reference price without taxes	$/kWh	0.308	0.246	0.273	0.329	0.343	0.320
C2	Total reference price at counterfactual tax rates	$/kWh	0.349	0.287	0.314	0.375	0.396	0.369
D.	**End-user price**							
D1	Applied tariff, excluding all taxes	$/kWh	0.258	0.257	0.258	0.289	0.290	0.273
D2	Applied tariff, including all taxes	$/kWh	0.258	0.257	0.258	0.289	0.290	0.273
E.	**Price gap**							
E1	Price gap, excluding all taxes (C1 − D1)	$/kWh	0.049	−0.011	0.016	0.040	0.053	0.047
E2	Price gap, with fuel at counterfactual tax rates (C2 − D2)	$/kWh	0.091	0.030	0.057	0.086	0.105	0.096
F.	**Consumption**							
F1	Volume	1,000 GWh	0.49	0.48	0.50	0.49	0.58	0.63
G.	**Fiscal cost**							
G1	Fiscal cost, excluding all taxes (E1 * F1)	$, millions	24.12	(5.17)	7.80	19.56	31.05	29.63
G2	Fiscal cost, with fuel at counterfactual tax rates (E2 * F1)	$, millions	44.26	14.74	28.49	42.42	61.58	60.45
G3	Fiscal expenditures (F1 * (E2 − E1))	$, millions	20.14	19.91	20.69	22.86	30.53	30.82

Source: Coordinating Body of Interconnected System (OC).
a. All prices and costs are expressed in current U.S. dollars. GWh = gigawatt-hour; kWh = kilowatt-hour.
b. Generation cost + transmission and distribution margins: indexed tariff.
c. Electricity sector is exempted from VAT (value added tax).

Table B.36 Dominican Republic—BTD (Low Voltage with Capacity)

		Unit[a]	2008	2009	2010	2011	2012	2013
A.	**Supply cost[b]**	$/kWh	0.200	0.159	0.177	0.214	0.223	0.208
B1.	**Taxes, excluding transfers (VAT)[c]**	$/kWh	0.000	0.000	0.000	0.000	0.000	0.000
B2.	**Taxes (counterfactual): standard VAT**	$/kWh	0.025	0.026	0.027	0.030	0.034	0.032
C.	**Total reference benchmark price**							
C1	Total reference price without taxes	$/kWh	0.200	0.159	0.177	0.214	0.223	0.208
C2	Total reference price at counterfactual tax rates	$/kWh	0.225	0.185	0.204	0.244	0.257	0.240

table continues next page

Table B.36 Dominican Republic—BTD (Low Voltage with Capacity) *(continued)*

	Unit[a]	2008	2009	2010	2011	2012	2013
D.	**End-user price**						
D1	Applied tariff, excluding all taxes	$/kWh	0.159	0.162	0.166	0.187	0.187	0.176
D2	Applied tariff, including all taxes	$/kWh	0.159	0.162	0.166	0.187	0.187	0.176
E.	**Price gap**						
E1	Price gap, excluding all taxes (C1 − D1)	$/kWh	0.041	−0.003	0.011	0.027	0.036	0.032
E2	Price gap, with fuel at counterfactual tax rates (C2 − D2)	$/kWh	0.066	0.023	0.038	0.057	0.070	0.063
F.	**Consumption**						
F1	Volume	1,000 GWh	0.23	0.29	0.28	0.29	0.34	0.37
G.	**Fiscal cost**						
G1	Fiscal cost, excluding all taxes (E1 * F1)	$, millions	9.54	(0.79)	3.09	7.76	12.22	11.55
G2	Fiscal cost, with fuel at counterfactual tax rates (E2 * F1)	$, millions	15.49	6.64	10.56	16.36	23.71	23.16
G3	Fiscal expenditures (F1 * (E2 − E1))	$, millions	5.95	7.43	7.47	8.60	11.49	11.61

Source: Coordinating Body of Interconnected System (OC).
a. All prices and costs are expressed in current U.S. dollars. GWh = gigawatt-hour; kWh = kilowatt-hour.
b. Generation cost + transmission and distribution margins: indexed tariff.
c. Electricity sector is exempted from VAT (value added tax).

Table B.37 Dominican Republic—BTH (Low Voltage with Time of Use)

	Unit[a]	2008	2009	2010	2011	2012	2013
A.	**Supply cost[b]**	$/kWh	0.196	0.157	0.174	0.210	0.604	0.204
B1.	**Taxes, excluding transfers (VAT)[c]**	$/kWh	0.000	0.000	0.000	0.000	0.000	0.000
B2.	**Taxes (counterfactual): standard VAT**	$/kWh	0.025	0.026	0.026	0.029	0.033	0.031
C.	**Total reference benchmark price**						
C1	Total reference price without taxes	$/kWh	0.196	0.157	0.174	0.210	0.604	0.204
C2	Total reference price at counterfactual tax rates	$/kWh	0.222	0.182	0.200	0.239	0.637	0.235
D.	**End-user price**						
D1	Applied tariff, excluding all taxes	$/kWh	0.159	0.160	0.164	0.184	0.185	0.174
D2	Applied tariff, including all taxes	$/kWh	0.159	0.160	0.164	0.184	0.185	0.174
E.	**Price gap**						
E1	Price gap, excluding all taxes (C1 − D1)	$/kWh	0.037	−0.003	0.010	0.026	0.419	0.030
E2	Price gap, with fuel at counterfactual tax rates (C2 − D2)	$/kWh	0.063	0.023	0.037	0.055	0.452	0.061
F.	**Consumption**						
F1	Volume	1,000 GWh	0.00	0.00	0.00	0.00	0.00	0.00
G.	**Fiscal cost**						
G1	Fiscal cost, excluding all taxes (E1 * F1)	$, millions	0.09	(0.01)	0.03	0.09	1.81	0.14
G2	Fiscal cost, with fuel at counterfactual tax rates (E2 * F1)	$, millions	0.15	0.06	0.11	0.20	1.95	0.28
G3	Fiscal expenditures (F1 * (E2 − E1))	$, millions	0.06	0.07	0.08	0.11	0.14	0.14

Source: Coordinating Body of Interconnected System (OC).
a. All prices and costs are expressed in current U.S. dollars. GWh = gigawatt-hour; kWh = kilowatt-hour.
b. Generation cost + transmission and distribution margins: indexed tariff.
c. Electricity sector is exempted from VAT (value added tax).

Table B.38 Dominican Republic—MTD1 (Medium Voltage with Capacity)

		Unit[a]	2008	2009	2010	2011	2012	2013
A.	Supply cost[b]	$/kWh	0.200	0.159	0.177	0.214	0.223	0.207
B1.	Taxes, excluding transfers (VAT)[c]	$/kWh	0.000	0.000	0.000	0.000	0.000	0.000
B2.	Taxes (counterfactual): standard VAT	$/kWh	0.027	0.028	0.028	0.032	0.036	0.034
C.	Total reference benchmark price							
C1	Total reference price without taxes	$/kWh	0.200	0.159	0.177	0.214	0.223	0.207
C2	Total reference price at counterfactual tax rates	$/kWh	0.227	0.187	0.205	0.245	0.259	0.241
D.	End-user price							
D1	Applied tariff, excluding all taxes	$/kWh	0.168	0.172	0.176	0.198	0.199	0.187
D2	Applied tariff, including all taxes	$/kWh	0.168	0.172	0.176	0.198	0.199	0.187
E.	Price gap							
E1	Price gap, excluding all taxes (C1 − D1)	$/kWh	0.032	−0.012	0.001	0.016	0.025	0.020
E2	Price gap, with fuel at counterfactual tax rates (C2 − D2)	$/kWh	0.058	0.015	0.029	0.047	0.060	0.054
F.	Consumption							
F1	Volume	1,000 GWh	1.41	1.60	1.64	1.60	1.90	2.04
G.	Fiscal cost							
G1	Fiscal cost, excluding all taxes (E1 * F1)	$, millions	44.36	(19.99)	1.76	25.15	46.79	41.40
G2	Fiscal cost, with fuel at counterfactual tax rates (E2 * F1)	$, millions	82.17	24.15	47.90	75.92	114.60	109.87
G3	Fiscal expenditures (F1 * (E2 − E1))	$, millions	37.81	44.15	46.14	50.77	67.81	68.47

Source: Coordinating Body of Interconnected System (OC).
a. All prices and costs are expressed in current U.S. dollars. GWh = gigawatt-hour; kWh = kilowatt-hour.
b. Generation cost + transmission and distribution margins: indexed tariff.
c. Electricity sector is exempted from VAT (value added tax).

Table B.39 Dominican Republic—MTD2 (Medium Voltage, Large)

		Unit[a]	2008	2009	2010	2011	2012	2013
A.	Supply cost[b]	$/kWh	0.200	0.159	0.177	0.214	0.223	0.207
B1.	Taxes, excluding transfers (VAT)[c]	$/kWh	0.000	0.000	0.000	0.000	0.000	0.000
B2.	Taxes (counterfactual): standard VAT	$/kWh	0.025	0.026	0.027	0.030	0.034	0.032
C.	Total reference benchmark price							
C1	Total reference price without taxes	$/kWh	0.200	0.159	0.177	0.214	0.223	0.207
C2	Total reference price at counterfactual tax rates	$/kWh	0.225	0.186	0.204	0.244	0.257	0.239
D.	End-user price							
D1	Applied tariff, excluding all taxes	$/kWh	0.159	0.163	0.166	0.187	0.188	0.177
D2	Applied tariff, including all taxes	$/kWh	0.159	0.163	0.166	0.187	0.188	0.177
E.	Price gap							
E1	Price gap, excluding all taxes (C1 − D1)	$/kWh	0.041	−0.003	0.011	0.027	0.036	0.031
E2	Price gap, with fuel at counterfactual tax rates (C2 − D2)	$/kWh	0.066	0.023	0.037	0.057	0.069	0.062
F.	Consumption							
F1	Volume	1,000 GWh	0.78	0.91	0.93	0.96	1.14	1.22

table continues next page

Table B.39 Dominican Republic—MTD2 (Medium Voltage, Large) *(continued)*

		Unit[a]	2008	2009	2010	2011	2012	2013
G.	**Fiscal cost**							
G1	Fiscal cost, excluding all taxes (E1 * F1)	$, millions	31.68	(2.77)	10.00	25.69	40.58	37.46
G2	Fiscal cost, with fuel at counterfactual tax rates (E2 * F1)	$, millions	51.45	20.91	34.86	54.51	79.09	76.34
G3	Fiscal expenditures (F1 * (E2 − E1))	$, millions	19.77	23.69	24.85	28.82	38.50	38.88

Source: Coordinating Body of Interconnected System (OC).
a. All prices and costs are expressed in current U.S. dollars. GWh = gigawatt-hour; kWh = kilowatt-hour.
b. Generation cost + transmission and distribution margins: indexed tariff.
c. Electricity sector is exempted from VAT (value added tax).

Table B.40 Dominican Republic—MTH (Medium Voltage, Time of Use)

		Unit[a]	2008	2009	2010	2011	2012	2013
A.	**Supply cost[b]**	$/kWh	0.196	0.157	0.174	0.210	0.219	0.204
B1.	**Taxes, excluding transfers (VAT)[c]**	$/kWh	0.000	0.000	0.000	0.000	0.000	0.000
B2.	**Taxes (counterfactual): standard VAT**	$/kWh	0.025	0.026	0.026	0.029	0.033	0.031
C.	**Total reference benchmark price**							
C1	Total reference price without taxes	$/kWh	0.196	0.157	0.174	0.210	0.219	0.204
C2	Total reference price at counterfactual tax rates	$/kWh	0.221	0.182	0.200	0.239	0.253	0.235
D.	**End-user price**							
D1	Applied tariff, excluding all taxes	$/kWh	0.156	0.160	0.164	0.184	0.185	0.174
D2	Applied tariff, including all taxes	$/kWh	0.156	0.160	0.164	0.184	0.185	0.174
E.	**Price gap**							
E1	Price gap, excluding all taxes (C1 − D1)	$/kWh	0.040	−0.003	0.010	0.026	0.035	0.030
E2	Price gap, with fuel at counterfactual tax rates (C2 − D2)	$/kWh	0.065	0.023	0.037	0.055	0.068	0.061
F.	**Consumption**							
F1	Volume	1,000 GWh	0.03	0.04	0.04	0.05	0.06	0.06
G.	**Fiscal cost**							
G1	Fiscal cost, excluding all taxes (E1 * F1)	$, millions	1.29	(0.12)	0.46	1.25	2.01	1.87
G2	Fiscal cost, with fuel at counterfactual tax rates (E2 * F1)	$, millions	2.11	0.85	1.61	2.68	3.92	3.80
G3	Fiscal expenditures (F1 * (E2 − E1))	$, millions	0.81	0.97	1.15	1.44	1.92	1.93

Source: Coordinating Body of Interconnected System (OC).
a. All prices and costs are expressed in current U.S. dollars. GWh = gigawatt-hour; kWh = kilowatt-hour.
b. Generation cost + transmission and distribution margins: indexed tariff.
c. Electricity sector is exempted from VAT (value added tax).

Table B.41 El Salvador—Residential (Less Than 99 kWh)

		Unit[a]	2008	2009	2010	2011	2012
A.	**Supply cost[b]**	$/kWh	0.196	0.198	0.200	0.233	0.253
B1.	**Taxes, excluding transfers (VAT)[c]**	$/kWh	0.024	0.031	0.032	0.035	0.039
B2.	**Taxes (counterfactual): standard VAT**	$/kWh	0.024	0.031	0.032	0.035	0.039
C.	**Total reference benchmark price**						
C1	Total reference price without taxes	$/kWh	0.196	0.198	0.200	0.233	0.253
C2	Total reference price at counterfactual tax rates	$/kWh	0.220	0.229	0.232	0.269	0.292

table continues next page

Table B.41 El Salvador—Residential (Less Than 99 kWh) *(continued)*

		Unit[a]	2008	2009	2010	2011	2012
D.	**End-user price**						
D1	Applied tariff, excluding all taxes	$/kWh	0.107	0.114	0.113	0.115	0.120
D2	Applied tariff, including all taxes	$/kWh	0.131	0.145	0.144	0.150	0.159
E.	**Price gap**						
E1	Price gap, excluding all taxes (C1 − D1)	$/kWh	0.089	0.084	0.088	0.119	0.133
E2	Price gap, with fuel at counterfactual tax rates (C2 − D2)	$/kWh	0.089	0.084	0.088	0.119	0.133
F.	**Consumption**						
F1	Volume	1,000 GWh	0.52	0.55	0.58	0.62	0.64
G.	**Fiscal cost**						
G1	Fiscal cost, excluding all taxes (E1 * F1)	$, millions	46.02	45.81	50.61	73.23	84.89
G2	Fiscal cost, with fuel at counterfactual tax rates (E2 * F1)	$, millions	46.02	45.81	50.61	73.23	84.89
G3	Fiscal expenditures (F1 * (E2 − E1))	$, millions	0.00	0.00	0.00	0.00	0.00

Sources: Superintendency of Electricity and Telecommunications (regulator) and National Energy Council.
a. All prices and costs are expressed in current U.S. dollars. GWh = gigawatt-hour; kWh = kilowatt-hour.
b. Generation cost (short-term operational marginal cost) + transmission and distribution margins (efficient tariff schedules).
c. VAT (value added tax) is 13 percent of the applied tariff.

Table B.42 El Salvador—Residential (between 100 kWh and 199 kWh)

		Unit[a]	2008	2009	2010	2011	2012
A.	**Supply cost**[b]	$/kWh	0.180	0.181	0.184	0.223	0.236
B1.	**Taxes, excluding transfers (VAT)**[c]	$/kWh	0.019	0.025	0.026	0.029	0.032
B2.	**Taxes (counterfactual): standard VAT**	$/kWh	0.019	0.025	0.026	0.029	0.032
C.	**Total reference benchmark price**						
C1	Total reference price without taxes	$/kWh	0.180	0.181	0.184	0.223	0.236
C2	Total reference price at counterfactual tax rates	$/kWh	0.200	0.206	0.209	0.252	0.268
D.	**End-user price**						
D1	Applied tariff, excluding all taxes	$/kWh	0.152	0.199	0.189	0.204	0.204
D2	Applied tariff, including all taxes	$/kWh	0.172	0.224	0.215	0.233	0.236
E.	**Price gap**						
E1	Price gap, excluding all taxes (C1 − D1)	$/kWh	0.028	−0.018	−0.005	0.019	0.032
E2	Price gap, with fuel at counterfactual tax rates (C2 − D2)	$/kWh	0.028	−0.018	−0.005	0.019	0.032
F.	**Consumption**						
F1	Volume	1,000 GWh	0.51	0.50	0.49	0.48	0.52
G.	**Fiscal cost**						
G1	Fiscal cost, excluding all taxes (E1 * F1)	$, millions	14.49	(8.92)	(2.47)	8.99	16.55
G2	Fiscal cost, with fuel at counterfactual tax rates (E2 * F1)	$, millions	14.49	(8.92)	(2.47)	8.99	16.55
G3	Fiscal expenditures (F1 * (E2 − E1))	$, millions	0.00	0.00	0.00	0.00	0.00

Sources: Superintendency of Electricity and Telecommunications (regulator) and National Energy Council.
a. All prices and costs are expressed in current U.S. dollars. GWh = gigawatt-hour; kWh = kilowatt-hour.
b. Generation cost (short-term operational marginal cost) + transmission and distribution margins (efficient tariff schedules).
c. VAT (value added tax) is 13 percent of the applied tariff.

Table B.43 El Salvador—Residential (More Than 200 kWh)

		Unit[a]	2008	2009	2010	2011	2012
A.	Supply cost[b]	$/kWh	0.183	0.183	0.186	0.225	0.238
B1.	Taxes, excluding transfers (VAT)[c]	$/kWh	0.020	0.025	0.026	0.029	0.032
B2.	Taxes (counterfactual): standard VAT	$/kWh	0.020	0.025	0.026	0.029	0.032
C.	Total reference benchmark price						
C1	Total reference price without taxes	$/kWh	0.183	0.183	0.186	0.225	0.238
C2	Total reference price at counterfactual tax rates	$/kWh	0.202	0.208	0.212	0.254	0.270
D.	End-user price						
D1	Applied tariff, excluding all taxes	$/kWh	0.154	0.201	0.197	0.225	0.257
D2	Applied tariff, including all taxes	$/kWh	0.174	0.226	0.223	0.254	0.289
E.	Price gap						
E1	Price gap, excluding all taxes (C1 − D1)	$/kWh	0.029	−0.018	−0.012	0.000	−0.019
E2	Price gap, with fuel at counterfactual tax rates (C2 − D2)	$/kWh	0.029	−0.018	−0.012	0.000	−0.019
F.	Consumption						
F1	Volume	1,000 GWh	0.61	0.58	0.55	0.51	0.50
G.	Fiscal cost						
G1	Fiscal cost, excluding all taxes (E1 * F1)	$, millions	17.53	(10.45)	(6.42)	0.16	(9.26)
G2	Fiscal cost, with fuel at counterfactual tax rates (E2 * F1)	$, millions	17.53	(10.45)	(6.42)	0.16	(9.26)
G3	Fiscal expenditures (F1 * (E2 − E1))	$, millions	0.00	0.00	0.00	0.00	0.00

Sources: Superintendency of Electricity and Telecommunications (regulator) and National Energy Council.
a. All prices and costs are expressed in current U.S. dollars. GWh = gigawatt-hour; kWh = kilowatt-hour.
b. Generation cost (short-term operational marginal cost) + transmission and distribution margins (efficient tariff schedules).
c. VAT (value added tax) is 13 percent of the applied tariff.

Table B.44 El Salvador—Small Demands, General

		Unit[a]	2008	2009	2010	2011	2012
A.	Supply cost[b]	$/kWh	0.163	0.162	0.165	0.204	0.217
B1.	Taxes, excluding transfers (VAT)[c]	$/kWh	0.018	0.024	0.024	0.028	0.031
B2.	Taxes (counterfactual): standard VAT	$/kWh	0.018	0.024	0.024	0.028	0.031
C.	Total reference benchmark price						
C1	Total reference price without taxes	$/kWh	0.163	0.162	0.165	0.204	0.217
C2	Total reference price at counterfactual tax rates	$/kWh	0.181	0.186	0.189	0.232	0.247
D.	End-user price						
D1	Applied tariff, excluding all taxes	$/kWh	0.140	0.185	0.187	0.213	0.237
D2	Applied tariff, including all taxes	$/kWh	0.159	0.208	0.212	0.240	0.268
E.	Price gap						
E1	Price gap, excluding all taxes (C1 − D1)	$/kWh	0.022	−0.022	−0.022	−0.009	−0.021
E2	Price gap, with fuel at counterfactual tax rates (C2 − D2)	$/kWh	0.022	−0.022	−0.022	−0.009	−0.021
F.	Consumption						
F1	Volume	1,000 GWh	0.42	0.42	0.41	0.41	0.43

table continues next page

Table B.44 El Salvador—Small Demands, General *(continued)*

		Unit[a]	2008	2009	2010	2011	2012
G.	**Fiscal cost**						
G1	Fiscal cost, excluding all taxes (E1 * F1)	$, millions	9.26	(9.22)	(9.25)	(3.52)	(8.88)
G2	Fiscal cost, with fuel at counterfactual tax rates (E2 * F1)	$, millions	9.26	(9.22)	(9.25)	(3.52)	(8.88)
G3	Fiscal expenditures (F1 * (E2 – E1))	$, millions	0.00	0.00	0.00	0.00	0.00

Sources: Superintendency of Electricity and Telecommunications (regulator) and National Energy Council.
a. All prices and costs are expressed in current U.S. dollars. GWh = gigawatt-hour; kWh = kilowatt-hour.
b. Generation cost (short-term operational marginal cost) + transmission and distribution margins (efficient tariff schedules).
c. VAT (value added tax) is 13 percent of the applied tariff.

Table B.45 El Salvador—Small Demands, Public Lighting

		Unit[a]	2008	2009	2010	2011	2012
A.	**Supply cost**[b]	$/kWh	0.171	0.172	0.174	0.213	0.226
B1.	**Taxes, excluding transfers (VAT)**[c]	$/kWh	0.018	0.024	0.024	0.028	0.031
B2.	**Taxes (counterfactual): standard VAT**	$/kWh	0.018	0.024	0.024	0.028	0.031
C.	**Total reference benchmark price**						
C1	Total reference price without taxes	$/kWh	0.171	0.172	0.174	0.213	0.226
C2	Total reference price at counterfactual tax rates	$/kWh	0.190	0.195	0.198	0.241	0.257
D.	**End-user price**						
D1	Applied tariff, excluding all taxes	$/kWh	0.140	0.182	0.186	0.214	0.241
D2	Applied tariff, including all taxes	$/kWh	0.159	0.206	0.210	0.241	0.272
E.	**Price gap**						
E1	Price gap, excluding all taxes (C1 – D1)	$/kWh	0.031	−0.011	−0.012	−0.001	−0.015
E2	Price gap, with fuel at counterfactual tax rates (C2 – D2)	$/kWh	0.031	−0.011	−0.012	−0.001	−0.015
F.	**Consumption**						
F1	Volume	1,000 GWh	0.11	0.12	0.13	0.13	0.13
G.	**Fiscal cost**						
G1	Fiscal cost, excluding all taxes (E1 * F1)	$, millions	3.45	(1.32)	(1.53)	(0.10)	(2.06)
G2	Fiscal cost, with fuel at counterfactual tax rates (E2 * F1)	$, millions	3.45	(1.32)	(1.53)	(0.10)	(2.06)
G3	Fiscal expenditures (F1 * (E2 – E1))	$, millions	0.00	0.00	0.00	0.00	0.00

Sources: Superintendency of Electricity and Telecommunications (regulator) and National Energy Council.
a. All prices and costs are expressed in current U.S. dollars. GWh = gigawatt-hour; kWh = kilowatt-hour.
b. Generation cost (short-term operational marginal cost) + transmission and distribution margins (efficient tariff schedules).
c. VAT (value added tax) is 13 percent of the applied tariff.

Table B.46 El Salvador—Medium Demands

		Unit[a]	2008	2009	2010	2011	2012
A.	**Supply cost**[b]	$/kWh	0.162	0.161	0.164	0.203	0.226
B1.	**Taxes, excluding transfers (VAT)**[c]	$/kWh	0.021	0.026	0.026	0.030	0.033
B2.	**Taxes (counterfactual): standard VAT**	$/kWh	0.021	0.026	0.026	0.030	0.033
C.	**Total reference benchmark price**						
C1	Total reference price without taxes	$/kWh	0.162	0.161	0.164	0.203	0.226

table continues next page

Table B.46 El Salvador—Medium Demands (continued)

		Unit[a]	2008	2009	2010	2011	2012
C2	Total reference price at counterfactual tax rates	$/kWh	0.183	0.187	0.190	0.233	0.259
D.	**End-user price**						
D1	Applied tariff, excluding all taxes	$/kWh	0.161	0.203	0.203	0.231	0.254
D2	Applied tariff, including all taxes	$/kWh	0.182	0.229	0.229	0.261	0.287
E.	**Price gap**						
E1	Price gap, excluding all taxes (C1 − D1)	$/kWh	0.001	−0.042	−0.039	−0.028	−0.028
E2	Price gap, with fuel at counterfactual tax rates (C2 − D2)	$/kWh	0.001	−0.042	−0.039	−0.028	−0.028
F.	**Consumption**						
F1	Volume	1,000 GWh	0.38	0.38	0.38	0.38	0.40
G.	**Fiscal cost**						
G1	Fiscal cost, excluding all taxes (E1 * F1)	$, millions	0.36	(16.23)	(14.79)	(10.67)	(10.98)
G2	Fiscal cost, with fuel at counterfactual tax rates (E2 * F1)	$, millions	0.36	(16.23)	(14.79)	(10.67)	(10.98)
G3	Fiscal expenditures (F1 * (E2 − E1))	$, millions	0.00	0.00	0.00	0.00	0.00

Sources: Superintendency of Electricity and Telecommunications (regulator) and National Energy Council.
a. All prices and costs are expressed in current U.S. dollars. GWh = gigawatt-hour; kWh = kilowatt-hour.
b. Generation cost (short-term operational marginal cost) + transmission and distribution margins (efficient tariff schedules).
c. VAT (value added tax) is 13 percent of the applied tariff.

Table B.47 El Salvador—Large Demands

		Unit[a]	2008	2009	2010	2011	2012
A.	**Supply cost[b]**	$/kWh	0.162	0.161	0.164	0.203	0.226
B1.	**Taxes, excluding transfers (VAT)[c]**	$/kWh	0.020	0.025	0.025	0.028	0.031
B2.	**Taxes (counterfactual): standard VAT**	$/kWh	0.020	0.025	0.025	0.028	0.031
C.	**Total reference benchmark price**						
C1	Total reference price without taxes	$/kWh	0.162	0.161	0.164	0.203	0.226
C2	Total reference price at counterfactual tax rates	$/kWh	0.181	0.186	0.189	0.231	0.257
D.	**End-user price**						
D1	Applied tariff, excluding all taxes	$/kWh	0.151	0.193	0.189	0.216	0.239
D2	Applied tariff, including all taxes	$/kWh	0.170	0.218	0.213	0.244	0.270
E.	**Price gap**						
E1	Price gap, excluding all taxes (C1 − D1)	$/kWh	0.011	−0.032	−0.024	−0.013	−0.013
E2	Price gap, with fuel at counterfactual tax rates (C2 − D2)	$/kWh	0.011	−0.032	−0.024	−0.013	−0.013
F.	**Consumption**						
F1	Volume	1,000 GWh	1.99	1.97	2.03	2.14	2.25
G.	**Fiscal cost**						
G1	Fiscal cost, excluding all taxes (E1 * F1)	$, millions	22.07	(62.86)	(49.53)	(28.40)	(28.66)
G2	Fiscal cost, with fuel at counterfactual tax rates (E2 * F1)	$, millions	22.07	(62.86)	(49.53)	(28.40)	(28.66)
G3	Fiscal expenditures (F1 * (E2 − E1))	$, millions	0.00	0.00	0.00	0.00	0.00

Sources: Superintendency of Electricity and Telecommunications (regulator) and National Energy Council.
a. All prices and costs are expressed in current U.S. dollars. GWh = gigawatt-hour; kWh = kilowatt-hour.
b. Generation cost (short-term operational marginal cost) + transmission and distribution margins (efficient tariff schedules).
c. VAT (value added tax) is 13 percent of the applied tariff.

Table B.48 El Salvador—Low Tension

		Unit[a]	2008	2009	2010	2011	2012
A.	Supply cost[b]	$/kWh	0.162	0.161	0.164	0.203	0.226
B1.	Taxes, excluding transfers (VAT)[c]	$/kWh	0.020	0.026	0.027	0.030	0.033
B2.	Taxes (counterfactual): standard VAT	$/kWh	0.020	0.026	0.027	0.030	0.033
C.	Total reference benchmark price						
C1	Total reference price without taxes	$/kWh	0.162	0.161	0.164	0.203	0.226
C2	Total reference price at counterfactual tax rates	$/kWh	0.182	0.187	0.191	0.233	0.260
D.	End-user price						
D1	Applied tariff, excluding all taxes	$/kWh	0.155	0.200	0.206	0.233	0.258
D2	Applied tariff, including all taxes	$/kWh	0.175	0.226	0.233	0.263	0.291
E.	Price gap						
E1	Price gap, excluding all taxes (C1 − D1)	$/kWh	0.007	−0.039	−0.042	−0.030	−0.032
E2	Price gap, with fuel at counterfactual tax rates (C2 − D2)	$/kWh	0.007	−0.039	−0.042	−0.030	−0.032
F.	Consumption						
F1	Volume	1,000 GWh	2.27	2.26	2.24	2.22	2.29
G.	Fiscal cost						
G1	Fiscal cost, excluding all taxes (E1 * F1)	$, millions	15.55	(89.22)	(93.59)	(66.75)	(72.19)
G2	Fiscal cost, with fuel at counterfactual tax rates (E2 * F1)	$, millions	15.55	(89.22)	(93.59)	(66.75)	(72.19)
G3	Fiscal expenditures (F1 * (E2 − E1))	$, millions	0.00	0.00	0.00	0.00	0.00

Sources: Superintendency of Electricity and Telecommunications (regulator) and National Energy Council.

a. All prices and costs are expressed in current U.S. dollars. GWh = gigawatt-hour; kWh = kilowatt-hour.

b. Generation cost (short-term operational marginal cost) + transmission and distribution margins (efficient tariff schedules).

c. VAT (value added tax) is 13 percent of the applied tariff.

Table B.49 El Salvador—Medium Tension

		Unit[a]	2008	2009	2010	2011	2012
A.	Supply cost[b]	$/kWh	0.129	0.127	0.130	0.168	0.180
B1.	Taxes, excluding transfers (VAT)[c]	$/kWh	0.016	0.021	0.021	0.024	0.027
B2.	Taxes (counterfactual): standard VAT	$/kWh	0.016	0.021	0.021	0.024	0.027
C.	Total reference benchmark price						
C1	Total reference price without taxes	$/kWh	0.129	0.127	0.130	0.168	0.180
C2	Total reference price at counterfactual tax rates	$/kWh	0.145	0.148	0.151	0.192	0.206
D.	End-user price						
D1	Applied tariff, excluding all taxes	$/kWh	0.121	0.158	0.160	0.183	0.206
D2	Applied tariff, including all taxes	$/kWh	0.137	0.179	0.180	0.206	0.232
E.	Price gap						
E1	Price gap, excluding all taxes (C1 − D1)	$/kWh	0.008	−0.031	−0.030	−0.014	−0.026
E2	Price gap, with fuel at counterfactual tax rates (C2 − D2)	$/kWh	0.008	−0.031	−0.030	−0.014	−0.026
F.	Consumption						
F1	Volume	1,000 GWh	2.27	2.26	2.32	2.45	2.56

table continues next page

Table B.49 El Salvador—Medium Tension *(continued)*

	Unit[a]	2008	2009	2010	2011	2012
G.	**Fiscal cost**					
G1	Fiscal cost, excluding all taxes (E1 * F1)	$, millions	18.10	(71.08)	(69.09)	(34.35)	(66.67)
G2	Fiscal cost, with fuel at counterfactual tax rates (E2 * F1)	$, millions	18.10	(71.08)	(69.09)	(34.35)	(66.67)
G3	Fiscal expenditures (F1 * (E2 − E1))	$, millions	0.00	0.00	0.00	0.00	0.00

Sources: Superintendency of Electricity and Telecommunications (regulator) and National Energy Council.
a. All prices and costs are expressed in current U.S. dollars. GWh = gigawatt-hour; kWh = kilowatt-hour.
b. Generation cost (short-term operational marginal cost) + transmission and distribution margins (efficient tariff schedules).
c. VAT (value added tax) is 13 percent of the applied tariff.

Table B.50 El Salvador—High Tension

		Unit[a]	2008	2009	2010	2011	2012
A.	**Supply cost[b]**	$/kWh	0.129	0.127	0.130	0.168	0.180
B1.	**Taxes, excluding transfers (VAT)[c]**	$/kWh	0.013	0.017	0.017	0.021	0.023
B2.	**Taxes (counterfactual): standard VAT**	$/kWh	0.013	0.017	0.017	0.021	0.023
C.	**Total reference benchmark price**						
C1	Total reference price without taxes	$/kWh	0.129	0.127	0.130	0.168	0.180
C2	Total reference price at counterfactual tax rates	$/kWh	0.142	0.144	0.147	0.189	0.203
D.	**End-user price**						
D1	Applied tariff, excluding all taxes	$/kWh	0.101	0.130	0.134	0.160	0.178
D2	Applied tariff, including all taxes	$/kWh	0.114	0.147	0.151	0.181	0.202
E.	**Price gap**						
E1	Price gap, excluding all taxes (C1 − D1)	$/kWh	0.028	−0.003	−0.004	0.008	0.001
E2	Price gap, with fuel at counterfactual tax rates (C2 − D2)	$/kWh	0.028	−0.003	−0.004	0.008	0.001
F.	**Consumption**						
F1	Volume	1,000 GWh	4.95	4.96	5.01	5.11	5.31
G.	**Fiscal cost**						
G1	Fiscal cost, excluding all taxes (E1 * F1)	$, millions	136.39	(14.04)	(18.39)	43.28	6.73
G2	Fiscal cost, with fuel at counterfactual tax rates (E2 * F1)	$, millions	136.39	(14.04)	(18.39)	43.28	6.73
G3	Fiscal expenditures (F1 * (E2 − E1))	$, millions	0.00	0.00	0.00	0.00	0.00

Sources: Superintendency of Electricity and Telecommunications (regulator) and National Energy Council.
a. All prices and costs are expressed in current U.S. dollars. GWh = gigawatt-hour; kWh = kilowatt-hour.
b. Generation cost (short-term operational marginal cost) + transmission and distribution margins (efficient tariff schedules).
c. VAT (value added tax) is 13 percent of the applied tariff.

Table B.51 Haiti—Residential

		Unit[a]	2010	2011	2012	2013
A.	**Supply cost[b]**	$/kWh	0.255	0.264	0.270	0.282
B1.	**Taxes, excluding transfers (VAT)[c]**	$/kWh	0.000	0.000	0.000	0.000
B2.	**Taxes (counterfactual): standard VAT**	$/kWh	0.000	0.000	0.000	0.000
C.	**Total reference benchmark price**					
C1	Total reference price without taxes	$/kWh	0.255	0.264	0.270	0.282
C2	Total reference price at counterfactual tax rates	$/kWh	0.255	0.264	0.270	0.282

table continues next page

Energy Pricing Policies for Inclusive Growth in Latin America and the Caribbean
http://dx.doi.org/10.1596/978-1-4648-1111-1

Table B.51 Haiti—Residential *(continued)*

		Unit[a]	2010	2011	2012	2013
D.	**End-user price**					
D1	Applied tariff, excluding all taxes	$/kWh	0.190	0.186	0.180	0.174
D2	Applied tariff, including all taxes	$/kWh	0.190	0.186	0.180	0.174
E.	**Price gap**					
E1	Price gap, excluding all taxes (C1 − D1)	$/kWh	0.066	0.078	0.090	0.109
E2	Price gap, with fuel at counterfactual tax rates (C2 − D2)	$/kWh	0.066	0.078	0.090	0.109
F.	**Consumption**					
F1	Volume	1,000 GWh	0.03	0.09	0.25	0.24
G.	**Fiscal cost**					
G1	Fiscal cost, excluding all taxes (E1 * F1)	$, millions	1.93	6.93	22.38	25.99
G2	Fiscal cost, with fuel at counterfactual tax rates (E2 * F1)	$, millions	1.93	6.93	22.38	25.99
G3	Fiscal expenditures (F1 * (E2 − E1))	$, millions	0.00	0.00	0.00	0.00

Sources: Superintendency of Electricity and Telecommunications (regulator) and National Energy Council.
a. All prices and costs are expressed in current U.S. dollars. GWh = gigawatt-hour; kWh = kilowatt-hour.
b. Generation cost (system average generation cost) + transmission and distribution margins ($.05/liter).
c. Electricity sector is exempted from VAT (value added tax).

Table B.52 Haiti—Commercial

		Unit[a]	2010	2011	2012	2013
A.	**Supply cost**[b]	$/kWh	0.255	0.264	0.270	0.282
B1.	**Taxes, excluding transfers (VAT)**[c]	$/kWh	0.000	0.000	0.000	0.000
B2.	**Taxes (counterfactual): standard VAT**	$/kWh	0.000	0.000	0.000	0.000
C.	**Total reference benchmark price**					
C1	Total reference price without taxes	$/kWh	0.255	0.264	0.270	0.282
C2	Total reference price at counterfactual tax rates	$/kWh	0.255	0.264	0.270	0.282
D.	**End-user price**					
D1	Applied tariff, excluding all taxes	$/kWh	0.329	0.324	0.313	0.302
D2	Applied tariff, including all taxes	$/kWh	0.329	0.324	0.313	0.302
E.	**Price gap**					
E1	Price gap, excluding all taxes (C1 − D1)	$/kWh	−0.074	−0.059	−0.042	−0.019
E2	Price gap, with fuel at counterfactual tax rates (C2 − D2)	$/kWh	−0.074	−0.059	−0.042	−0.019
F.	**Consumption**					
F1	Volume	1,000 GWh	0.01	0.02	0.04	0.05
G.	**Fiscal cost**					
G1	Fiscal cost, excluding all taxes (E1 * F1)	$, millions	(0.53)	(1.18)	(1.81)	(0.88)
G2	Fiscal cost, with fuel at counterfactual tax rates (E2 * F1)	$, millions	(0.53)	(1.18)	(1.81)	(0.88)
G3	Fiscal expenditures (F1 * (E2 − E1))	$, millions	0.00	0.00	0.00	0.00

Sources: Superintendency of Electricity and Telecommunications (regulator) and National Energy Council.
a. All prices and costs are expressed in current U.S. dollars. GWh = gigawatt-hour; kWh = kilowatt-hour.
b. Generation cost (system average generation cost) + transmission and distribution margins ($.05/liter).
c. Electricity sector is exempted from VAT (value added tax).

Table B.53 Haiti—Industrial (Low Tension)

		Unit[a]	2010	2011	2012	2013
A.	Supply cost[b]	$/kWh	0.255	0.264	0.270	0.282
B1.	Taxes, excluding transfers (VAT)[c]	$/kWh	0.000	0.000	0.000	0.000
B2.	Taxes (counterfactual): standard VAT	$/kWh	0.000	0.000	0.000	0.000
C.	Total reference benchmark price					
C1	Total reference price without taxes	$/kWh	0.255	0.264	0.270	0.282
C2	Total reference price at counterfactual tax rates	$/kWh	0.255	0.264	0.270	0.282
D.	End-user price					
D1	Applied tariff, excluding all taxes	$/kWh	0.351	0.345	0.333	0.321
D2	Applied tariff, including all taxes	$/kWh	0.351	0.345	0.333	0.321
E.	Price gap					
E1	Price gap, excluding all taxes (C1 − D1)	$/kWh	−0.096	−0.081	−0.063	−0.039
E2	Price gap, with fuel at counterfactual tax rates (C2 − D2)	$/kWh	−0.096	−0.081	−0.063	−0.039
F.	Consumption					
F1	Volume	1,000 GWh	0.01	0.02	0.02	0.03
G.	Fiscal cost					
G1	Fiscal cost, excluding all taxes (E1 * F1)	$, millions	(0.48)	(1.49)	(1.56)	(0.98)
G2	Fiscal cost, with fuel at counterfactual tax rates (E2 * F1)	$, millions	(0.48)	(1.49)	(1.56)	(0.98)
G3	Fiscal expenditures (F1 * (E2 − E1))	$, millions	0.00	0.00	0.00	0.00

Sources: Superintendency of Electricity and Telecommunications (regulator) and National Energy Council.
a. All prices and costs are expressed in current U.S. dollars. GWh = gigawatt-hour; kWh = kilowatt-hour.
b. Generation cost (system average generation cost) + transmission and distribution margins ($.05/liter).
c. Electricity sector is exempted from VAT (value added tax).

Table B.54 Haiti—Industrial (Medium Tension)

		Unit[a]	2010	2011	2012	2013
A.	Supply cost[b]	$/kWh	0.255	0.264	0.270	0.282
B1.	Taxes, excluding transfers (VAT)[c]	$/kWh	0.000	0.000	0.000	0.000
B2.	Taxes (counterfactual): standard VAT	$/kWh	0.000	0.000	0.000	0.000
C.	Total reference benchmark price					
C1	Total reference price without taxes	$/kWh	0.255	0.264	0.270	0.282
C2	Total reference price at counterfactual tax rates	$/kWh	0.255	0.264	0.270	0.282
D.	End-user price					
D1	Applied tariff, excluding all taxes	$/kWh	0.334	0.328	0.317	0.306
D2	Applied tariff, including all taxes	$/kWh	0.334	0.328	0.317	0.306
E.	Price gap					
E1	Price gap, excluding all taxes (C1 − D1)	$/kWh	−0.079	−0.064	−0.047	−0.024
E2	Price gap, with fuel at counterfactual tax rates (C2 − D2)	$/kWh	−0.079	−0.064	−0.047	−0.024
F.	Consumption					
F1	Volume	1,000 GWh	0.01	0.05	0.07	0.06
G.	Fiscal cost					
G1	Fiscal cost, excluding all taxes (E1 * F1)	$, millions	(0.99)	(3.01)	(3.05)	(1.50)
G2	Fiscal cost, with fuel at counterfactual tax rates (E2 * F1)	$, millions	(0.99)	(3.01)	(3.05)	(1.50)
G3	Fiscal expenditures (F1 * (E2 − E1))	$, millions	0.00	0.00	0.00	0.00

Sources: Superintendency of Electricity and Telecommunications (regulator) and National Energy Council.
a. All prices and costs are expressed in current U.S. dollars. GWh = gigawatt-hour; kWh = kilowatt-hour.
b. Generation cost (system average generation cost) + transmission and distribution margins ($.05/liter).
c. Electricity sector is exempted from VAT (value added tax).

Table B.55 Haiti—Public Lighting

		Unit[a]	2010	2011	2012	2013
A.	Supply cost[b]	$/kWh	0.255	0.264	0.270	0.282
B1.	Taxes, excluding transfers (VAT)[c]	$/kWh	0.000	0.000	0.000	0.000
B2.	Taxes (counterfactual): standard VAT	$/kWh	0.000	0.000	0.000	0.000
C.	Total reference benchmark price					
C1	Total reference price without taxes	$/kWh	0.255	0.264	0.270	0.282
C2	Total reference price at counterfactual tax rates	$/kWh	0.255	0.264	0.270	0.282
D.	End-user price					
D1	Applied tariff, excluding all taxes	$/kWh	0.355	0.348	0.337	0.325
D2	Applied tariff, including all taxes	$/kWh	0.355	0.348	0.337	0.325
E.	Price gap					
E1	Price gap, excluding all taxes (C1 − D1)	$/kWh	−0.099	−0.084	−0.066	−0.042
E2	Price gap, with fuel at counterfactual tax rates (C2 − D2)	$/kWh	−0.099	−0.084	−0.066	−0.042
F.	Consumption					
F1	Volume	1,000 GWh	0.00	0.02	0.04	0.05
G.	Fiscal cost					
G1	Fiscal cost, excluding all taxes (E1 * F1)	$, millions	(0.40)	(1.67)	(2.51)	(2.33)
G2	Fiscal cost, with fuel at counterfactual tax rates (E2 * F1)	$, millions	(0.40)	(1.67)	(2.51)	(2.33)
G3	Fiscal expenditures (F1 * (E2 − E1))	$, millions	0.00	0.00	0.00	0.00

Sources: Superintendency of Electricity and Telecommunications (regulator) and National Energy Council.
a. All prices and costs are expressed in current U.S. dollars. GWh = gigawatt-hour; kWh = kilowatt-hour.
b. Generation cost (system average generation cost) + transmission and distribution margins ($.05/liter).
c. Electricity sector is exempted from VAT (value added tax).

Table B.56 Haiti—Public Organizations

		Unit[a]	2010	2011	2012	2013
A.	Supply cost[b]	$/kWh	0.255	0.264	0.270	0.282
B1.	Taxes, excluding transfers (VAT)[c]	$/kWh	0.000	0.000	0.000	0.000
B2.	Taxes (counterfactual): standard VAT	$/kWh	0.000	0.000	0.000	0.000
C.	Total reference benchmark price					
C1	Total reference price without taxes	$/kWh	0.255	0.264	0.270	0.282
C2	Total reference price at counterfactual tax rates	$/kWh	0.255	0.264	0.270	0.282
D.	End-user price					
D1	Applied tariff, excluding all taxes	$/kWh	0.355	0.348	0.337	0.325
D2	Applied tariff, including all taxes	$/kWh	0.355	0.348	0.337	0.325
E.	Price gap					
E1	Price gap, excluding all taxes (C1 − D1)	$/kWh	−0.099	−0.084	−0.066	−0.042
E2	Price gap, with fuel at counterfactual tax rates (C2 − D2)	$/kWh	−0.099	−0.084	−0.066	−0.042
F.	Consumption					
F1	Volume	1,000 GWh	0.00	0.01	0.02	0.02
G.	Fiscal cost					
G1	Fiscal cost, excluding all taxes (E1 * F1)	$, millions	(0.32)	(0.96)	(1.02)	(0.77)
G2	Fiscal cost, with fuel at counterfactual tax rates (E2 * F1)	$, millions	(0.32)	(0.96)	(1.02)	(0.77)
G3	Fiscal expenditures (F1 * (E2 − E1))	$, millions	0.00	0.00	0.00	0.00

Sources: Superintendency of Electricity and Telecommunications (regulator) and National Energy Council.
a. All prices and costs are expressed in current U.S. dollars. GWh = gigawatt-hour; kWh = kilowatt-hour.
b. Generation cost (system average generation cost) + transmission and distribution margins ($.05/liter).
c. Electricity sector is exempted from VAT (value added tax).

Table B.57 Haiti—Autonomous Public Organizations

		Unit[a]	2010	2011	2012	2013
A.	**Supply cost[b]**	$/kWh	0.255	0.264	0.270	0.282
B1.	**Taxes, excluding transfers (VAT)[c]**	$/kWh	0.000	0.000	0.000	0.000
B2.	**Taxes (counterfactual): standard VAT**	$/kWh	0.000	0.000	0.000	0.000
C.	**Total reference benchmark price**					
C1	Total reference price without taxes	$/kWh	0.255	0.264	0.270	0.282
C2	Total reference price at counterfactual tax rates	$/kWh	0.255	0.264	0.270	0.282
D.	**End-user price**					
D1	Applied tariff, excluding all taxes	$/kWh	0.355	0.348	0.337	0.325
D2	Applied tariff, including all taxes	$/kWh	0.355	0.348	0.337	0.325
E.	**Price gap**					
E1	Price gap, excluding all taxes (C1 − D1)	$/kWh	−0.099	−0.084	−0.066	−0.042
E2	Price gap, with fuel at counterfactual tax rates (C2 − D2)	$/kWh	−0.099	−0.084	−0.066	−0.042
F.	**Consumption**					
F1	Volume	1,000 GWh	0.00	0.01	0.01	0.01
G.	**Fiscal cost**					
G1	Fiscal cost, excluding all taxes (E1 * F1)	$, millions	(0.30)	(1.10)	(0.72)	(0.39)
G2	Fiscal cost, with fuel at counterfactual tax rates (E2 * F1)	$, millions	(0.30)	(1.10)	(0.72)	(0.39)
G3	Fiscal expenditures (F1 * (E2 − E1))	$, millions	0.00	0.00	0.00	0.00

Sources: Superindency of Electricity and Telecommunications (regulator) and National Energy Council.
a. All prices and costs are expressed in current U.S. dollars. GWh = gigawatt-hour; kWh = kilowatt-hour.
b. Generation cost (system average generation cost) + transmission and distribution margins ($.05/liter).
c. Electricity sector is exempted from VAT (value added tax).

Table B.58 Haiti—Low Tension

		Unit[a]	2010	2011	2012	2013
A.	**Supply cost[b]**	$/kWh	0.255	0.264	0.270	0.282
B1.	**Taxes, excluding transfers (VAT)[c]**	$/kWh	0.000	0.000	0.000	0.000
B2.	**Taxes (counterfactual): standard VAT**	$/kWh	0.000	0.000	0.000	0.000
C.	**Total reference benchmark price**					
C1	Total reference price without taxes	$/kWh	0.255	0.264	0.270	0.282
C2	Total reference price at counterfactual tax rates	$/kWh	0.255	0.264	0.270	0.282
D.	**End-user price**					
D1	Applied tariff, excluding all taxes	$/kWh	0.322	0.317	0.306	0.295
D2	Applied tariff, including all taxes	$/kWh	0.322	0.317	0.306	0.295
E.	**Price gap**					
E1	Price gap, excluding all taxes (C1 − D1)	$/kWh	−0.067	−0.052	−0.035	−0.013
E2	Price gap, with fuel at counterfactual tax rates (C2 − D2)	$/kWh	−0.067	−0.052	−0.035	−0.013
F.	**Consumption**					
F1	Volume	1,000 GWh	0.05	0.17	0.38	0.39
G.	**Fiscal cost**					
G1	Fiscal cost, excluding all taxes (E1 * F1)	$, millions	(3.47)	(9.01)	(13.46)	(5.04)
G2	Fiscal cost, with fuel at counterfactual tax rates (E2 * F1)	$, millions	(3.47)	(9.01)	(13.46)	(5.04)
G3	Fiscal expenditures (F1 * (E2 − E1))	$, millions	0.00	0.00	0.00	0.00

Sources: Superintendency of Electricity and Telecommunications (regulator) and National Energy Council.
a. All prices and costs are expressed in current U.S. dollars. GWh = gigawatt-hour; kWh = kilowatt-hour.
b. Generation cost (system average generation cost) + transmission and distribution margins ($.05/liter).
c. Electricity sector is exempted from VAT (value added tax).

Energy Pricing Policies for Inclusive Growth in Latin America and the Caribbean
http://dx.doi.org/10.1596/978-1-4648-1111-1

Table B.59 Haiti—Medium Tension

		Unit[a]	2010	2011	2012	2013
A.	Supply cost[b]	$/kWh	0.255	0.264	0.270	0.282
B1.	Taxes, excluding transfers (VAT)[c]	$/kWh	0.000	0.000	0.000	0.000
B2.	Taxes (counterfactual): standard VAT	$/kWh	0.000	0.000	0.000	0.000
C.	Total reference benchmark price					
C1	Total reference price without taxes	$/kWh	0.255	0.264	0.270	0.282
C2	Total reference price at counterfactual tax rates	$/kWh	0.255	0.264	0.270	0.282
D.	End-user price					
D1	Applied tariff, excluding all taxes	$/kWh	0.334	0.328	0.317	0.306
D2	Applied tariff, including all taxes	$/kWh	0.334	0.328	0.317	0.306
E.	Price gap					
E1	Price gap, excluding all taxes (C1 – D1)	$/kWh	−0.079	−0.064	−0.047	−0.024
E2	Price gap, with fuel at counterfactual tax rates (C2 – D2)	$/kWh	−0.079	−0.064	−0.047	−0.024
F.	Consumption					
F1	Volume	1,000 GWh	0.01	0.05	0.07	0.06
G.	Fiscal cost					
G1	Fiscal cost, excluding all taxes (E1 * F1)	$, millions	(0.99)	(3.01)	(3.05)	(1.50)
G2	Fiscal cost, with fuel at counterfactual tax rates (E2 * F1)	$, millions	(0.99)	(3.01)	(3.05)	(1.50)
G3	Fiscal expenditures (F1 * (E2 – E1))	$, millions	0.00	0.00	0.00	0.00

Sources: Superindency of Electricity and Telecommunications (regulator) and National Energy Council.
a. All prices and costs are expressed in current U.S. dollars. GWh = gigawatt-hour; kWh = kilowatt-hour.
b. Generation cost (system average generation cost) + transmission and distribution margins ($.05/liter).
c. Electricity sector is exempted from VAT (value added tax).

Table B.60 Honduras—Residential (Less Than 75 kWh)

		Unit[a]	2008	2009	2010	2011	2012	2013
A.	Supply cost[b]	$/kWh	0.178	0.147	0.163	0.199	0.214	0.221
B1.	Taxes, excluding transfers (VAT)[c]	$/kWh	0.000	0.000	0.000	0.000	0.000	0.000
B2.	Taxes (counterfactual): standard VAT	$/kWh	0.008	0.011	0.010	0.011	0.012	0.014
C.	Total reference benchmark price							
C1	Total reference price without taxes	$/kWh	0.178	0.147	0.163	0.199	0.214	0.221
C2	Total reference price at counterfactual tax rates	$/kWh	0.186	0.158	0.172	0.211	0.226	0.235
D.	End-user price							
D1	Applied tariff, excluding all taxes	$/kWh	0.070	0.093	0.081	0.094	0.100	0.095
D2	Applied tariff, including all taxes	$/kWh	0.070	0.093	0.081	0.094	0.100	0.095
E.	Price gap							
E1	Price gap, excluding all taxes (C1 – D1)	$/kWh	0.107	0.054	0.082	0.105	0.115	0.126
E2	Price gap, with fuel at counterfactual tax rates (C2 – D2)	$/kWh	0.116	0.065	0.092	0.117	0.127	0.140
F.	Consumption							
F1	Volume	1,000 GWh	0.13	0.14	0.15	0.16	0.18	0.20

table continues next page

Table B.60 Honduras—Residential (Less Than 75 kWh) *(continued)*

	Unit[a]	2008	2009	2010	2011	2012	2013
G.	**Fiscal cost**							
G1	Fiscal cost, excluding all taxes (E1 * F1)	$, millions	13.78	7.41	12.07	16.79	20.12	24.64
G2	Fiscal cost, with fuel at counterfactual tax rates (E2 * F1)	$, millions	14.86	8.94	13.49	18.58	22.22	27.44
G3	Fiscal expenditures (F1 * (E2 − E1))	$, millions	1.09	1.53	1.43	1.80	2.10	2.80

Source: National Power Utility (ENEE).
a. All prices and costs are expressed in current U.S. dollars. GWh = gigawatt-hour; kWh = kilowatt-hour.
b. Generation cost (system average generation cost) + transmission and distribution margins (system average transmission and distribution costs).
c. VAT (value added tax) is 12 percent of the applied tariff for residential consumption above 750 kWh; other categories are exempted from VAT.

Table B.61 Honduras—Residential (between 75 kWh and 150 kWh)

		Unit[a]	2008	2009	2010	2011	2012	2013
A.	**Supply cost**[b]	$/kWh	0.178	0.147	0.163	0.199	0.214	0.221
B1.	**Taxes, excluding transfers (VAT)**[c]	$/kWh	0.000	0.000	0.000	0.000	0.000	0.000
B2.	**Taxes (counterfactual): standard VAT**	$/kWh	0.009	0.010	0.011	0.012	0.013	0.016
C.	**Total reference benchmark price**							
C1	Total reference price without taxes	$/kWh	0.178	0.147	0.163	0.199	0.214	0.221
C2	Total reference price at counterfactual tax rates	$/kWh	0.187	0.157	0.173	0.212	0.228	0.237
D.	**End-user price**							
D1	Applied tariff, excluding all taxes	$/kWh	0.078	0.087	0.089	0.104	0.110	0.105
D2	Applied tariff, including all taxes	$/kWh	0.078	0.087	0.089	0.104	0.110	0.105
E.	**Price gap**							
E1	Price gap, excluding all taxes (C1 − D1)	$/kWh	0.100	0.060	0.074	0.096	0.104	0.116
E2	Price gap, with fuel at counterfactual tax rates (C2 − D2)	$/kWh	0.109	0.070	0.084	0.108	0.117	0.132
F.	**Consumption**							
F1	Volume	1,000 GWh	0.40	0.45	0.49	0.52	0.57	0.59
G.	**Fiscal cost**							
G1	Fiscal cost, excluding all taxes (E1 * F1)	$, millions	39.70	27.13	36.11	50.04	58.86	68.69
G2	Fiscal cost, with fuel at counterfactual tax rates (E2 * F1)	$, millions	43.40	31.87	41.37	56.56	66.34	78.02
G3	Fiscal expenditures (F1 * (E2 − E1))	$, millions	3.71	4.74	5.26	6.52	7.48	9.33

Source: National Power Utility (ENEE).
a. All prices and costs are expressed in current U.S. dollars. GWh = gigawatt-hour; kWh = kilowatt-hour.
b. Generation cost (system average generation cost) + transmission and distribution margins (system average transmission and distribution costs).
c. VAT (value added tax) is 12 percent of the applied tariff for residential consumption above 750 kWh; other categories are exempted from VAT.

Table B.62 Honduras—Residential (between 150 kWh and 300 kWh)

		Unit[a]	2008	2009	2010	2011	2012	2013
A.	**Supply cost**[b]	$/kWh	0.178	0.147	0.163	0.199	0.214	0.221
B1.	**Taxes, excluding transfers (VAT)**[c]	$/kWh	0.000	0.000	0.000	0.000	0.000	0.000
B2.	**Taxes (counterfactual): standard VAT**	$/kWh	0.012	0.012	0.014	0.016	0.017	0.020
C.	**Total reference benchmark price**							
C1	Total reference price without taxes	$/kWh	0.178	0.147	0.163	0.199	0.214	0.221
C2	Total reference price at counterfactual tax rates	$/kWh	0.189	0.159	0.176	0.215	0.231	0.241

table continues next page

Table B.62 Honduras—Residential (between 150 kWh and 300 kWh) *(continued)*

		Unit[a]	2008	2009	2010	2011	2012	2013
D.	**End-user price**							
D1	Applied tariff, excluding all taxes	$/kWh	0.098	0.099	0.113	0.131	0.139	0.133
D2	Applied tariff, including all taxes	$/kWh	0.098	0.099	0.113	0.131	0.139	0.133
E.	**Price gap**							
E1	Price gap, excluding all taxes (C1 − D1)	$/kWh	0.079	0.048	0.050	0.068	0.075	0.088
E2	Price gap, with fuel at counterfactual tax rates (C2 − D2)	$/kWh	0.091	0.060	0.064	0.084	0.092	0.108
F.	**Consumption**							
F1	Volume	1,000 GWh	0.77	0.73	0.72	0.71	0.68	0.69
G.	**Fiscal cost**							
G1	Fiscal cost, excluding all taxes (E1 * F1)	$, millions	60.58	35.20	36.19	48.24	51.35	60.58
G2	Fiscal cost, with fuel at counterfactual tax rates (E2 * F1)	$, millions	69.61	43.84	45.98	59.39	62.80	74.31
G3	Fiscal expenditures (F1 * (E2 − E1))	$, millions	9.04	8.64	9.79	11.15	11.44	13.72

Source: National Power Utility (ENEE).

a. All prices and costs are expressed in current U.S. dollars. GWh = gigawatt-hour; kWh = kilowatt-hour.

b. Generation cost (system average generation cost) + transmission and distribution margins (system average transmission and distribution costs).

c. VAT (value added tax) is 12 percent of the applied tariff for residential consumption above 750 kWh; other categories are exempted from VAT.

Table B.63 Honduras—Residential (between 300 kWh and 750 kWh)

		Unit[a]	2008	2009	2010	2011	2012	2013
A.	**Supply cost[b]**	$/kWh	0.178	0.147	0.163	0.199	0.214	0.221
B1.	**Taxes, excluding transfers (VAT)[c]**	$/kWh	0.000	0.000	0.000	0.000	0.000	0.000
B2.	**Taxes (counterfactual): standard VAT**	$/kWh	0.014	0.014	0.017	0.019	0.021	0.025
C.	**Total reference benchmark price**							
C1	Total reference price without taxes	$/kWh	0.178	0.147	0.163	0.199	0.214	0.221
C2	Total reference price at counterfactual tax rates	$/kWh	0.192	0.161	0.179	0.219	0.235	0.246
D.	**End-user price**							
D1	Applied tariff, excluding all taxes	$/kWh	0.121	0.120	0.139	0.162	0.172	0.164
D2	Applied tariff, including all taxes	$/kWh	0.121	0.120	0.139	0.162	0.172	0.164
E.	**Price gap**							
E1	Price gap, excluding all taxes (C1 − D1)	$/kWh	0.057	0.027	0.024	0.038	0.043	0.057
E2	Price gap, with fuel at counterfactual tax rates (C2 − D2)	$/kWh	0.071	0.041	0.041	0.057	0.063	0.082
F.	**Consumption**							
F1	Volume	1,000 GWh	0.59	0.58	0.57	0.56	0.53	0.54
G.	**Fiscal cost**							
G1	Fiscal cost, excluding all taxes (E1 * F1)	$, millions	33.74	15.58	13.70	20.89	22.67	30.80
G2	Fiscal cost, with fuel at counterfactual tax rates (E2 * F1)	$, millions	42.33	23.87	23.26	31.66	33.66	44.13
G3	Fiscal expenditures (F1 * (E2 − E1))	$, millions	8.59	8.30	9.56	10.77	10.99	13.33

Source: National Power Utility (ENEE).

a. All prices and costs are expressed in current U.S. dollars. GWh = gigawatt-hour; kWh = kilowatt-hour.

b. Generation cost (system average generation cost) + transmission and distribution margins (system average transmission and distribution costs).

c. VAT (value added tax) is 12 percent of the applied tariff for residential consumption above 750 kWh; other categories are exempted from VAT.

Table B.64 Honduras—Residential (More Than 750 kWh)

		Unit[a]	2008	2009	2010	2011	2012	2013
A.	Supply cost[b]	$/kWh	0.178	0.147	0.163	0.199	0.214	0.221
B1.	Taxes, excluding transfers (VAT)[c]	$/kWh	0.018	0.019	0.021	0.025	0.026	0.025
B2.	Taxes (counterfactual): standard VAT	$/kWh	0.018	0.019	0.021	0.025	0.026	0.031
C.	Total reference benchmark price							
C1	Total reference price without taxes	$/kWh	0.178	0.147	0.163	0.199	0.214	0.221
C2	Total reference price at counterfactual tax rates	$/kWh	0.196	0.166	0.184	0.224	0.240	0.252
D.	End-user price							
D1	Applied tariff, excluding all taxes	$/kWh	0.153	0.156	0.175	0.204	0.217	0.207
D2	Applied tariff, including all taxes	$/kWh	0.171	0.175	0.197	0.229	0.243	0.232
E.	Price gap							
E1	Price gap, excluding all taxes (C1 − D1)	$/kWh	0.025	−0.009	−0.013	−0.005	−0.003	0.014
E2	Price gap, with fuel at counterfactual tax rates (C2 − D2)	$/kWh	0.025	−0.009	−0.013	−0.005	−0.003	0.020
F.	Consumption							
F1	Volume	1,000 GWh	0.24	0.25	0.24	0.21	0.20	0.20
G.	Fiscal cost							
G1	Fiscal cost, excluding all taxes (E1 * F1)	$, millions	6.05	(2.26)	(3.01)	(1.03)	(0.53)	2.76
G2	Fiscal cost, with fuel at counterfactual tax rates (E2 * F1)	$, millions	6.05	(2.26)	(3.01)	(1.03)	(0.53)	4.01
G3	Fiscal expenditures (F1 * (E2 − E1))	$, millions	0.00	0.00	0.00	0.00	0.00	1.25

Source: National Power Utility (ENEE).
a. All prices and costs are expressed in current U.S. dollars. GWh = gigawatt-hour; kWh = kilowatt-hour.
b. Generation cost (system average generation cost) + transmission and distribution margins (system average transmission and distribution costs).
c. VAT (value added tax) is 12 percent of the applied tariff for residential consumption above 750 kWh; other categories are exempted from VAT.

Table B.65 Honduras—Commercial

		Unit[a]	2008	2009	2010	2011	2012	2013
A.	Supply cost[b]	$/kWh	0.178	0.147	0.163	0.199	0.214	0.221
B1.	Taxes, excluding transfers (VAT)[c]	$/kWh	0.000	0.000	0.000	0.000	0.000	0.000
B2.	Taxes (counterfactual): standard VAT	$/kWh	0.018	0.018	0.020	0.024	0.024	0.029
C.	Total reference benchmark price							
C1	Total reference price without taxes	$/kWh	0.178	0.147	0.163	0.199	0.214	0.221
C2	Total reference price at counterfactual tax rates	$/kWh	0.195	0.165	0.182	0.223	0.238	0.250
D.	End-user price							
D1	Applied tariff, excluding all taxes	$/kWh	0.148	0.148	0.163	0.196	0.199	0.190
D2	Applied tariff, including all taxes	$/kWh	0.148	0.148	0.163	0.196	0.199	0.190
E.	Price gap							
E1	Price gap, excluding all taxes (C1 − D1)	$/kWh	0.029	−0.001	−0.001	0.003	0.015	0.031
E2	Price gap, with fuel at counterfactual tax rates (C2 − D2)	$/kWh	0.047	0.016	0.019	0.027	0.039	0.059
F.	Consumption							
F1	Volume	1,000 GWh	1.27	1.26	1.28	1.28	1.33	1.38

table continues next page

Table B.65 Honduras—Commercial *(continued)*

		Unit[a]	2008	2009	2010	2011	2012	2013
G.	**Fiscal cost**							
G1	Fiscal cost, excluding all taxes (E1 * F1)	$, millions	37.17	(1.71)	(0.65)	4.07	19.94	42.60
G2	Fiscal cost, with fuel at counterfactual tax rates (E2 * F1)	$, millions	59.74	20.72	24.37	34.24	51.67	81.92
G3	Fiscal expenditures (F1 * (E2 − E1))	$, millions	22.58	22.43	25.03	30.17	31.73	39.32

Source: National Power Utility (ENEE).
a. All prices and costs are expressed in current U.S. dollars. GWh = gigawatt-hour; kWh = kilowatt-hour.
b. Generation cost (system average generation cost) + transmission and distribution margins (system average transmission and distribution costs).
c. VAT (value added tax) is 12 percent of the applied tariff for residential consumption above 750 kWh; other categories are exempted from VAT.

Table B.66 Honduras—Industrial

		Unit[a]	2008	2009	2010	2011	2012	2013
A.	**Supply cost[b]**	$/kWh	0.178	0.147	0.163	0.199	0.214	0.221
B1.	**Taxes, excluding transfers (VAT)[c]**	$/kWh	0.000	0.000	0.000	0.000	0.000	0.000
B2.	**Taxes (counterfactual): standard VAT**	$/kWh	0.018	0.018	0.020	0.023	0.024	0.029
C.	**Total reference benchmark price**							
C1	Total reference price without taxes	$/kWh	0.178	0.147	0.163	0.199	0.214	0.221
C2	Total reference price at counterfactual tax rates	$/kWh	0.196	0.165	0.183	0.223	0.238	0.250
D.	**End-user price**							
D1	Applied tariff, excluding all taxes	$/kWh	0.153	0.151	0.165	0.195	0.199	0.192
D2	Applied tariff, including all taxes	$/kWh	0.153	0.151	0.165	0.195	0.199	0.192
E.	**Price gap**							
E1	Price gap, excluding all taxes (C1 − D1)	$/kWh	0.024	−0.004	−0.002	0.004	0.015	0.029
E2	Price gap, with fuel at counterfactual tax rates (C2 − D2)	$/kWh	0.043	0.014	0.017	0.027	0.039	0.058
F.	**Consumption**							
F1	Volume	1,000 GWh	1.39	1.24	1.27	1.38	1.43	1.49
G.	**Fiscal cost**							
G1	Fiscal cost, excluding all taxes (E1 * F1)	$, millions	33.74	(5.48)	(2.94)	5.49	21.72	42.99
G2	Fiscal cost, with fuel at counterfactual tax rates (E2 * F1)	$, millions	59.28	16.96	22.16	37.78	55.79	85.91
G3	Fiscal expenditures (F1 * (E2 − E1))	$, millions	25.55	22.44	25.10	32.29	34.08	42.92

Source: National Power Utility (ENEE).
a. All prices and costs are expressed in current U.S. dollars. GWh = gigawatt-hour; kWh = kilowatt-hour.
b. Generation cost (system average generation cost) + transmission and distribution margins (system average transmission and distribution costs).
c. VAT (value added tax) is 12 percent of the applied tariff for residential consumption above 750 kWh; other categories are exempted from VAT.

Table B.67 Honduras—Public Sector

		Unit[a]	2008	2009	2010	2011	2012	2013
A.	**Supply cost[b]**	$/kWh	0.178	0.147	0.163	0.199	0.214	0.221
B1.	**Taxes, excluding transfers (VAT)[c]**	$/kWh	0.000	0.000	0.000	0.000	0.000	0.000
B2.	**Taxes (counterfactual): standard VAT**	$/kWh	0.021	0.023	0.025	0.030	0.031	0.037
C.	**Total reference benchmark price**							
C1	Total reference price without taxes	$/kWh	0.178	0.147	0.163	0.199	0.214	0.221

table continues next page

Table B.67 Honduras—Public Sector (continued)

		Unit[a]	2008	2009	2010	2011	2012	2013
C2	Total reference price at counterfactual tax rates	$/kWh	0.199	0.169	0.188	0.229	0.245	0.258
D.	**End-user price**							
D1	Applied tariff, excluding all taxes	$/kWh	0.178	0.188	0.207	0.250	0.255	0.245
D2	Applied tariff, including all taxes	$/kWh	0.178	0.188	0.207	0.250	0.255	0.245
E.	**Price gap**							
E1	Price gap, excluding all taxes (C1 − D1)	$/kWh	0.000	−0.042	−0.045	−0.051	−0.041	−0.024
E2	Price gap, with fuel at counterfactual tax rates (C2 − D2)	$/kWh	0.021	−0.019	−0.020	−0.021	−0.010	0.012
F.	**Consumption**							
F1	Volume	1,000 GWh	0.26	0.27	0.26	0.27	0.27	0.28
G.	**Fiscal cost**							
G1	Fiscal cost, excluding all taxes (E1 * F1)	$, millions	(0.04)	(11.07)	(11.52)	(13.54)	(11.11)	(6.89)
G2	Fiscal cost, with fuel at counterfactual tax rates (E2 * F1)	$, millions	5.46	(5.05)	(5.09)	(5.51)	(2.72)	3.52
G3	Fiscal expenditures (F1 * (E2 − E1))	$, millions	5.50	6.02	6.43	8.03	8.39	10.41

Source: National Power Utility (ENEE).
a. All prices and costs are expressed in current U.S. dollars. GWh = gigawatt-hour; kWh = kilowatt-hour.
b. Generation cost (system average generation cost) + transmission and distribution margins (system average transmission and distribution costs).
c. VAT (value added tax) is 12 percent of the applied tariff for residential consumption above 750 kWh; other categories are exempted from VAT.

Table B.68 Honduras—Public Lighting

		Unit[a]	2008	2009	2010	2011	2012	2013
A.	**Supply cost[b]**	$/kWh	0.178	0.147	0.163	0.199	0.214	0.221
B1.	**Taxes, excluding transfers (VAT)[c]**	$/kWh	0.000	0.000	0.000	0.000	0.000	0.000
B2.	**Taxes (counterfactual): standard VAT**	$/kWh	0.016	0.017	0.018	0.021	0.022	0.026
C.	**Total reference benchmark price**							
C1	Total reference price without taxes	$/kWh	0.178	0.147	0.163	0.199	0.214	0.221
C2	Total reference price at counterfactual tax rates	$/kWh	0.194	0.164	0.181	0.221	0.236	0.247
D.	**End-user price**							
D1	Applied tariff, excluding all taxes	$/kWh	0.133	0.140	0.150	0.177	0.184	0.175
D2	Applied tariff, including all taxes	$/kWh	0.133	0.140	0.150	0.177	0.184	0.175
E.	**Price gap**							
E1	Price gap, excluding all taxes (C1 − D1)	$/kWh	0.044	0.007	0.012	0.022	0.030	0.046
E2	Price gap, with fuel at counterfactual tax rates (C2 − D2)	$/kWh	0.060	0.024	0.030	0.044	0.052	0.072
F.	**Consumption**							
F1	Volume	1,000 GWh	0.12	0.12	0.12	0.12	0.13	0.13
G.	**Fiscal cost**							
G1	Fiscal cost, excluding all taxes (E1 * F1)	$, millions	5.52	0.84	1.54	2.79	3.79	5.79
G2	Fiscal cost, with fuel at counterfactual tax rates (E2 * F1)	$, millions	7.51	2.94	3.79	5.43	6.56	9.08
G3	Fiscal expenditures (F1 * (E2 − E1))	$, millions	1.99	2.10	2.25	2.65	2.77	3.29

Source: National Power Utility (ENEE).
a. All prices and costs are expressed in current U.S. dollars. GWh = gigawatt-hour; kWh = kilowatt-hour.
b. Generation cost (system average generation cost) + transmission and distribution margins (system average transmission and distribution costs).
c. VAT (value added tax) is 12 percent of the applied tariff for residential consumption above 750 kWh; other categories are exempted from VAT.

Table B.69 Mexico—Residential

		Unit[a]	2008	2009	2010	2011	2012
A.	Supply cost[b]	$/kWh	0.282	0.220	0.221	0.225	0.216
B1.	Taxes, excluding transfers (VAT)[c]	$/kWh	0.014	0.012	0.014	0.015	0.014
B2.	Taxes (counterfactual): standard VAT	$/kWh	0.014	0.012	0.014	0.015	0.014
C.	Total reference benchmark price						
C1	Total reference price without taxes	$/kWh	0.282	0.220	0.221	0.225	0.216
C2	Total reference price at counterfactual tax rates	$/kWh	0.296	0.232	0.235	0.240	0.230
D.	End-user price						
D1	Applied tariff, excluding all taxes	$/kWh	0.093	0.077	0.088	0.094	0.089
D2	Applied tariff, including all taxes	$/kWh	0.107	0.089	0.102	0.109	0.103
E.	Price gap						
E1	Price gap, excluding all taxes (C1 − D1)	$/kWh	0.189	0.143	0.133	0.130	0.127
E2	Price gap, with fuel at counterfactual tax rates (C2 − D2)	$/kWh	0.189	0.143	0.133	0.130	0.127
F.	Consumption						
F1	Volume	1,000 GWh	40.85	42.14	46.89	51.75	52.01
G.	Fiscal cost						
G1	Fiscal cost, excluding all taxes (E1 * F1)	$, millions	7,717.14	6,011.66	6,235.08	6,748.21	6,600.13
G2	Fiscal cost, with fuel at counterfactual tax rates (E2 * F1)	$, millions	7,717.14	6,011.66	6,235.08	6,748.21	6,600.13
G3	Fiscal expenditures (F1 * (E2 − E1))	$, millions	0.00	0.00	0.00	0.00	0.00

Source: Secretariat of Finance and Public Credit (SHCP).
a. All prices and costs are expressed in current U.S. dollars. GWh = gigawatt-hour; kWh = kilowatt-hour.
b. Generation cost + transmission and distribution margins: average cost of electricity.
c. VAT (value added tax) is 16 percent of the applied tariff.

Table B.70 Mexico—Commercial

		Unit[a]	2008	2009	2010	2011	2012
A.	Supply cost[b]	$/kWh	0.297	0.228	0.194	0.199	0.191
B1.	Taxes, excluding transfers (VAT)[c]	$/kWh	0.035	0.029	0.033	0.035	0.035
B2.	Taxes (counterfactual): standard VAT	$/kWh	0.035	0.029	0.033	0.035	0.035
C.	Total reference benchmark price						
C1	Total reference price without taxes	$/kWh	0.297	0.228	0.194	0.199	0.191
C2	Total reference price at counterfactual tax rates	$/kWh	0.332	0.256	0.227	0.234	0.226
D.	End-user price						
D1	Applied tariff, excluding all taxes	$/kWh	0.233	0.178	0.204	0.219	0.221
D2	Applied tariff, including all taxes	$/kWh	0.267	0.207	0.237	0.255	0.256
E.	Price gap						
E1	Price gap, excluding all taxes (C1 − D1)	$/kWh	0.065	0.050	−0.010	−0.021	−0.030
E2	Price gap, with fuel at counterfactual tax rates (C2 − D2)	$/kWh	0.065	0.050	−0.010	−0.021	−0.030
F.	Consumption						
F1	Volume	1,000 GWh	9.68	9.63	11.94	13.59	13.92

table continues next page

Table B.70 Mexico—Commercial *(continued)*

	Unit[a]	2008	2009	2010	2011	2012
G.	**Fiscal cost**						
G1	Fiscal cost, excluding all taxes (E1 * F1)	$, millions	627.52	478.94	(121.34)	(279.63)	(422.14)
G2	Fiscal cost, with fuel at counterfactual tax rates (E2 * F1)	$, millions	627.52	478.94	(121.34)	(279.63)	(422.14)
G3	Fiscal expenditures (F1 * (E2 − E1))	$, millions	0.00	0.00	0.00	0.00	0.00

Source: Secretariat of Finance and Public Credit (SHCP).
a. All prices and costs are expressed in current U.S. dollars. GWh = gigawatt-hour; kWh = kilowatt-hour.
b. Generation cost + transmission and distribution margins: average cost of electricity.
c. VAT (value added tax) is 16 percent of the applied tariff.

Table B.71 Mexico—Commercial (Temporary)

		Unit[a]	2008	2009	2010	2011	2012
A.	**Supply cost[b]**	$/kWh	0.330	0.266	0.256	0.246	0.238
B1.	**Taxes, excluding transfers (VAT)[c]**	$/kWh	0.057	0.048	0.050	0.055	0.058
B2.	**Taxes (counterfactual): standard VAT**	$/kWh	0.057	0.048	0.050	0.055	0.058
C.	**Total reference benchmark price**						
C1	Total reference price without taxes	$/kWh	0.330	0.266	0.256	0.246	0.238
C2	Total reference price at counterfactual tax rates	$/kWh	0.386	0.314	0.306	0.301	0.296
D.	**End-user price**						
D1	Applied tariff, excluding all taxes	$/kWh	0.377	0.297	0.314	0.343	0.360
D2	Applied tariff, including all taxes	$/kWh	0.433	0.345	0.365	0.398	0.417
E.	**Price gap**						
E1	Price gap, excluding all taxes (C1 − D1)	$/kWh	−0.047	−0.031	−0.059	−0.097	−0.121
E2	Price gap, with fuel at counterfactual tax rates (C2 − D2)	$/kWh	−0.047	−0.031	−0.059	−0.097	−0.121
F.	**Consumption**						
F1	Volume	1,000 GWh	0.01	0.01	0.02	0.02	0.02
G.	**Fiscal cost**						
G1	Fiscal cost, excluding all taxes (E1 * F1)	$, millions	(0.63)	(0.38)	(0.95)	(2.01)	(2.01)
G2	Fiscal cost, with fuel at counterfactual tax rates (E2 * F1)	$, millions	(0.63)	(0.38)	(0.95)	(2.01)	(2.01)
G3	Fiscal expenditures (F1 * (E2 − E1))	$, millions	0.00	0.00	0.00	0.00	(0.00)

Source: Secretariat of Finance and Public Credit (SHCP).
a. All prices and costs are expressed in current U.S. dollars. GWh = gigawatt-hour; kWh = kilowatt-hour.
b. Generation cost + transmission and distribution margins: average cost of electricity.
c. VAT (value added tax) is 16 percent of the applied tariff.

Table B.72 Mexico—Industrial

		Unit[a]	2008	2009	2010	2011	2012
A.	**Supply cost[b]**	$/kWh	0.140	0.106	0.108	0.112	0.109
B1.	**Taxes, excluding transfers (VAT)[c]**	$/kWh	0.019	0.014	0.017	0.018	0.018
B2.	**Taxes (counterfactual): standard VAT**	$/kWh	0.019	0.014	0.017	0.018	0.018
C.	**Total reference benchmark price**						
C1	Total reference price without taxes	$/kWh	0.140	0.106	0.108	0.112	0.109

table continues next page

Table B.72 Mexico—Industrial *(continued)*

		Unit[a]	2008	2009	2010	2011	2012
C2	Total reference price at counterfactual tax rates	$/kWh	0.159	0.119	0.124	0.130	0.128
D.	**End-user price**						
D1	Applied tariff, excluding all taxes	$/kWh	0.125	0.085	0.103	0.116	0.114
D2	Applied tariff, including all taxes	$/kWh	0.144	0.099	0.120	0.134	0.133
E.	**Price gap**						
E1	Price gap, excluding all taxes (C1 − D1)	$/kWh	0.015	0.021	0.004	−0.004	−0.005
E2	Price gap, with fuel at counterfactual tax rates (C2 − D2)	$/kWh	0.015	0.021	0.004	−0.004	−0.005
F.	**Consumption**						
F1	Volume	1,000 GWh	89.78	84.32	102.95	116.54	121.34
G.	**Fiscal cost**						
G1	Fiscal cost, excluding all taxes (E1 * F1)	$, millions	1,376.12	1,746.53	457.08	(425.90)	(615.50)
G2	Fiscal cost, with fuel at counterfactual tax rates (E2 * F1)	$, millions	1,376.12	1,746.53	457.08	(425.90)	(615.50)
G3	Fiscal expenditures (F1 * (E2 − E1))	$, millions	0.00	0.00	0.00	0.00	0.00

Source: Secretariat of Finance and Public Credit (SHCP).
a. All prices and costs are expressed in current U.S. dollars. GWh = gigawatt-hour; kWh = kilowatt-hour.
b. Generation cost + transmission and distribution margins: average cost of electricity.
c. VAT (value added tax) is 16 percent of the applied tariff.

Table B.73 Mexico—Public Lighting

		Unit[a]	2008	2009	2010	2011	2012
A.	**Supply cost[b]**	$/kWh	0.247	0.189	0.174	0.179	0.173
B1.	**Taxes, excluding transfers (VAT)[c]**	$/kWh	0.025	0.023	0.027	0.029	0.030
B2.	**Taxes (counterfactual): standard VAT**	$/kWh	0.025	0.023	0.027	0.029	0.030
C.	**Total reference benchmark price**						
C1	Total reference price without taxes	$/kWh	0.247	0.189	0.174	0.179	0.173
C2	Total reference price at counterfactual tax rates	$/kWh	0.272	0.213	0.201	0.208	0.203
D.	**End-user price**						
D1	Applied tariff, excluding all taxes	$/kWh	0.168	0.146	0.168	0.182	0.185
D2	Applied tariff, including all taxes	$/kWh	0.193	0.170	0.195	0.212	0.215
E.	**Price gap**						
E1	Price gap, excluding all taxes (C1 − D1)	$/kWh	0.079	0.043	0.006	−0.004	−0.012
E2	Price gap, with fuel at counterfactual tax rates (C2 − D2)	$/kWh	0.079	0.043	0.006	−0.004	−0.012
F.	**Consumption**						
F1	Volume	1,000 GWh	3.70	3.80	4.52	4.87	4.86
G.	**Fiscal cost**						
G1	Fiscal cost, excluding all taxes (E1 * F1)	$, millions	292.03	164.39	27.18	(18.02)	(58.48)
G2	Fiscal cost, with fuel at counterfactual tax rates (E2 * F1)	$, millions	292.03	164.39	27.18	(18.02)	(58.48)
G3	Fiscal expenditures (F1 * (E2 − E1))	$, millions	0.00	0.00	0.00	0.00	0.00

Source: Secretariat of Finance and Public Credit (SHCP).
a. All prices and costs are expressed in current U.S. dollars. GWh = gigawatt-hour; kWh = kilowatt-hour.
b. Generation cost + transmission and distribution margins: average cost of electricity.
c. VAT (value added tax) is 16 percent of the applied tariff.

Table B.74 Mexico—Water Pumping

		Unit[a]	2008	2009	2010	2011	2012
A.	Supply cost[b]	$/kWh	0.230	0.172	0.169	0.173	0.167
B1.	Taxes, excluding transfers (VAT)[c]	$/kWh	0.018	0.016	0.018	0.019	0.019
B2.	Taxes (counterfactual): standard VAT	$/kWh	0.018	0.016	0.018	0.019	0.019
C.	Total reference benchmark price						
C1	Total reference price without taxes	$/kWh	0.230	0.172	0.169	0.173	0.167
C2	Total reference price at counterfactual tax rates	$/kWh	0.248	0.188	0.187	0.192	0.186
D.	End-user price						
D1	Applied tariff, excluding all taxes	$/kWh	0.117	0.099	0.114	0.121	0.120
D2	Applied tariff, including all taxes	$/kWh	0.134	0.115	0.132	0.140	0.140
E.	Price gap						
E1	Price gap, excluding all taxes (C1 − D1)	$/kWh	0.113	0.072	0.055	0.052	0.047
E2	Price gap, with fuel at counterfactual tax rates (C2 − D2)	$/kWh	0.113	0.072	0.055	0.052	0.047
F.	Consumption						
F1	Volume	1,000 GWh	1.51	2.22	2.72	3.20	3.51
G.	Fiscal cost						
G1	Fiscal cost, excluding all taxes (E1 * F1)	$, millions	171.48	159.98	149.12	166.36	164.00
G2	Fiscal cost, with fuel at counterfactual tax rates (E2 * F1)	$, millions	171.48	159.98	149.12	166.36	164.00
G3	Fiscal expenditures (F1 * (E2 − E1))	$, millions	0.00	0.00	0.00	0.00	0.00

Source: Secretariat of Finance and Public Credit (SHCP).
a. All prices and costs are expressed in current U.S. dollars. GWh = gigawatt-hour; kWh = kilowatt-hour.
b. Generation cost + transmission and distribution margins: average cost of electricity.
c. VAT (value added tax) is 16 percent of the applied tariff.

Table B.75 Mexico—Agriculture

		Unit[a]	2008	2009	2010	2011	2012
A.	Supply cost[b]	$/kWh	0.164	0.123	0.134	0.136	0.132
B1.	Taxes, excluding transfers (VAT)[c]	$/kWh	0.007	0.005	0.006	0.007	0.007
B2.	Taxes (counterfactual): standard VAT	$/kWh	0.007	0.005	0.006	0.007	0.007
C.	Total reference benchmark price						
C1	Total reference price without taxes	$/kWh	0.164	0.123	0.134	0.136	0.132
C2	Total reference price at counterfactual tax rates	$/kWh	0.171	0.128	0.140	0.143	0.139
D.	End-user price						
D1	Applied tariff, excluding all taxes	$/kWh	0.046	0.030	0.039	0.044	0.044
D2	Applied tariff, including all taxes	$/kWh	0.052	0.035	0.045	0.051	0.051
E.	Price gap						
E1	Price gap, excluding all taxes (C1 − D1)	$/kWh	0.119	0.093	0.095	0.092	0.088
E2	Price gap, with fuel at counterfactual tax rates (C2 − D2)	$/kWh	0.119	0.093	0.095	0.092	0.088
F.	Consumption						
F1	Volume	1,000 GWh	8.04	9.24	8.59	10.97	10.82

table continues next page

Table B.75 Mexico—Agriculture *(continued)*

	Unit[a]	2008	2009	2010	2011	2012
G.	**Fiscal cost**					
G1	Fiscal cost, excluding all taxes (E1 * F1)	$, millions	955.48	856.03	814.87	1,006.52	950.51
G2	Fiscal cost, with fuel at counterfactual tax rates (E2 * F1)	$, millions	955.48	856.03	814.87	1,006.52	950.51
G3	Fiscal expenditures (F1 * (E2 − E1))	$, millions	0.00	0.00	0.00	0.00	0.00

Source: Secretariat of Finance and Public Credit (SHCP).
a. All prices and costs are expressed in current U.S. dollars. GWh = gigawatt-hour; kWh = kilowatt-hour.
b. Generation cost + transmission and distribution margins: average cost of electricity.
c. VAT (value added tax) is 16 percent of the applied tariff.

Table B.76 Mexico—Medium Tension

		Unit[a]	2008	2009	2010	2011	2012
A.	**Supply cost[b]**	$/kWh	0.159	0.121	0.118	0.122	0.120
B1.	**Taxes, excluding transfers (VAT)[c]**	$/kWh	0.019	0.014	0.017	0.018	0.018
B2.	**Taxes (counterfactual): standard VAT**	$/kWh	0.019	0.014	0.017	0.018	0.018
C.	**Total reference benchmark price**						
C1	Total reference price without taxes	$/kWh	0.159	0.121	0.118	0.122	0.120
C2	Total reference price at counterfactual tax rates	$/kWh	0.178	0.134	0.135	0.141	0.138
D.	**End-user price**						
D1	Applied tariff, excluding all taxes	$/kWh	0.125	0.084	0.104	0.115	0.115
D2	Applied tariff, including all taxes	$/kWh	0.144	0.098	0.121	0.134	0.133
E.	**Price gap**						
E1	Price gap, excluding all taxes (C1 − D1)	$/kWh	0.034	0.036	0.014	0.007	0.005
E2	Price gap, with fuel at counterfactual tax rates (C2 − D2)	$/kWh	0.034	0.036	0.014	0.007	0.005
F.	**Consumption**						
F1	Volume	1,000 GWh	62.39	61.76	73.73	84.33	86.60
G.	**Fiscal cost**						
G1	Fiscal cost, excluding all taxes (E1 * F1)	$, millions	2,093.93	2,234.47	1,023.08	595.56	437.14
G2	Fiscal cost, with fuel at counterfactual tax rates (E2 * F1)	$, millions	2,093.93	2,234.47	1,023.08	595.56	437.14
G3	Fiscal expenditures (F1 * (E2 − E1))	$, millions	0.00	0.00	0.00	0.00	0.00

Source: Secretariat of Finance and Public Credit (SHCP).
a. All prices and costs are expressed in current U.S. dollars. GWh = gigawatt-hour; kWh = kilowatt-hour.
b. Generation cost + transmission and distribution margins: average cost of electricity.
c. VAT (value added tax) is 16 percent of the applied tariff.

Table B.77 Mexico—Industrial (High Tension)

		Unit[a]	2008	2009	2010	2011	2012
A.	**Supply cost[b]**	$/kWh	0.106	0.078	0.089	0.093	0.093
B1.	**Taxes, excluding transfers (VAT)[c]**	$/kWh	0.016	0.011	0.014	0.016	0.015
B2.	**Taxes (counterfactual): standard VAT**	$/kWh	0.016	0.011	0.014	0.016	0.015
C.	**Total reference benchmark price**						
C1	Total reference price without taxes	$/kWh	0.106	0.078	0.089	0.093	0.093
C2	Total reference price at counterfactual tax rates	$/kWh	0.122	0.089	0.103	0.109	0.108

table continues next page

Table B.77 Mexico—Industrial (High Tension) *(continued)*

		Unit[a]	2008	2009	2010	2011	2012
D.	**End-user price**						
D1	Applied tariff, excluding all taxes	$/kWh	0.106	0.070	0.087	0.098	0.097
D2	Applied tariff, including all taxes	$/kWh	0.122	0.082	0.101	0.114	0.112
E.	**Price gap**						
E1	Price gap, excluding all taxes (C1 − D1)	$/kWh	0.000	0.007	0.002	−0.005	−0.004
E2	Price gap, with fuel at counterfactual tax rates (C2 − D2)	$/kWh	0.000	0.007	0.002	−0.005	−0.004
F.	**Consumption**						
F1	Volume	1,000 GWh	35.36	31.73	37.73	43.11	45.51
G.	**Fiscal cost**						
G1	Fiscal cost, excluding all taxes (E1 * F1)	$, millions	9.01	231.22	67.46	(195.36)	(185.37)
G2	Fiscal cost, with fuel at counterfactual tax rates (E2 * F1)	$, millions	9.01	231.22	67.46	(195.36)	(185.37)
G3	Fiscal expenditures (F1 * (E2 − E1))	$, millions	0.00	0.00	0.00	0.00	0.00

Source: Secretariat of Finance and Public Credit (SHCP).
a. All prices and costs are expressed in current U.S. dollars. GWh = gigawatt-hour; kWh = kilowatt-hour.
b. Generation cost + transmission and distribution margins: average cost of electricity.
c. VAT (value added tax) is 16 percent of the applied tariff.

Table B.78 Peru—Residential

		Unit[a]	2010	2011	2012	2013
A.	**Supply cost[b]**	$/kWh	0.060	0.050	0.056	0.051
B1.	**Taxes, excluding transfers (VAT)[c]**	$/kWh	0.011	0.009	0.010	0.009
B2.	**Taxes (counterfactual): standard VAT**	$/kWh	0.011	0.009	0.010	0.009
C.	**Total reference benchmark price**					
C1	Total reference price without taxes	$/kWh	0.060	0.050	0.056	0.051
C2	Total reference price at counterfactual tax rates	$/kWh	0.072	0.059	0.066	0.060
D.	**End-user price**					
D1	Applied tariff, excluding all taxes	$/kWh	0.018	0.020	0.030	0.025
D2	Applied tariff, including all taxes	$/kWh	0.021	0.023	0.035	0.029
E.	**Price gap**					
E1	Price gap, excluding all taxes (C1 − D1)	$/kWh	0.043	0.030	0.026	0.026
E2	Price gap, with fuel at counterfactual tax rates (C2 − D2)	$/kWh	0.051	0.036	0.031	0.031
F.	**Consumption**					
F1	Volume	1,000 GWh	7.04	7.53	7.95	8.37
G.	**Fiscal cost**					
G1	Fiscal cost, excluding all taxes (E1 * F1)	$, millions	300.90	229.27	207.13	216.47
G2	Fiscal cost, with fuel at counterfactual tax rates (E2 * F1)	$, millions	358.07	270.53	244.41	255.44
G3	Fiscal expenditures (F1 * (E2 − E1))	$, millions	57.17	41.27	37.28	38.97

Source: Committee for Economic Operation of the National Interconnected System (COES).
a. All prices and costs are expressed in current U.S. dollars. GWh = gigawatt-hour; kWh = kilowatt-hour.
b. Generation cost (natural gas spot prices and system averages) + transmission and distribution margins (system average for transmission, aggregate value of tension levels for distribution).
c. VAT (value added tax) is 18 percent of the applied tariff.

Table B.79 Peru—Commercial

		Unit[a]	2010	2011	2012	2013
A.	Supply cost[b]	$/kWh	0.060	0.050	0.056	0.051
B1.	Taxes, excluding transfers (VAT)[c]	$/kWh	0.011	0.009	0.010	0.009
B2.	Taxes (counterfactual): standard VAT	$/kWh	0.011	0.009	0.010	0.009
C.	Total reference benchmark price					
C1	Total reference price without taxes	$/kWh	0.060	0.050	0.056	0.051
C2	Total reference price at counterfactual tax rates	$/kWh	0.072	0.059	0.066	0.060
D.	End-user price					
D1	Applied tariff, excluding all taxes	$/kWh	0.018	0.020	0.030	0.025
D2	Applied tariff, including all taxes	$/kWh	0.021	0.023	0.035	0.029
E.	Price gap					
E1	Price gap, excluding all taxes (C1 − D1)	$/kWh	0.043	0.030	0.026	0.026
E2	Price gap, with fuel at counterfactual tax rates (C2 − D2)	$/kWh	0.051	0.036	0.031	0.031
F.	Consumption					
F1	Volume	1,000 GWh	5.82	6.48	6.96	7.30
G.	Fiscal cost					
G1	Fiscal cost, excluding all taxes (E1 * F1)	$, millions	248.69	197.36	181.31	188.90
G2	Fiscal cost, with fuel at counterfactual tax rates (E2 * F1)	$, millions	295.95	232.88	213.94	222.90
G3	Fiscal expenditures (F1 * (E2 − E1))	$, millions	47.25	35.52	32.64	34.00

Source: Committee for Economic Operation of the National Interconnected System (COES).
a. All prices and costs are expressed in current U.S. dollars. GWh = gigawatt-hour; kWh = kilowatt-hour.
b. Generation cost (natural gas spot prices and system averages) + transmission and distribution margins (system average for transmission, aggregate value of tension levels for distribution).
c. VAT (value added tax) is 18 percent of the applied tariff.

Table B.80 Peru—Industrial

		Unit[a]	2010	2011	2012	2013
A.	Supply cost[b]	$/kWh	0.060	0.050	0.056	0.051
B1.	Taxes, excluding transfers (VAT)[c]	$/kWh	0.011	0.009	0.010	0.009
B2.	Taxes (counterfactual): standard VAT	$/kWh	0.011	0.009	0.010	0.009
C.	Total reference benchmark price					
C1	Total reference price without taxes	$/kWh	0.060	0.050	0.056	0.051
C2	Total reference price at counterfactual tax rates	$/kWh	0.072	0.059	0.066	0.060
D.	End-user price					
D1	Applied tariff, excluding all taxes	$/kWh	0.018	0.020	0.030	0.025
D2	Applied tariff, including all taxes	$/kWh	0.021	0.023	0.035	0.029
E.	Price gap					
E1	Price gap, excluding all taxes (C1 − D1)	$/kWh	0.043	0.030	0.026	0.026
E2	Price gap, with fuel at counterfactual tax rates (C2 − D2)	$/kWh	0.051	0.036	0.031	0.031
F.	Consumption					
F1	Volume	1,000 GWh	16.00	17.02	17.92	19.20
G.	Fiscal cost					
G1	Fiscal cost, excluding all taxes (E1 * F1)	$, millions	683.93	518.02	466.67	496.58
G2	Fiscal cost, with fuel at counterfactual tax rates (E2 * F1)	$, millions	813.88	611.27	550.68	585.97
G3	Fiscal expenditures (F1 * (E2 − E1))	$, millions	129.95	93.24	84.00	89.38

Source: Committee for Economic Operation of the National Interconnected System (COES).
a. All prices and costs are expressed in current U.S. dollars. GWh = gigawatt-hour; kWh = kilowatt-hour.
b. Generation cost (natural gas spot prices and system averages) + transmission and distribution margins (system average for transmission, aggregate value of tension levels for distribution).
c. VAT (value added tax) is 18 percent of the applied tariff.

Table B.81 Peru—Public Lighting

	Unit[a]	2010	2011	2012	2013
A.	**Supply cost[b]**	$/kWh	0.060	0.050	0.056	0.051
B1.	**Taxes, excluding transfers (VAT)[c]**	$/kWh	0.011	0.009	0.010	0.009
B2.	**Taxes (counterfactual): standard VAT**	$/kWh	0.011	0.009	0.010	0.009
C.	**Total reference benchmark price**					
C1	Total reference price without taxes	$/kWh	0.060	0.050	0.056	0.051
C2	Total reference price at counterfactual tax rates	$/kWh	0.072	0.059	0.066	0.060
D.	**End-user price**					
D1	Applied tariff, excluding all taxes	$/kWh	0.018	0.020	0.030	0.025
D2	Applied tariff, including all taxes	$/kWh	0.021	0.023	0.035	0.029
E.	**Price gap**					
E1	Price gap, excluding all taxes (C1 − D1)	$/kWh	0.043	0.030	0.026	0.026
E2	Price gap, with fuel at counterfactual tax rates (C2 − D2)	$/kWh	0.051	0.036	0.031	0.031
F.	**Consumption**					
F1	Volume	1,000 GWh	0.71	0.75	0.79	0.85
G.	**Fiscal cost**					
G1	Fiscal cost, excluding all taxes (E1 * F1)	$, millions	30.46	22.90	20.55	22.10
G2	Fiscal cost, with fuel at counterfactual tax rates (E2 * F1)	$, millions	36.24	27.02	24.25	26.08
G3	Fiscal expenditures (F1 * (E2 − E1))	$, millions	5.79	4.12	3.70	3.98

Source: Committee for Economic Operation of the National Interconnected System (COES).
a. All prices and costs are expressed in current U.S. dollars. GWh = gigawatt-hour; kWh = kilowatt-hour.
b. Generation cost (natural gas spot prices and system averages) + transmission and distribution margins (system average for transmission, aggregate value of tension levels for distribution).
c. VAT (value added tax) is 18 percent of the applied tariff.

Table B.82 Peru—Low Tension

	Unit[a]	2010	2011	2012	2013
A.	**Supply cost[b]**	$/kWh	0.060	0.050	0.056	0.051
B1.	**Taxes, excluding transfers (VAT)[c]**	$/kWh	0.011	0.009	0.010	0.009
B2.	**Taxes (counterfactual): standard VAT**	$/kWh	0.011	0.009	0.010	0.009
C.	**Total reference benchmark price**					
C1	Total reference price without taxes	$/kWh	0.060	0.050	0.056	0.051
C2	Total reference price at counterfactual tax rates	$/kWh	0.072	0.059	0.066	0.060
D.	**End-user price**					
D1	Applied tariff, excluding all taxes	$/kWh	0.018	0.020	0.030	0.025
D2	Applied tariff, including all taxes	$/kWh	0.021	0.023	0.035	0.029
E.	**Price gap**					
E1	Price gap, excluding all taxes (C1 − D1)	$/kWh	0.043	0.030	0.026	0.026
E2	Price gap, with fuel at counterfactual tax rates (C2 − D2)	$/kWh	0.051	0.036	0.031	0.031
F.	**Consumption**					
F1	Volume	1,000 GWh	10.53	11.24	11.82	12.39
G.	**Fiscal cost**					
G1	Fiscal cost, excluding all taxes (E1 * F1)	$, millions	450.33	342.00	307.96	320.60
G2	Fiscal cost, with fuel at counterfactual tax rates (E2 * F1)	$, millions	535.90	403.56	363.39	378.31
G3	Fiscal expenditures (F1 * (E2 − E1))	$, millions	85.56	61.56	55.43	57.71

Source: Committee for Economic Operation of the National Interconnected System (COES).
a. All prices and costs are expressed in current U.S. dollars. GWh = gigawatt-hour; kWh = kilowatt-hour.
b. Generation cost (natural gas spot prices and system averages) + transmission and distribution margins (system average for transmission, aggregate value of tension levels for distribution).
c. VAT (value added tax) is 18 percent of the applied tariff.

Table B.83 Peru—Medium Tension

		Unit[a]	2010	2011	2012	2013
A.	Supply cost[b]	$/kWh	0.060	0.050	0.056	0.051
B1.	Taxes, excluding transfers (VAT)[c]	$/kWh	0.011	0.009	0.010	0.009
B2.	Taxes (counterfactual): standard VAT	$/kWh	0.011	0.009	0.010	0.009
C.	Total reference benchmark price					
C1	Total reference price without taxes	$/kWh	0.060	0.050	0.056	0.051
C2	Total reference price at counterfactual tax rates	$/kWh	0.072	0.059	0.066	0.060
D.	End-user price					
D1	Applied tariff, excluding all taxes	$/kWh	0.018	0.020	0.030	0.025
D2	Applied tariff, including all taxes	$/kWh	0.021	0.023	0.035	0.029
E.	Price gap					
E1	Price gap, excluding all taxes (C1 − D1)	$/kWh	0.043	0.030	0.026	0.026
E2	Price gap, with fuel at counterfactual tax rates (C2 − D2)	$/kWh	0.051	0.036	0.031	0.031
F.	Consumption					
F1	Volume	1,000 GWh	9.25	10.32	10.84	10.86
G.	Fiscal cost					
G1	Fiscal cost, excluding all taxes (E1 * F1)	$, millions	395.65	314.13	282.33	280.84
G2	Fiscal cost, with fuel at counterfactual tax rates (E2 * F1)	$, millions	470.82	370.68	333.15	331.40
G3	Fiscal expenditures (F1 * (E2 − E1))	$, millions	75.17	56.54	50.82	50.55

Source: Committee for Economic Operation of the National Interconnected System (COES).
a. All prices and costs are expressed in current U.S. dollars. GWh = gigawatt-hour; kWh = kilowatt-hour.
b. Generation cost (natural gas spot prices and system averages) + transmission and distribution margins (system average for transmission, aggregate value of tension levels for distribution).
c. VAT (value added tax) is 18 percent of the applied tariff.

Table B.84 Peru—High Tension

		Unit[a]	2010	2011	2012	2013
A.	Supply cost[b]	$/kWh	0.060	0.050	0.056	0.051
B1.	Taxes, excluding transfers (VAT)[c]	$/kWh	0.011	0.009	0.010	0.009
B2.	Taxes (counterfactual): standard VAT	$/kWh	0.011	0.009	0.010	0.009
C.	Total reference benchmark price					
C1	Total reference price without taxes	$/kWh	0.060	0.050	0.056	0.051
C2	Total reference price at counterfactual tax rates	$/kWh	0.072	0.059	0.066	0.060
D.	End-user price					
D1	Applied tariff, excluding all taxes	$/kWh	0.018	0.020	0.030	0.025
D2	Applied tariff, including all taxes	$/kWh	0.021	0.023	0.035	0.029
E.	Price gap					
E1	Price gap, excluding all taxes (C1 − D1)	$/kWh	0.043	0.030	0.026	0.026
E2	Price gap, with fuel at counterfactual tax rates (C2 − D2)	$/kWh	0.051	0.036	0.031	0.031
F.	Consumption					
F1	Volume	1,000 GWh	1.48	1.86	2.14	2.98
G.	Fiscal cost					
G1	Fiscal cost, excluding all taxes (E1 * F1)	$, millions	63.42	56.76	55.80	77.05
G2	Fiscal cost, with fuel at counterfactual tax rates (E2 * F1)	$, millions	75.47	66.97	65.85	90.92
G3	Fiscal expenditures (F1 * (E2 − E1))	$, millions	12.05	10.22	10.04	13.87

Source: Committee for Economic Operation of the National Interconnected System (COES).
a. All prices and costs are expressed in current U.S. dollars. GWh = gigawatt-hour; kWh = kilowatt-hour.
b. Generation cost (natural gas spot prices and system averages) + transmission and distribution margins (system average for transmission, aggregate value of tension levels for distribution).
c. VAT (value added tax) is 18 percent of the applied tariff.

Table B.85 Peru—Very High Tension

		Unit[a]	2010	2011	2012	2013
A.	Supply cost[b]	$/kWh	0.060	0.050	0.056	0.051
B1.	Taxes, excluding transfers (VAT)[c]	$/kWh	0.011	0.009	0.010	0.009
B2.	Taxes (counterfactual): standard VAT	$/kWh	0.011	0.009	0.010	0.009
C.	Total reference benchmark price					
C1	Total reference price without taxes	$/kWh	0.060	0.050	0.056	0.051
C2	Total reference price at counterfactual tax rates	$/kWh	0.072	0.059	0.066	0.060
D.	End-user price					
D1	Applied tariff, excluding all taxes	$/kWh	0.018	0.020	0.030	0.025
D2	Applied tariff, including all taxes	$/kWh	0.021	0.023	0.035	0.029
E.	Price gap					
E1	Price gap, excluding all taxes (C1 − D1)	$/kWh	0.043	0.030	0.026	0.026
E2	Price gap, with fuel at counterfactual tax rates (C2 − D2)	$/kWh	0.051	0.036	0.031	0.031
F.	Consumption					
F1	Volume	1,000 GWh	8.29	8.37	8.81	9.49
G.	Fiscal cost					
G1	Fiscal cost, excluding all taxes (E1 * F1)	$, millions	354.57	254.66	229.56	245.56
G2	Fiscal cost, with fuel at counterfactual tax rates (E2 * F1)	$, millions	421.94	300.50	270.89	289.76
G3	Fiscal expenditures (F1 * (E2 − E1))	$, millions	67.37	45.84	41.32	44.20

Source: Committee for Economic Operation of the National Interconnected System (COES).
a. All prices and costs are expressed in current U.S. dollars. GWh = gigawatt-hour; kWh = kilowatt-hour.
b. Generation cost (natural gas spot prices and system averages) + transmission and distribution margins (system average for transmission, aggregate value of tension levels for distribution).
c. VAT (value added tax) is 18 percent of the applied tariff.

Price Shocks

Figure C.1 Sectoral Impact of a US$.25 per Liter Price Shock in Refined Oil Products, Selected LAC Countries

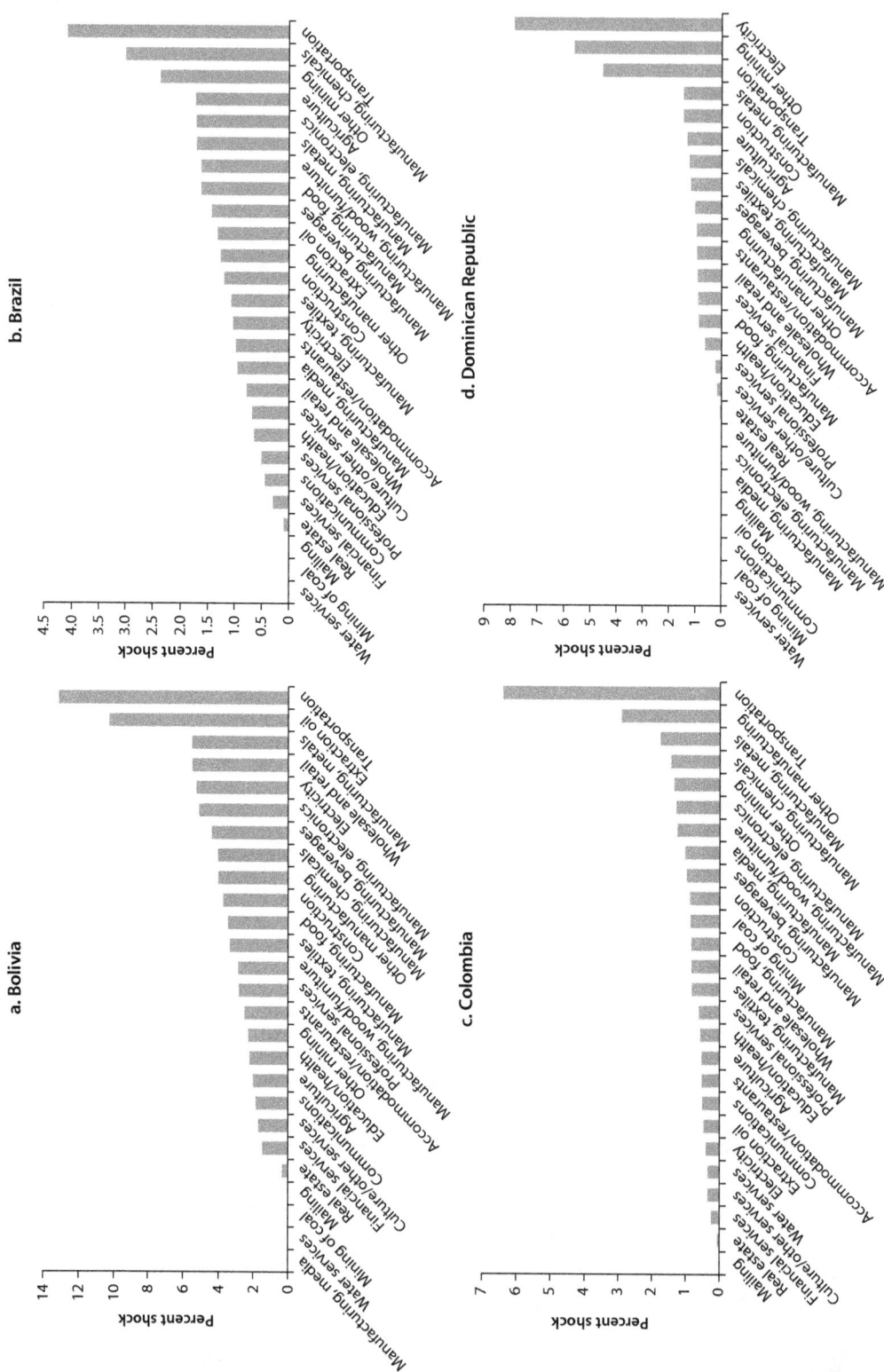

a. Bolivia

b. Brazil

c. Colombia

d. Dominican Republic

figure continues next page

Figure C.1 Sectoral Impact of a US$.25 per Liter Price Shock in Refined Oil Products, Selected LAC Countries (continued)

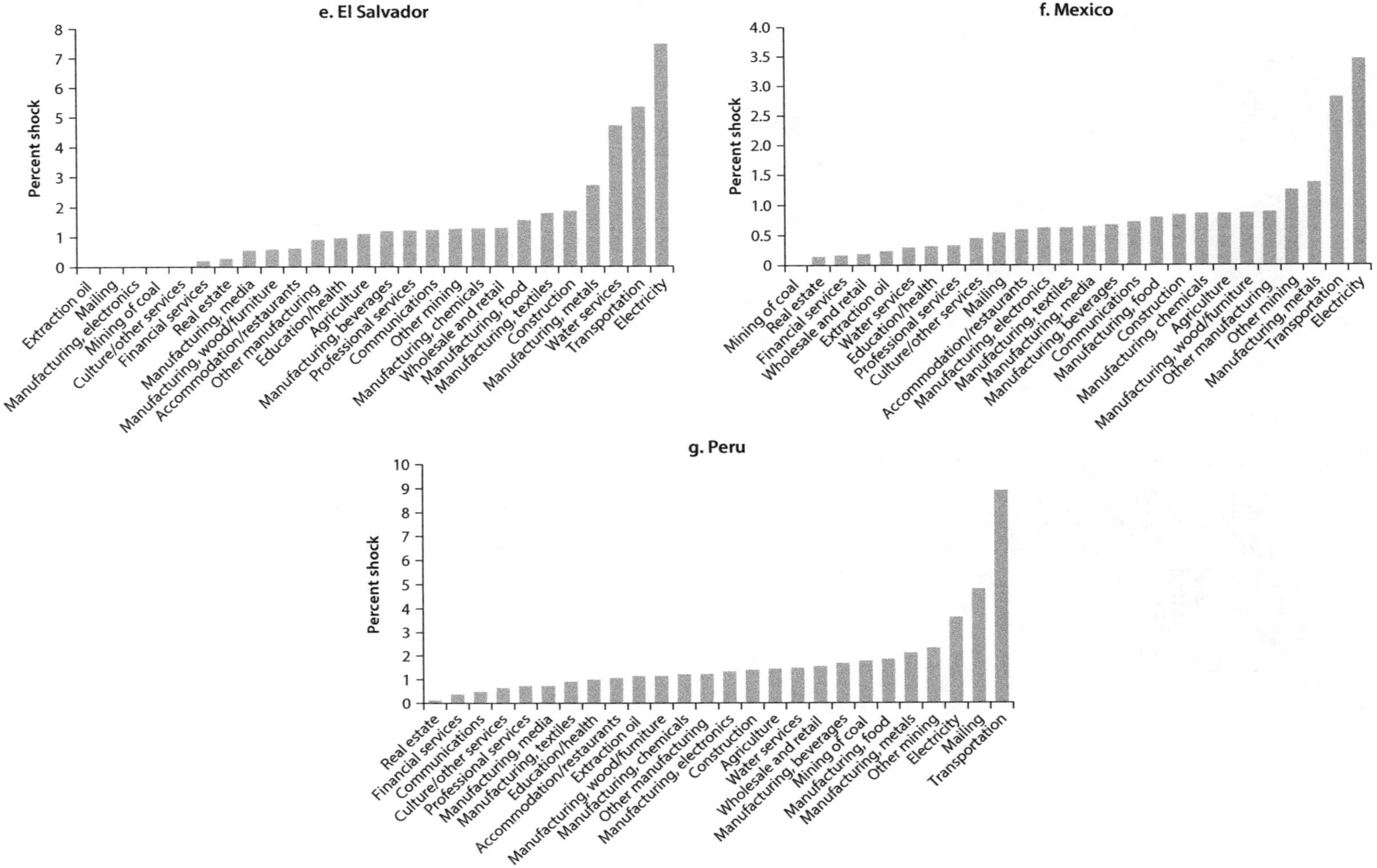

e. El Salvador

f. Mexico

g. Peru

Source: Calculations based on each country's input-output (IO) matrixes.
Note: The graphs plot the percentage change in prices by sector and by country of a US$.25 per liter price shock in gasoline and diesel.

Figure C.2 Sectoral Impact of a US$.05 per Kilowatt-Hour Price Shock in Electricity, Selected LAC Countries

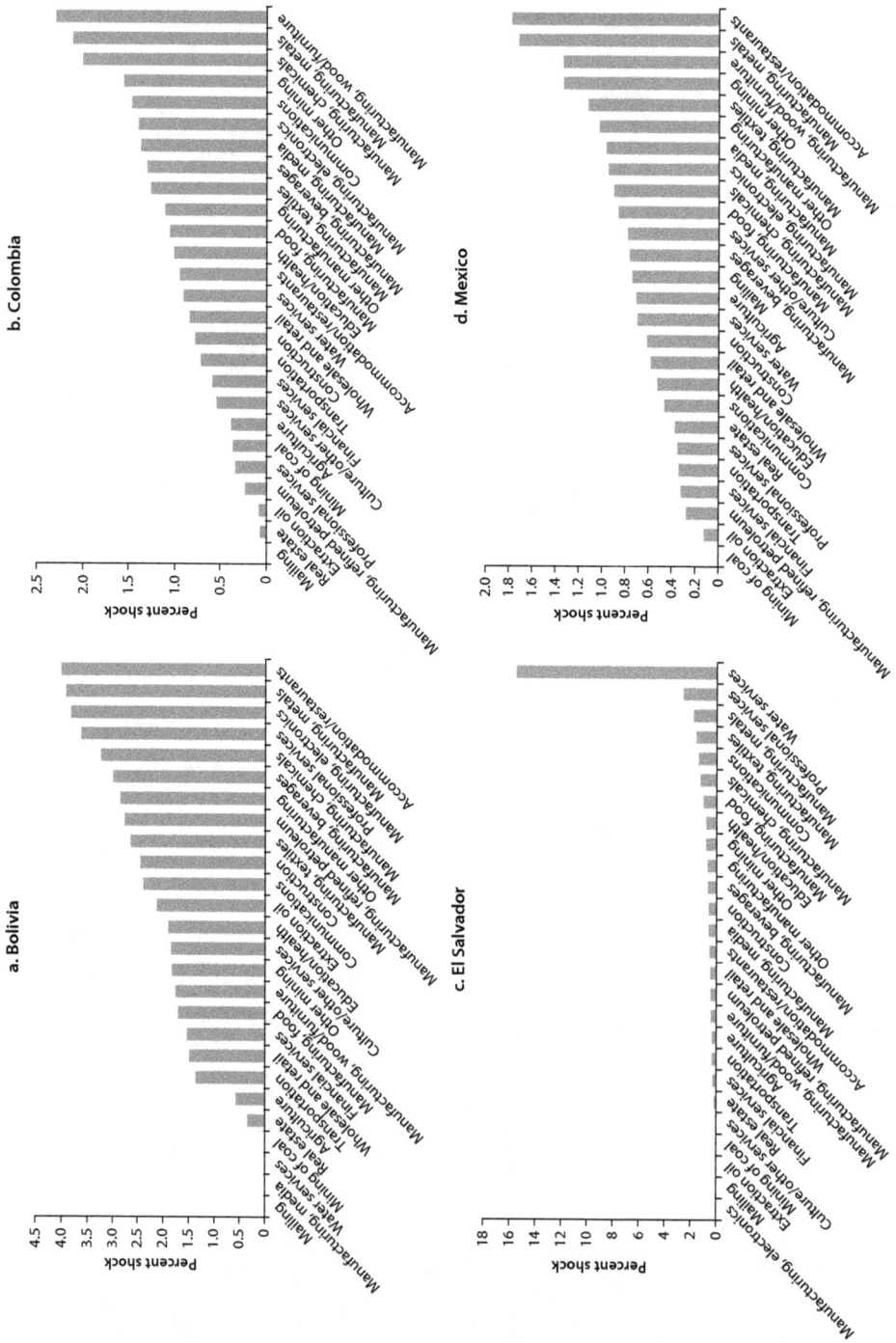

a. Bolivia

b. Colombia

c. El Salvador

d. Mexico

figure continues next page

Figure C.2 Sectoral Impact of a US$.05 per Kilowatt-Hour Price Shock in Electricity, Selected LAC Countries *(continued)*

e. Peru

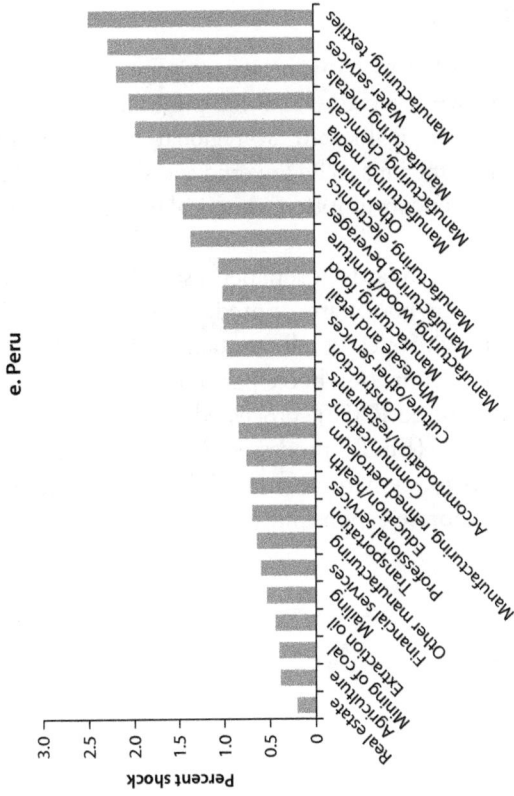

Source: Calculations based on each country's input-output (IO) matrixes.

Note: The graphs plot the percentage change in prices by sector and by country of a US$.05 per kilowatt-hour price shock in electricity.

Environmental Benefits Statement

The World Bank Group is committed to reducing its environmental footprint. In support of this commitment, we leverage electronic publishing options and print-on-demand technology, which is located in regional hubs worldwide. Together, these initiatives enable print runs to be lowered and shipping distances decreased, resulting in reduced paper consumption, chemical use, greenhouse gas emissions, and waste.

We follow the recommended standards for paper use set by the Green Press Initiative. The majority of our books are printed on Forest Stewardship Council (FSC)–certified paper, with nearly all containing 50–100 percent recycled content. The recycled fiber in our book paper is either unbleached or bleached using totally chlorine-free (TCF), processed chlorine–free (PCF), or enhanced elemental chlorine–free (EECF) processes.

More information about the Bank's environmental philosophy can be found at http://www.worldbank.org/corporateresponsibility.

www.ingramcontent.com/pod-product-compliance
Lightning Source LLC
Chambersburg PA
CBHW080419270326
41929CB00018B/3091

* 9 7 8 1 4 6 4 8 1 1 1 1 1 *